24 0543170 X

AF606268

Very Low-Mass Stars and Brown Dwarfs

This volume provides a state-of-the-art review of our current knowledge of brown dwarfs and very low-mass stars. The hunt for and study of these elusive objects is currently one of the most dynamic areas of research in astronomy for two reasons. Brown dwarfs bridge the gap between stars and planets. They could be as numerous as stars in galaxies and contribute to the 'dark matter' of the Universe. This volume presents review articles from a team of international authorities who gathered at a conference in La Palma to assess the spectacular progress that has been made in this field in the last few years.

This volume reviews both the latest observations and theory to provide an essential reference to one of the most exciting fields in contemporary astronomy.

CAMBRIDGE CONTEMPORARY ASTROPHYSICS

Series editors
José Franco, Steven M. Kahn, Andrew R. King and Barry F. Madore

Titles available in this series

Gravitational Dynamics, *edited by O. Lahav, E. Terlevich and R. J. Terlevich* (ISBN 0 521 56327 5)

High-sensitivity Radio Astronomy, *edited by N. Jackson and R. J. Davis* (ISBN 0 521 57350 5)

Relativistic Astrophysics, *edited by B. J. T. Jones and D. Marković* (ISBN 0 521 62113 5)

Advances in Stellar Evolution, *edited by R. T. Rood and A. Renzini* (ISBN 0 521 59184 8)

Relativistic Gravitation and Gravitational Radiation, *edited by J.-A. Marck and J.-P. Lasota* (ISBN 0 521 59065 5)

Instrumentation for Large Telescopes, *edited by J. M. Rodríguez Espinosa, A. Herrero and F. Sánchez* (ISBN 0 521 58291 1)

Stellar Astrophysics for the Local Group, *edited by A. Aparicio, A. Herrero and F. Sánchez* (ISBN 0 521 63255 2)

Nuclear and Particle Astrophysics, *edited by J. G. Hirsch and D. Page* (ISBN 0 521 63010 X)

Theory of Black Hole Accretion Discs, *edited by M. A. Abramowicz, G. Björnsson and J. E. Pringle* (ISBN 0 521 62362 6)

Interstellar Turbulence, *edited by J. Franco and A. Carramiñana* (ISBN 0 521 65131 X)

Globular Clusters, *edited by C. Martínez Roger, I. Pérez Fournón and F. Sánchez* (ISBN 0 521 77058 0)

The Formation of Galactic Bulges, *edited by C. M. Carollo, H. C. Ferguson and R. F. G. Wyse* (ISBN 0 521 66334 2)

Very Low-Mass Stars and Brown Dwarfs

Edited by
R. REBOLO
Instituto de Astrofísica de Canarias, Tenerife, Spain
M. R. ZAPATERO-OSORIO
Instituto de Astrofísica de Canarias, Tenerife, Spain

CAMBRIDGE
UNIVERSITY PRESS

PUBLISHED BY THE PRESS SYNDICATE OF THE UNIVERSITY OF CAMBRIDGE
The Pitt Building, Trumpington Street, Cambridge CB2 1RP, United Kingdom

CAMBRIDGE UNIVERSITY PRESS
The Edinburgh Building, Cambridge CB2 2RU, UK http://www.cup.cam.ac.uk
40 West 20th Street, New York, NY 10011-4211, USA http://www.cup.org
10 Stamford Road, Oakleigh, Melbourne 3166, Australia
Ruiz de Alarcón 13, 28014 Madrid, Spain

First published 2000

Printed in the United States of America

10/12 pt. Typeset in LaTeX by the author

A catalog record for this book is available from the British Library.

Library of Congress Cataloging-in-Publication Data is available

ISBN 0 521 66335 0 hardback

Contents

III. Convection, Rotation and Activity

Preface

The first of the "Three-Island" Euroconferences on *Stellar Clusters and Associations* was dedicated to *Very Low-Mass Stars and Brown Dwarfs.* It was held in the island of La Palma (May 11–15, 1998) where the Observatory of Roque de los Muchachos is located. These series of Euroconferences, an initiative led by Roberto Pallavicini (coordinator), Thierry Montmerle and Rafael Rebolo, are aimed to cover a very broad range of astrophysical problems where research on Stellar Clusters and Associations is crucial. In the first Euroconference, we reviewed, in a beautiful location, problems related to the formation, evolution and characterization of objects at the bottom of the Main Sequence and beyond. The first discoveries of brown dwarfs in 1995 have been followed by numerous detections in stellar clusters and in the solar vicinity. The drastic increase in the number of known examples of these fascinating objects, which suggests they are indeed very numerous in the Galaxy, has allowed a better comparison with theoretical predictions and a better and faster development of our knowledge about their physical conditions.

Some of the questions addressed in the papers compiled in this volume and delivered by active researchers in the field are: how very low-mass stars and brown dwarfs form, how many there are in the Galaxy, how they evolve, what the physical conditions of their atmospheres and interiors are, how magnetic activity develops in fully convective objects, if they generate magnetic fields, if brown dwarfs are chromospherically active and show coronae. Most of these problems remain unsolved but there is little doubt that in the coming years there will be a lot of progress in this field.

The Editors
Instituto de Astrofísica de Canarias
January, 2000

Acknowledgements

We thank the financial support of the European Union through their Euroconference Programme. Other financial and logistical support were provided by the Instituto de Astrofísica de Canarias and the "Cabildo de La Palma" (local authorities of La Palma Island).

The Scientific Organizing Committee consisted of: Gibor Basri (USA), Álvaro Giménez (Spain), Richard Jameson (UK), James Liebert (USA), Antonio Magazù (Italy), Thierry Montmerle (France), Roberto Pallavicini (Italy), Rafael Rebolo (IAC), Jurgen Schmitt (Germany), John Stauffer (USA), Hans Zinnecker (Germany).

This Euroconference would have not been possible without the hard work of the LOC members: Monica Murphy, Judith de Araoz, Beatriz Mederos and the generous support of many other people at the Instituto de Astrofísica de Canarias.

I.
SEARCHES IN CLUSTERS, STELLAR ASSOCIATIONS AND THE FIELD

Open Clusters After HIPPARCOS

By JEAN- CLAUDE MERMILLIOD

Institut d'Astronomie, Université de Lausanne, Switzerland

The current distance determinations from the HIPPARCOS satellite for nearby open clusters are reviewed and compared with ground based estimates in various photometric systems. Photometrically determined distances of open clusters are in generally good agreement with HIPPARCOS distances, when the chemical composition is properly taken into account, except for the Pleiades cluster, for which the difference between the HIPPARCOS and ground-based distance moduli still amounts to 0.3–0.4 mag if the metallicity value is assumed to be [Fe/H] = −0.03. Evidence from the (M_V, β) diagrams are somewhat contradictory and the sense of metallicity effects is not clear. In the (M_V, $(V-I)_{Ko}$) and (M_V, $(V-I)_{Co}$) diagrams, several clusters (Praesepe, α Persei, NGC 6475, IC 2391 and IC 2602) define very closely the same colour-magnitude relation, while the Pleiades are again too low. However, extensive check and analysis of HIPPARCOS observations do not reveal any failure on the satellite or reduction side that could explain this situation.

---

1. Introduction

The distances of the Hyades and other nearby open clusters are fundamental values to fix the zero point of the distance scale and absolute magnitude calibrations. Most important are the distance moduli of open clusters containing cepheid variables, usually estimated by the well known ZAMS fitting method. Therefore much hope was placed in the results of the HIPPARCOS satellite to determine distances for nearby open clusters that are independent of any previous knowledge on these objects, and especially of any determination of their chemical composition. Indeed, these results should, in principle, offer the best opportunity to get information on the effects of chemical composition.

In practice, and this is not to surprise us, the situation is not as simple as desired. If the distance modulus of the Hyades (Perryman et al. 1998) corresponds rather well to the average value of most previous determinations ($m - M = 3.33 \pm 0.01$), the distance moduli of nearby open clusters are determined with much larger errors, of the order of 0.05 - 0.20 mag. In addition, the true chemical composition is only known with large uncertainties, and the relative positions of the main sequences of various clusters are sometimes in contradiction with what is expected. If one take into account the errors resulting from the uncertainties on

- the distance moduli estimated from HIPPARCOS parallaxes,
- the chemical composition,
- the transformation from theoretical parameters to observational units,

the determination of cluster parameters has not progressed so much. Therefore, open clusters are not able to help very much in fixing the zero point of the period-luminosity relation for cepheids unless we do not use the Pleiades main sequence as a reference, but the ZAMS obtained from a combination of those clusters that have the best distance determinations.

2. HIPPARCOS distances

The mean cluster parallaxes cannot be computed without caution. As was already explained before the satellite launch (Lindegren 1988), the estimation of the mean para-

TABLE 1. Cluster mean astrometric parameters

Cluster	N stars	N meas	π	σ_π	dist	+	−	$(m-M)_o$	±
Coma Ber	30	1563	11.49	0.21	87.0	1.6	1.6	4.70	0.04
Pleiades	54	2158	8.46	0.22	118.2	3.2	3.0	5.36	0.06
IC 2391	11	807	6.85	0.22	146.0	4.8	4.5	5.82	0.07
IC 2602	23	1766	6.58	0.16	152.0	3.8	3.6	5.91	0.05
Praesepe	26	1126	5.54	0.31	180.5	10.7	9.6	6.28	0.13
NGC 2451	12	908	5.30	0.19	188.7	7.0	6.5	6.38	0.08
α Per	46	2198	5.25	0.19	190.5	7.2	6.7	6.40	0.08
NGC 6475	22	772	3.57	0.30	280.1	25.7	21.7	7.24	0.19
NGC 2516	14	947	2.89	0.21	346.0	27.1	23.4	7.70	0.16

llax or proper motion of a cluster observed by HIPPARCOS must take into account the observation mode of the satellite. This is due to the fact that stars within a small area in the sky have frequently been observed in the same field of view of the satellite. Consequently, one may expect correlations between measurements obtained for stars separated by a few degrees, or with a separation being a multiple of the basic angle between the two fields of view.

The consequence is that, when averaging the parallaxes or proper motions for n stars, the improvement factor does not follow the expected $1/\sqrt{n}$ law and will not be asymptotically better than $\sqrt{\rho}$, where ρ is the mean positive correlation between data. In the case of clusters, the improvement was estimated to about $n^{-0.35}$ (Lindegren 1988). The straight average of individual parallaxes would then not be an optimal estimate of the mean cluster parallax, and moreover its standard error would be underestimated.

The proper way to take these correlations into account is to go back to the Great Circle (RGC) level, to calibrate the correlations between the RGC abscissae, so that the full co-variance matrix between observations allows to find the optimal astrometric parameters. The method adopted by Robichon et al. (1998) is similar to that of van Leeuwen & Evans (1998) with the exception that the calibration of correlation coefficients has been done on each RGC. This has been done using the theoretical formulae of Lindegren (1988) to which harmonics were added through the use of cosine transform.

The quantities of interest are the mean parallax π_0 and the mean proper motion $\mu_{\alpha_0} \cos\delta_0$, μ_{δ_0} of each cluster center and the position α_i, δ_i of each cluster member i. Such a method is explained in more details in Robichon et al. (1998) and the new mean parallaxes and distance moduli for open clusters closer than 500 pc have been computed in this way. Table 1 summarizes the results for the nine clusters nearer than 350 pc which have errors on their distance moduli less than 0.20 mag.

3. Metallicity effects

To check if the first order assumption that the position of a cluster sequence in the colour-magnitude diagram is mainly governed by the metallicity, the magnitude differences, in absolute magnitude, between the sequence of Hyades - Praesepe and that of the other cluster have been computed. The results are summarized in Fig. 1 and the data are listed in Table 2. Error bars on magnitudes are the distance-modulus uncertainties from Table 1 and the error bars on [Fe/H] are taken at 0.05, if nothing else has been found in the original publications. The curve is the model prediction of VandenBerg & Poll

TABLE 2. Magnitude differences and metallicities

Cluster	Diff	Err	[Fe/H]	Err
Praesepe	0.00	0.13	+0.14	0.05
Coma Ber	−0.22	0.04	−0.07	0.04
Pleiades	−0.42	0.06	−0.03	0.05
IC 2391	−0.37	0.07	−0.10	0.05
IC 2602	−0.22	0.05	−0.20	0.05
α Per	−0.07	0.08	+0.06	0.05
NGC 6475	−0.13	0.19	+0.07	0.09
NGC 7092	−0.15	0.20	+0.01	0.03
NGC 2516	−0.67	0.16	−0.28	0.05

(1989)

$$\Delta M_V = -[\text{Fe/H}] * (1.444 + 0.362\ [\text{Fe/H}])$$

and nicely matches the observed trend.

Apart for the three clusters with errors on the distance moduli between 0.15 et 0.20, the weakest point is the metallicity determination. Some are spectroscopic, as discussed by Cayrel (1990) and Pinsonneault et al. (1998), some are photometric, mainly from the *uvby* system. The [Fe/H/] values adopted are from Pinsonneault et al. (1998) for the Pleiades and Praesepe, from Cayrel (1990) for Coma Ber and α Persei, from Strobel (1991) for NGC 2516, NGC 7092 and IC 2391, from Lyngå (1987) for IC 2602.

The three discrepant points are the Pleiades, which appears too metal-rich for its distance, Coma Berenices, which appears too metal-poor for its parallax, [Fe/H] $\sim$ 0 would be better, and IC 2602. This cluster has the most uncertain point because of the poorly known metallicity and the lack of stars on the Zero Age Main Sequence. Solar-type stars are still under contraction and are therefore too bright for their colours. It is thus possible that the estimated magnitude difference is not large enough. But, a more realistic estimate of the chemical composition would probably solve a large fraction of the discrepancy.

Good agreement is found between HIPPARCOS distances and previously determined distance moduli, mostly derived from ground-based photometric data, within the uncertainties due to the errors on the chemical composition, except for the Pleiades, where the HIPPARCOS distance is about 10% too small. To obtain such a distance (118 pc) from the ZAMS fitting method, one needs to adopt a rather lower value of [Fe/H] and/or a higher than normal value of the Helium abundance. Such abnormal values are so far not supported by spectroscopic analyses. Taken at face value, this distance presents a challenge to the ZAMS fitting method.

In many colour-magnitude diagrams, the Pleiades sequences is located 0.3-0.4 mag fainter than the Hyades-Praesepe sequence and gives sometimes the impression that the distance modulus is wrong.

4. Illustrations of the problem

Several photometric diagrams may be used to compare the HIPPARCOS distances of nearby open clusters and the values derived from ground-based photometric data.

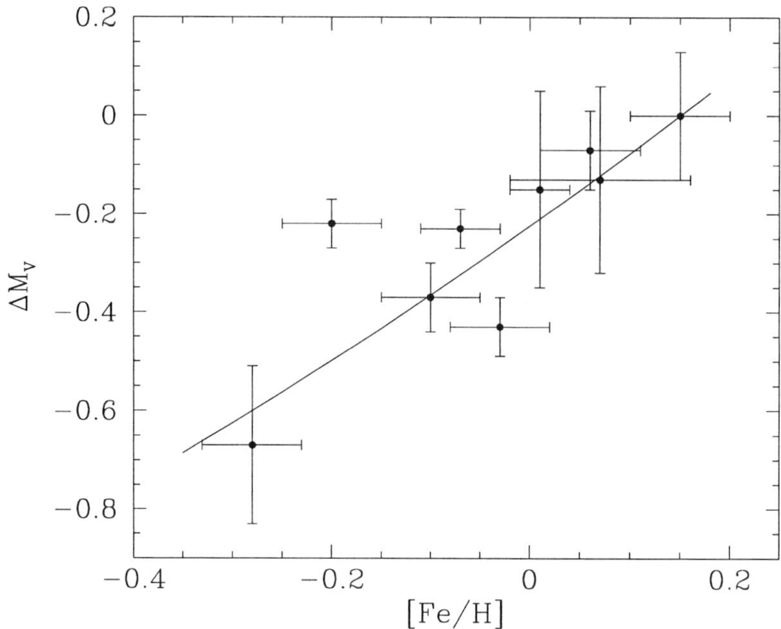

FIGURE 1. Position differences of cluster sequences relative to the Hyades - Praesepe sequence in function of metallicity.

4.1. *The (M_V, β) plane*

In the (M_V, β) plane, the sequences of the Pleiades and Praesepe clusters are obviously different (Fig. 2). Although there are evolutionary effects for $\beta > 2.7$ in the Pleiades-Praesepe diagram, the sequence for $\beta < 2.7$ should not be affected. The sequences of the Pleiades and IC 2391 are rather well superimposed (Fig. 3) all along the sequence, supporting the distance modulus of the Pleiades. However the (M_V, β) diagram for the Pleiades and α Persei is quite puzzling (Fig. 4). The B-type stars define the same sequence, while the solar-type stars are strikingly different. Both data sets are from Crawford (Crawford & Perry 1976; Crawford & Barnes 1974) and α Persei data have been used for the calibration of the $uvby\beta$ system. It happens that the solar-type star sequence of α Persei is nearly identical to that of Praesepe (Fig. 5)

At least these four diagrams offer contradictory evidences: the three open clusters, Pleiades, α Persei and IC 2391 define the same (M_V, β) relation for the B-type stars, there is clearly no evidence of systematic differences due to wrong distance modulus of the Pleiades. On the opposite, in the domain of A- and F-type stars, the Pleiades and IC 2391 show good agreement, while the Pleiades and α Persei differ notably. This behavior does not seem to be related with age because the age of IC 2391 is around 30 Myr, i.e. slightly younger than α Persei. If any effect is to be expected, the F-type stars of IC 2391 should be even more luminous and the effect should be even larger than that seen for α Persei, which is obviously not the case. Therefore, several questions arise from the examination of these (M_V, β) diagrams. Why does α Persei behaves this way? Is there any problem with the data sets? What are the real effects of the metallicity on the luminosity and β parameters? Does another parameter not yet taken into account play a decisive role?

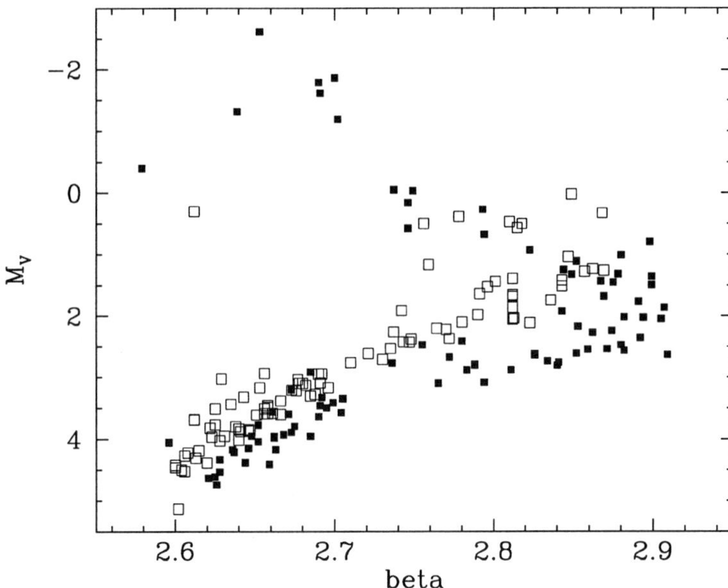

FIGURE 2. Comparison of the sequences for the Pleiades (filled squares) and Praesepe (open squares).

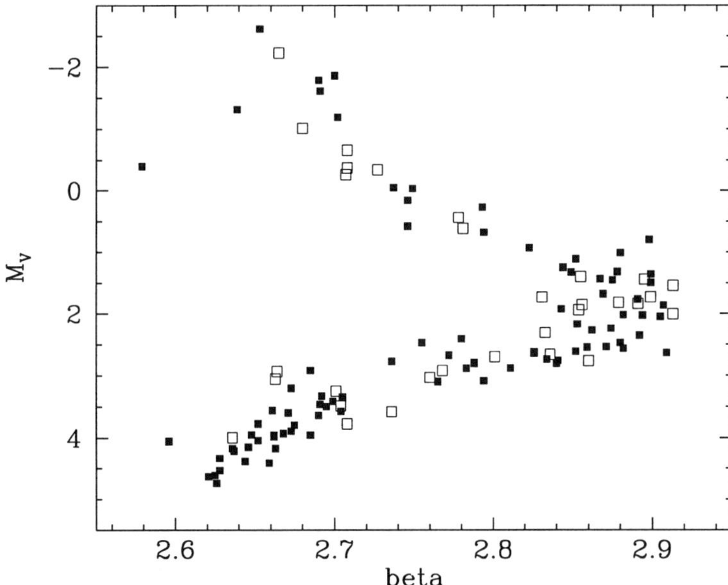

FIGURE 3. Comparison of the sequences for the Pleiades (filled squares) and IC 2391 (open squares).

4.2. *The Geneva photometry*

The diagrams (M_V, d) and (M_V, Δ) in the Geneva 7-colour photometric system for the Pleiades and Praesepe (Mermilliod et al. 1997) give the strong impression that the Pleiades sequence is globally displaced by 0.3 - 0.4 mag along the vertical axis, casting doubts on the Pleiades parallax.

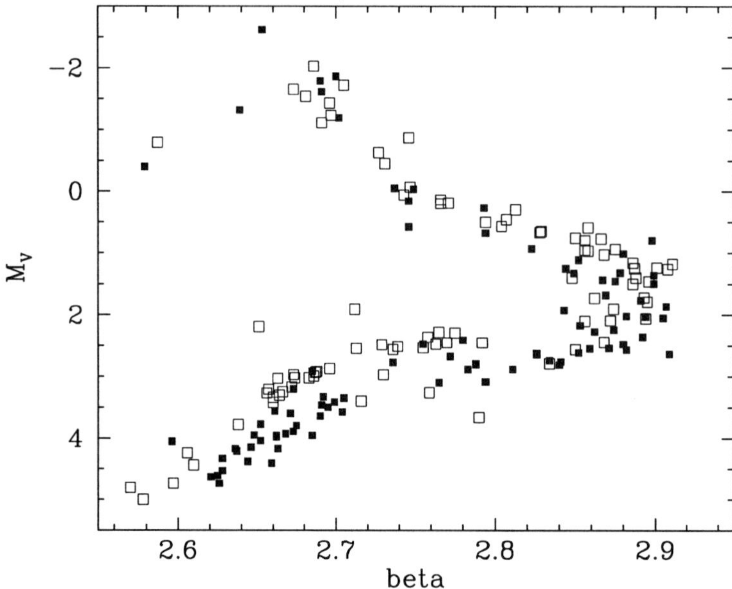

FIGURE 4. Comparison of the sequences for the Pleiades (filled squares) and α Persei (open squares).

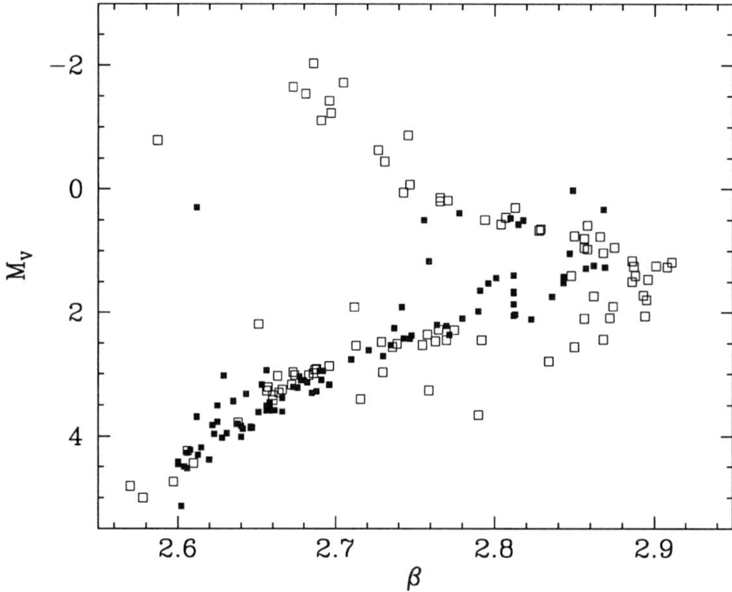

FIGURE 5. Comparison of the sequences for Praesepe (filled squares) and α Persei (open squares).

However, Grenon (1998) supports, from his analysis of Pleiades solar-type stars observed in the Geneva photometry, a value of [Fe/H] = -0.112 ± 0.025 for the Pleiades, the Hyades metallicity being taken at [Fe/H] = $+0.143 \pm 0.008$. A magnitude difference of $\Delta V = 0.38$ between the Hyades and Pleiades results from the calibration applied and is explained by the metallicity difference alone. This magnitude difference is in good agreement with HIPPARCOS results.

Does this example mean that the ZAMS fitting method is too crude and that it does not properly take into account all necessary parameters?

4.3. *The UBVRI system*

A very careful analysis of the ZAMS fitting method based on well calibrated new theoretical isochrones has been performed by Pinsonneault et al. (1998). By adopting the values of the chemical composition given by Cayrel (1990) and using simultaneously the $(V, B-V)$ and $(V, V-I)$ diagrams, they determine distance moduli in good agreement with those deduced form HIPPARCOS, except for the Coma Berenices and Pleiades clusters. In the first case, the problem is probably related to the old *VRI* photometry of Mendoza (1967), while for the Pleiades a conservative value of $m-M = 5.60$ is obtained, with [Fe/H] $= -0.03$. It should however be noticed that, except for the Hyades, for which the agreement is very good, the difference between ZAMS fitting and HIPPARCOS values are at least $\pm$ 0.1 mag. and reach 0.37 for the Pleiades. The minimum differences are about twice the errors quoted over the distance moduli derived from the main sequence fits.

They examined all possible reasons that could explain the observed difference for the Pleiades cluster and concluded that, with the spectroscopic data presently available, none was able to justify the discrepancy.

Therefore, on the basis of these results, Pinsonneault et al. (1998) concluded that there is a problem with the HIPPARCOS parallax of the Pleiades. They found an abnormally large correlation (ρ_{α}^{π}) between HIPPARCOS parallaxes and the right ascension. By taking the parallax for the null value of the correlation, they obtained $\pi = 7.46$ mas. Tests and simulations performed by F. Arenou and N. Robichon (Robichon et al. 1998) have shown that such a correlation does not induce an error on the average parallax of the order of that found for the Pleiades (1 mas) and that the correlation results from an asymmetry in the distribution of observations.

Finally, with Mermilliod et al. (1981) ZAMS extended with that of Schmidt-Kaler (1982) and using the formula given by VandenBerg & Poll (1989), a good fit is obtained for [Fe/H] $= -0.10$. The reliability of this value is supported by the fact that the sequence of Praesepe is well reproduced for the HIPPARCOS distance and [Fe/H] $= +0.15$, while that for NGC 2516 is also well matched for [Fe/H] $= -0.25$, the value quoted by Lyngå (1987), resulting from the analysis of *uvby* photometry. This result tends to support a value of the metallicity for the Pleiades lower than that usually accepted.

4.4. *Kron and Cousins VI photometry*

Because Pinsonneault et al. (1998) claims that the $(V-I)$ indices are less sensitive to metallicity effects than $(B-V)$ the available data for lower main-sequence stars produced by several groups can be used to check this assumption. Data are mostly in the Kron system (photoelectric and CCD) for the three northern clusters (Praesepe, Pleiades and α Persei), and in the Cousins system (CCD) for three southern objects (NGC 6475, IC 2391 and IC 2602). For the Pleiades, *VI* photoelectric and CCD photometry has been published mostly by Stauffer and collaborators. For α Persei, the *VI* photoelectric photometry has been obtained by Stauffer and collaborators, while *VI* CCD photometry has been published by Prosser. For Praesepe, the *VI* photometry come from several sources. The numerous bibliographic references to these data sets can be found in WEBDA, the Web site devoted to galactic open clusters (http://obswww.unige.ch/webda/) and are not reproduced here.

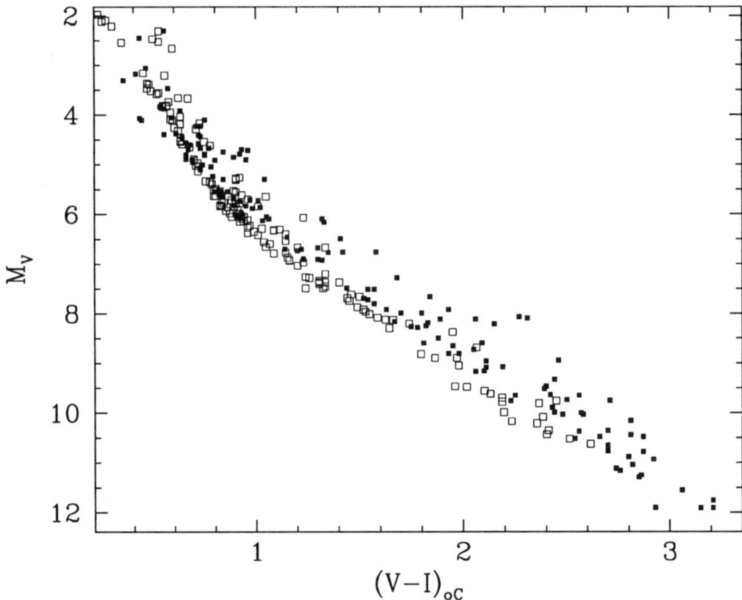

FIGURE 6. Comparison of the sequences for the combined data for IC 2391 and IC 2602 (filled squares) and Praesepe (open squares).

The reddening ratios given by Pinsonneault et al. (1998), $E(V-I)_{Kron} = 1.5\ E(B-V)$ and $A_V = 3.12\ E(B-V)$, have been used to correct the observed V magnitudes and $(V-I)$ colours for reddening and plot $(M_V,\ (V-I)_{Ko})$ diagrams. As could be expected, the sequences of Praesepe and α Persei do match well, while that for the Pleiades falls too low.

The $(V-I)_o$ colour indices in the Kron system have been transformed in the Cousins system with the relation given by Pinsonneault et al. (1998):

$$(V-I)_C = 0.227 + 0.9567\ (V-I)_K + 0.0128\ (V-I)_K^2 - 0.0053\ (V-I)_K^3$$

The new sequences have been compared with the sequences published for NGC 6475, IC 2391 and IC 2602. The faint cluster members of these three clusters have been mainly identified by their X-ray emission and the membership of many of them has been confirmed by radial velocity observations. The sequences of IC 2391 and IC 2602 define very precisely the same colour-magnitude relation according to the HIPPARCOS distances and these two clusters have been combined to improve the sequence definition.

Figure 6 presents the comparison of the combined sequence (IC 2391 + IC 2602) with that of Praesepe. The agreement is quite good. For $(V-I)_{Co} > 1.5$, stars represented by filled squares are slightly too bright because the lower main-sequence stars in these two young open clusters did not yet reach the Zero Age Main Sequence. Figure 7 presents the same comparison, but with α Persei. The agreement is also very nice. Both figures tend to confirm that the $(V-I)$ indices are less sensitive to metallicity effects that $(B-V)$, because the sequences define very closely the same colour-magnitude relation, while in the $(M_V,\ (B-V)_o)$ plane, these clusters define different sequences. The overall result for $(B-V)$ was presented in Fig. 1.

Figure 8 compares the sequence of Praesepe and NGC 6475 and both are nicely superposed. However, Figure 9 again shows that the Pleiades fall below the relation defined by Praesepe.

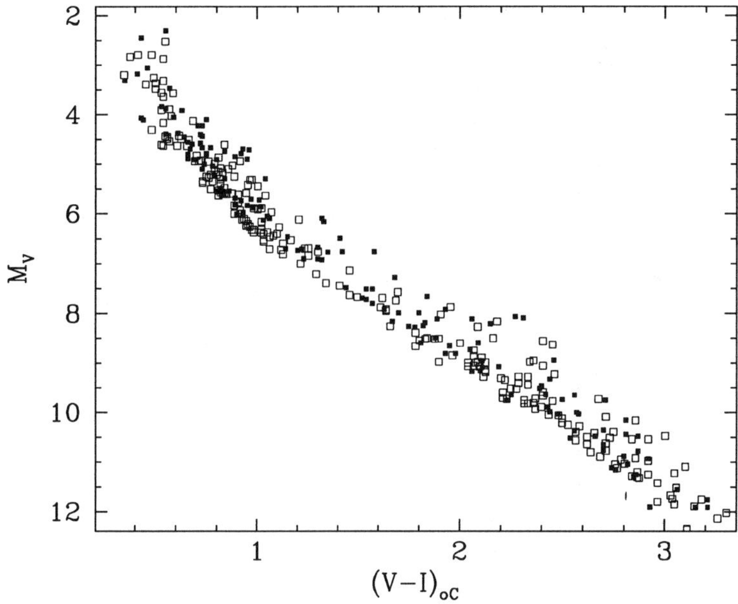

FIGURE 7. Comparison of the sequences for the combined data for IC 2391 and IC 2602 (filled squares) and α Persei (open squares).

4.5. *Conclusion of this section*

To conclude this section, I would summarize the situation as follows:

- The Pleiades distance estimated from the HIPPARCOS mean parallax does not agree with the value obtained from the main-sequence fitting method, if one assumes that [Fe/H] = −0.03,
- It also disagrees in photometric planes in which metallicity effects are not supposed to be predominant,
- No satellite or reduction anomaly can so far explain the discrepancy.

The correlation ρ_α^π observed is due to the fact that there are more observations made on one side of the Sun than on the other. If the 20 stars located at a distance less than 1° from the Pleiades center are considered, the correlation ρ_α^π disappears, by removing randomly measurements on the side with excess of observations, but the resulting parallax does not change. It however decreases to about 8 mas if only stars which have mutual distances larger than 2° are considered.

Although the HIPPARCOS distance of the Pleiades seems to be too short, there are reasons to believe that it is correct. If it is not correct, the causes of the problem have not yet been identified. We are working at the limit of precision of the data and comparisons of the various photometric data sets raise some questions, as mentioned above in the discussion of the β photometry of the α Persei cluster. In addition, the photoelectric *BVI* photometry in the Johnson system of stars in the northern nearby clusters is rather old. This system is now much less used than the Cousins system. Only few *VRI* data in the Kron or Cousins systems are available for the Coma Berenices open cluster. It should also be emphasized that the uncertainties on metallicity determinations are still rather large and a major source of scatter in the results.

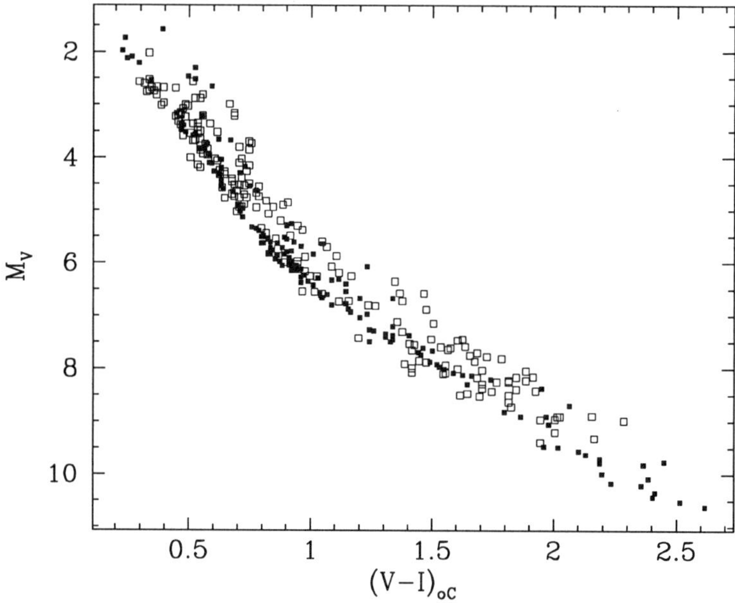

FIGURE 8. Comparison of the sequences for Praesepe (filled squares) and NGC 6475 (open squares).

5. Parameter determination

The determination of open cluster parameters would ideally require that the chemical composition is known with a precision sufficient to determine the reddening and distance moduli according to the appropriate reference sequence. However, this information is not known with a great accuracy, and there are still important differences between spectroscopic and photometric determinations. Fig. 1 shows that the metallicity plays an important role in the determination of the distance of an open cluster by the ZAMS fitting method. However, this fundamental parameter is not known for many clusters. Photometric systems like the DDO and Washington systems, designed to determine metallicity, mostly from the observations of red giants, have produced interesting results, but for a limited sample of open clusters. Revised calibrations of the DDO system (Piatti et al. 1993, Clariá et al. 1994) and of the Washington system (Geisler et al. 1991) now produce estimates of the metallicity that seem to be on the same scale. It would be important to obtain more observations in these two photometric systems to improve the number and reliability of metallicity determinations.

The method designed by Pinsonneault et al. (1998) is interesting, especially in connection with the present days common use of *BVI* CCD photometry. According to Alonso et al. (1966) the $(V-I)$ index is less sensitive than $(B-V)$ to the chemical composition. Therefore the distance modulus can be estimated from the $(V, V-I)$ diagram and chemical composition has to be adapted until the $(V, B-V)$ diagram gives the same value of the distance. The use of this method obviously requires that the reddening is known by some independent method, which does not appear to be so easy, especially when no $(U-B)$ colours are available.

In some papers, the four parameters, namely the reddening, distance, age and chemical composition, are determined at the same time from the $(V, B-V)$ diagram alone.

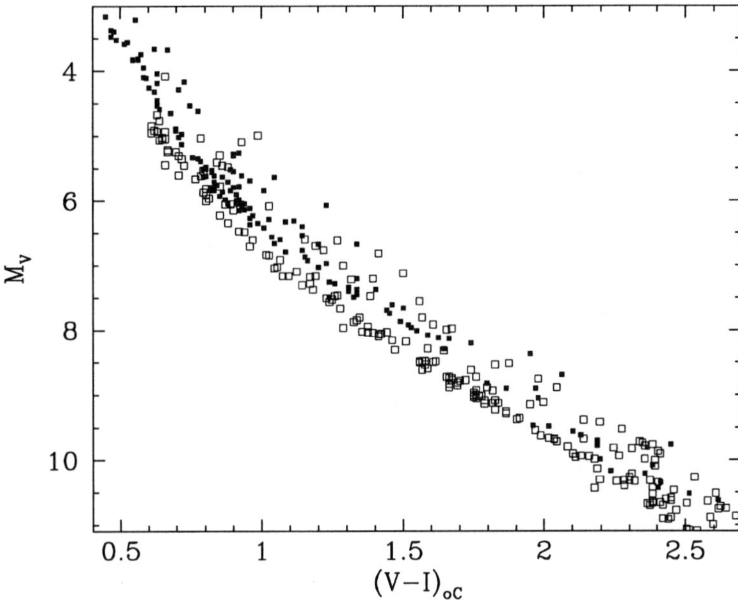

FIGURE 9. Comparison of the sequences for Praesepe (filled squares) and Pleiades (open squares).

Although this method may provide a good approximation of the true parameters, it seems to extract too much information from one plane only.

6. Age determination

In addition to the uncertainties on the chemical composition, the second problem in the determination of the ages of open clusters is that grid of models are computed at discrete values of Z and isochrones available from ftp servers are at most given for steps of 0.1 dex in log t. Therefore a valuable fit of the isochrones to the observed colour magnitude diagrams may be difficult to obtain. Isochrones with steps of 0.01 dex in log t are sometimes needed for clusters with very good photometry.

Furthermore the transformation of theoretical units ($T_{\rm eff}$) to photometric indices suffers from much uncertainties and is available mainly for solar composition and $UBV(RI)$ filters. Extension to other photometric systems would be valuable to permit the use of information provided by other photometric systems, Geneva, uvby, Walraven, Vilnius, for example, as well as reliable transformation for non-solar metallicity. The differences on age (and other parameter) estimates arise for a large part from the different methods used to transform the theoretical isochrones to observational planes. It would be very important to finally build a unique set of tables and use them solely.

With one set of isochrones, the number of degrees of freedom to realize the best fit is small if the cluster photometry is reasonably precise. As shown by Figures 10 and 11, the upper main sequence becomes narrower because it is more and more vertical. Therefore, if one uses both an isochrone and the upper binary limit produced by a shift of 0.75 mag of the same isochrone, the best age determination is that which fits at the same time the left and right envelopes, i.e. the single- and binary star sequences. Figures 10 and 11 show that a good agreement is found for the Pleiades at log t = 8.00 (100 Myr) and for α Persei at log t = 7.70 (50 Myr) for the isochrones computed with the

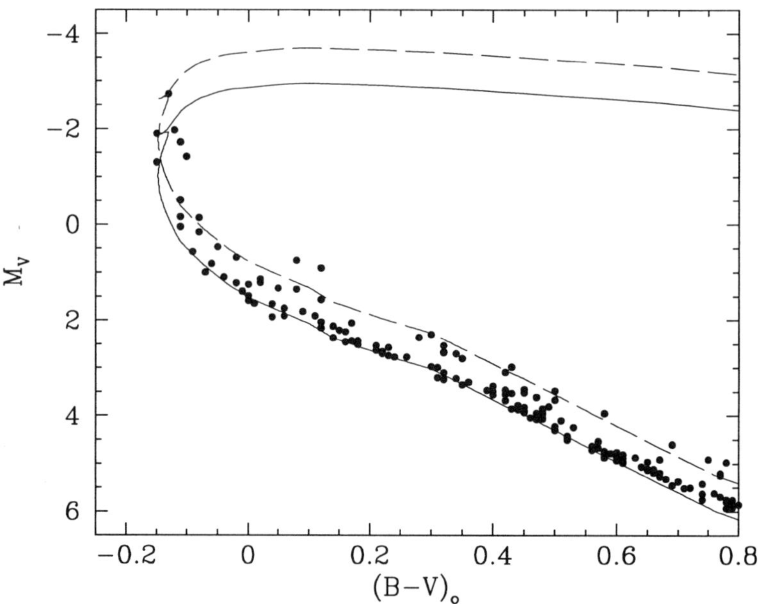

FIGURE 10. The isochrone for log t = 8.00 gives the best fit for the Pleiades.

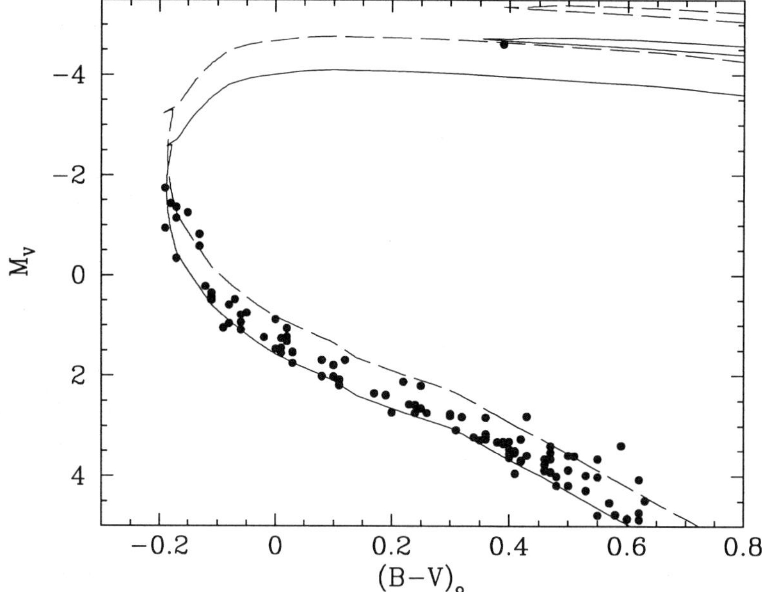

FIGURE 11. The isochrone for log t = 7.70 gives the best fit for α Persei.

models of Schaller et al. (1992). In both cases, the boundaries are well reproduced. It is however clear that stars with high rotational velocities or presenting Balmer lines in emission will most probably be found to the red of the main-sequence band defined by both sequences. This is obviously the case for the Pleiades. For α Persei, the 50 Myr isochrone also reproduces very well the position of the F5Ib supergiant α Persei itself.

7. Conclusions

The distances of six to seven open clusters can be considered as reliably determined both from the HIPPARCOS satellite and ground based measurements. They are the Hyades, Praesepe, α Persei, IC 2391 and IC 2602, and probably NGC 2516 and NGC 6475.

The distances of two clusters, the Pleiades and Coma Berenices, which are the two closest clusters after the Hyades presents some problems. The reason for Coma discrepancy may arise in the chemical composition or RI data of lesser quality. Modern VRI_K or VRI_C observations would be useful to settle this question and prove that HIPPARCOS distance is correct.

As concerns the Pleiades, the discrepant results from colour-magnitude diagrams in several photometric systems are more frequent that the agreements. This is evident in the (M_V, $(V-I)_{Ko}$) and (M_V, $(V-I)_{Co}$) diagrams comparing the nearby open clusters two by two. Because no failure in HIPPARCOS data has been found after careful examination of many aspects and properties of the observational data, the solution of the problem remains to be found.

REFERENCES

Alonso A., Arribas S., Martinez-Roger C. 1996 The empirical scale of temperatures of the low main sequence (F0V-K5V) *Astron. Astrophys.* **313**, 873–890.

Cayrel de Strobel G. 1990 Chemical abundances and ages of open clusters *Mem.S.A.It* **61**, 613–645.

Clariá J.J., Piatti A.E., Lapasset E. 1994 A revised effective temperature calibration for the DDO photometric system *Publ. Astron. Soc. Pacific* **106**, 436–443.

Crawford D.L., Barnes J.V. 1974 Four-color and Hβ photometry of open clusters. X. The α Persei cluster *Astron. J.* **79**, 687–697.

Crawford D.L., Perry C.L. 1976 Four-color and Hβ photometry of open clusters. XI. The Pleiades *Astron. J.* **81**, 419–426.

Geisler D., Clariá J.J., Minniti D. 1991 An improved metal abundance calibration for the Washington system *Astron. J.* **102**, 1836–1869.

Grenon M. 1998 *Private communication*

Lindegren L. 1988, A correlation study of simulated Hipparcos astrometry in *Scientific aspects of the Input Catalogue Preparation II* (eds J. Torra & C. Turon) pp. 179–188. CIRIT.

Lyngå G. 1987, Fifth catalogue of open cluster parameters *CDS, Strasbourg*

Mendoza E.E. 1967 Multicolor photometry of stellar aggregates *Bol. Tonantzintla Tacubaya* **4**, 149–196.

Mermilliod J.-C. 1981 Comparative studies of young open clusters. III. Empirical isochronous curves and the Zero Age Main Sequence *Astron. Astrophys.* **97**, 235–244.

Mermilliod J.-C., Turon C., Robichon N., Arenou F., Lebreton Y. 1997 The distance of the Pleiades and nearby clusters *ESA-SP* **402**, 643–650.

Perryman M.A.C., Brown A.G.A., Lebreton Y., Gómez A., Turon C., Cayrel G., Mermilliod J.-C., Robichon N., Kovalevsky J., Crifo F. 1998 The Hyades: distance, structure, dynamics, and age *Astron. Astrophys.* **331**, 81–120.

Piatti A.E., Clariá J.J., Minniti D. 1993, A revised abundance calibration for population I red giants *J. Astron. Astrophys.* **14**, 145–165.

Pinsonneault M.H., Stauffer J., Soderblom D.R., King J.R., Hanson R.B. 1998 The problem of HIPPARCOS distances to open clusters: I. Constraints from multicolor main sequence fitting *Astron. J.* **504**, 170–191.

Robichon N., Arenou F., Turon C., Mermilliod J.-C., Y. Lebreton 1997 Analysis of seven nearby open clusters using HIPPARCOS data *ESA-SP* **402**, 567–570.

ROBICHON N., ARENOU F., MERMILLIOD J.-C., TURON C. 1998 *in preparation*

SCHALLER G., SCHERER D., MEYNET G., MAEDER A. 1992 New grids of stellar models from 0.8 to 120 $M_\odot$ at Z = 0.020 and Z = 0.001 *Astron. Astrophys. Suppl. Ser.* **96**, 269–331.

SCHMIDT-KALER TH. 1982, *in Landolt & Börnstein* **2b**, 19

STROBEL A. 1991 Metallicities of open clusters *Astron. Nachr.* **312**, 177–196.

VANDENBERG D.A., POLL H.E. 1989 On the precise ZAMSs, the solar color, and pre-main-sequence lithium depletion *Astron. Astrophys.* **98**, 1451–1471.

VAN LEEUWEN F., EVANS D.W. 1998 On the use of HIPPARCOS intermediate astrometric data *Astron. Astrophys. Suppl. Ser.* **130**, 157–172.

Proper Motions of Very Low Mass Stars and Brown Dwarfs in Open Clusters

By NIGEL C. HAMBLY

Institute for Astronomy of the University of Edinburgh, Royal Observatory, Blackford Hill, Edinburgh, EH9 3HJ, UK

Open clusters are a rich source of very low mass stars and brown dwarfs of a single known metallicity, age, and distance. Proper motion surveys enable candidate members within these clusters to be identified with a reasonably high degree of confidence. The nearby clusters are therefore a challenging test–bed for the latest evolutionary models of these ellusive objects. In this talk, I will review the progress that has been made recently in pushing proper motion surveys through very low mass ranges into the substellar régime, and I examine the prospects for extending these surveys to other clusters and to lower masses.

---

1. Introduction

Open clusters provide the astronomer with a rich source of objects for studying stellar structure over the full mass range of stable, hydrogen burning stars; furthermore, stellar evolution can be studied as the higher mass stars evolve away from, and as the low mass stars contract onto, the main sequence. Moreover, open cluster studies of objects that have too low a mass to stabilise on the hydrogen burning main sequence (i.e. brown dwarfs) have recently come of age, so now it is possible to study the physics of coeval objects having masses ranging over three orders of magnitude (and luminosities over eight orders of magnitude). Properties of very low mass (VLM) stars being studied in open clusters include lithium evolution, angular momentum evolution, spotting and variability, choronal activity, the binary fraction, and, most fundamentally, the mass function. For a general introduction to this broad subject, the reader is refered to the series of Cambridge Cool Star workshop proceedings (e.g. Pallavicini & Dupree 1996 and references therein); Rebolo, Martín & Zapatero–Osorio (1998) and Micela, Pallavicini & Sciortino (1997).

The advantages of studying stars in open clusters are numerous. Open clusters present a single stellar population of known distance, age and metallicity, and such parameters can in some cases be determined accurately and independently. For example, parallax measurements yield distances for the nearer clusters (e.g. see the article by Mermilliod and references therein in these prodeedings); spectroscopy in conjunction with model atmospheric analysis yields metallicities (e.g. Chaffe, Carbon & Strom 1971); and stellar evolutionary models yield ages via main sequence turn–on/turn–off (Straniero, Chieffi & Limongi 1997). Relatively small solid–angle surveys cover all (or a large fraction) of the area of the clusters, and there are no Malmquist–type biases in object selection. Unresolved binarity is observed photometrically (e.g. Stauffer 1984), and for young clusters the mass function is the *initial* mass function. Finally, if we make the assumption that all Galactic disk stars originate in open clusters, then open cluster studies are relevant to the general field population whereas inferring ensemble properties from field studies is of course not so straight–forward due to dynamical mixing of populations and a combination of unresolved binarity and Malmquist bias with photometric parallax.

Naturally, there are disadvantages. The youngest open clusters have large and highly variable extinction; dynamical evolution leads to the lowest mass stars being scattered into the largest volume of space (e.g. Hambly *et al.* 1995b) and source confusion is a problem at the low latitudes presented by many clusters from our position within the Galactic disk.

Jones (1997) reviews cluster membership determinations from proper motion surveys. Because the rich, nearby open clusters like the Pleiades, Hyades and Praesepe show peculiar motions with respect to the general field populations, proper motions offer a way of detecting members — for typical proper motion vector–point diagrams, see e.g. Hambly *et al.* (1995a) for Praesepe and Hambly *et al.* (1993 – HHJ) for the Pleiades. The combination of photometric selection and proper motions yields candidate cluster members with high membership probability. Since old disk, nearby M–type dwarfs have generally large space motions, proper motion selection helps to weed out such contaminants which can easily be confused with cluster members on the basis of photometry alone. Finally, the technique of membership probabilities, in the modern implementation due to Sanders (1971), offers a statistical method for the accurate determination of luminosity and mass functions — see Jones (1997) and references therein.

Once again, there are disadvantages to the use of proper motions. For example, the reflex solar motion of general field stars in the direction of the Pleiades scatters them into the quadrant of the proper motion vector–point diagram occupied by cluster stars (e.g. Jones 1973) — in such a circumstance, additional membership criteria (e.g. radial velocities and spectral diagnostics like Hα equivalent width) can provide the necessary member/non–member discrimination. Also, beyond the peak in the cluster luminosity function the ratio of members to non–members falls dramatically while at the same time astrometric errors inevitably rise as centroiding errors increase due to object faintness (see later). In such circumstances, additional membership criteria become vital.

2. Astrometric errors as a funcion of magnitude

It is instructive to examine astrometric errors as a function of magnitude for objects imaged using two dimensional detectors (e.g. CCDs or digitised photographic plates). Using maximum information techniuqes, King (1983) and Irwin (1985) showed that provided the pixel size is negligible compared to the scale size of an unresolved image (i.e. provided the image profile is oversampled), the relative centroid error in either co-ordinate, in units of the scale size of the image, is to a good approximation equal to the relative error in the intensity:

$$\frac{\sigma_x}{a} \approx \frac{\sigma_I}{I} \tag{2.1}$$

where σ_x is the error in the centroid in either co-ordinate; a is the scale size of the image; I and σ_I are the intensity and intensity error in some arbitrary unit. Of course, in most imaging cameras the pixel scale is generally set such that the sampling is close to (but not under) the critical case to maximise areal coverage. Irwin (1985) discussed the effects of finite pixel size, and for Gaussian profiles found that centroids will be less accurate by a factor $(1 + p^2/6a^2)$ where p is the pixel size in the same units as a. For example, for $p \sim a$ (corresponding to critical sampling) the errors are only $\sim 20\%$ worse.

Figure 1 shows an example from an imaging survey of the Pleiades using I–band Schmidt plates measured on the SuperCOSMOS microdensitometer at Edinburgh (see later). The diagram shows the centroiding errors in one co-ordinate as a function of magnitude for two plates (of different depth) with respect to the mean positions defined by a set of 8 plates. The SuperCOSMOS image detection software uses an isophotal analysis as oppose to a full profile fitting technique (which corresponds to the maximum information analysis); however Irwin (1985) has shown that such a difference should yield centroiding errors only around 10% worse than the optimum case. Here, the pixel data

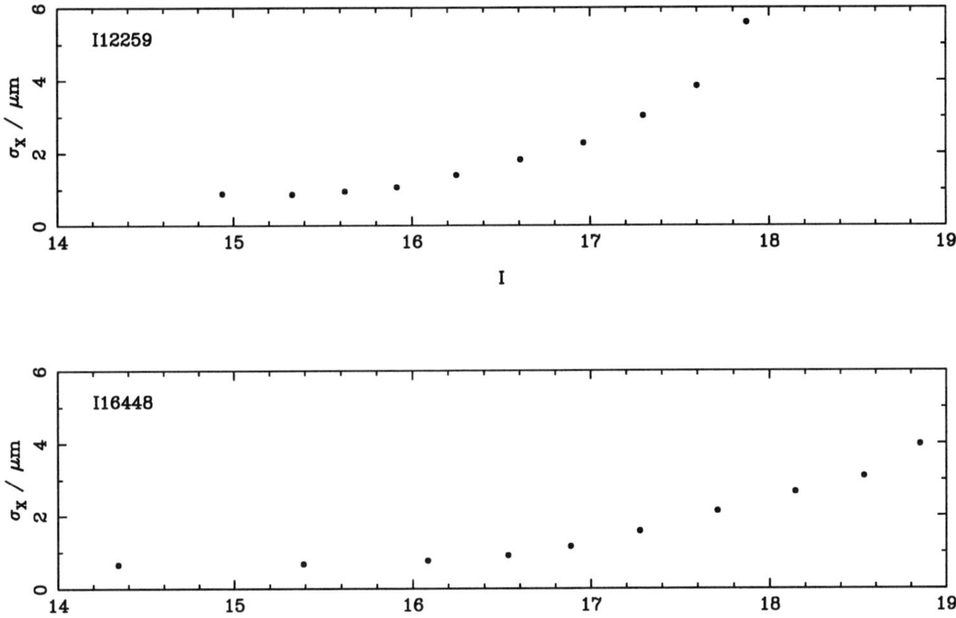

FIGURE 1. Astrometric errors, as a function of magnitude, for two I–band Schmidt plates measured with SuperCOSMOS (note that 1μm=67mas).

were thesholded above sky at 3σ (i.e. the signal–to–noise ratio is 3:1 per pixel) and a minimum of 6 connected pixels was required to define an image. The seeing on the plates is around 4 arcsec for the less sensitive plate (I12259) and around 2 arcsec for the better plate while the pixel size is 10μm or 0.67 arcsec at the plate scale of the 1.2 m UK Schmidt Telescope, so the images are oversampled by factors of ~ 1.5 to ~ 3. Bearing in mind the correction factors necessary to the idealised, maximum information result, the real measurements compare favourably with the theoretical predictions for objects at the plate limits.

As another example, consider the newly commissioned Wide Field Camera on the Isaac Newton Telescope at the Observatorio del Roque de los Muchachos on the Island of La Palma (`http://www.ast.cam.ac.uk/ING/WFCsur/`) This instrument has a pixel scale of 0.37 arcsec. A 600 sec exposure in 1 arsec seeing yields 10σ detections at I ~ 23.0 and Z ~ 21.2, for which the theoretical analysis yields $\sigma_x \sim 40$ milliarcsec. In 5 yr, $\sigma_\mu \sim 16$ milliarcsec yr^{-1}, noting the factor of 2 from $\sqrt{2}$ (2 independent measurements at 2 epochs) and $\sqrt{2}$ (2 independent co-ordinates for the total proper motion). Thus in 5 yr, a 3σ measurement of the Pleiades proper motion is attainable. It is interesting to note that PIZ1 (Cossburn *et al.* 1997) and Roque 4 (Zapatero–Osorio *et al.* 1997c) are $\sim 10\times$ brighter; i.e. there is an opportunity to get proper motions for Pleiads $10\times$ fainter than the currently faintest known members (which have $m \sim 0.05\mathrm{M}_\odot$).

3. Surveys for very low mass stars in nearby open clusters

Table 1 details some parameters for a selection of nearby open clusters and star formation regions from Allen (1973). It is not intended to be complete, nor is it representative of the most up–to–date measurements for some quantities (see, for example, the latest

TABLE 1. Some nearby OCs/star formation associations (Allen 1973)

Name	RA (1950) DEC	μ (mas yr^{-1})	π (mas)	age (10^8 yr)
Taurus–Auriga	04h26 +24	22	7	0.1
IC2602	10h44 −64	20	6	0.2
Alpha–Per	03h22 +49	30	6	0.8
Pleiades	03h44 +24	45	8	1
Ursa–Major	12h23 +57	> 100?	50	2
Praesepe	08h36 +20	40	6	4
Coma–Berenices	12h20 +26	100	13	5
Hyades	04h16 +15	100	22	6

Hipparcos results reported by Mermilliod elsewhere in these proceedings). Nonetheless, the point to stress is that, in general, proper motion measurements of cluster members are easier to make than parallax measurements, and that 2 epoch image data for the proper motion determinations are available *now* in the form of the whole–sky Schmidt survey plate collections.

For many clusters, exisiting proper motion surveys for members reach V $\sim$ 16 — for example, pioneering surveys done in the Pleiades, Praesepe, Orion and Taurus by Jones and co-workers — see Jones (1997) and references therein. Morgan *et al.* (1992) review the available Schmidt plate surveys in the northern and southern hemispheres. Glass copies of the survey originals are being distributed to many institutions, and the second epoch surveys are nearing completion. Several microdensitometers are currently engaged in digitising various subsets of these plates. Table 2 gives details and references, and compares the accuracies of the object parameters produced. The R and I–band Schmidt plates reach R $\sim$ 21 and I $\sim$ 18, and are therefore ideally suited to studying the very low mass main sequence of nearby clusters, as demonstrated by Leggett & Hawkins (1988) in the Hyades; HHJ in the Pleiades and Hambly *et al.* (1995a) in Praesepe. Centroiding accuracies of 1 to 2 μm coupled with epoch differences of 10 to 40 yr potentially yield proper motions at accuracies of order milliarcseconds per year. Figure 2 shows a mosaic of Schmidt survey fields from the first and second epoch Palomar sky surveys in the region of the Pleiades. Solid lines and three–figure numbers are the second epoch fields; dotted lines and four–figure numbers are first epoch. The dashed circles have radii 2 to 10° centred on the nominal cluster centre. The survey of HHJ used fields 0441 and 0031, but it is worth noting that cluster members will presumably exist out to the tidal radius which is currently estimated to be $\sim$ 6° (Pinfield 1998; Raboud & Mermilliod 1998). Clearly, multi–field studies using the survey plate collections will enable complete wide–field surveys of the Pleiades membership, and similar arguments hold for other clusters such as the Hyades, Praesepe and Alpha–Per. Note that the small overlap of the Palomar first epoch survey necessitates a machine that can accurately scan the full area of a 14 inch Schmidt plate. This was a problem for HHJ where the older COSMOS machine was employed. Note also that the ESO–R first epoch in the southern hemisphere has a similarly small overlap.

3.1. *SuperCOSMOS*

In terms of astrometric precision, SuperCOSMOS is the only machine with a demonstrated ability to centroid at accuracies of order 0.5μm (or 33 milliarcsec on 6×6° Schmidt plates). The SuperCOSMOS machine, which is described in Hambly *et al.* (1998) and

TABLE 2. Microdensitometers scanning the sky survey Schmidt plate collections. Accuracies quoted are for stellar images between 3 and 4 magnitudes above the plate limits.

Machine	**Location**	σ_x (arcsec)		σ_m (mag)		σ_μ (mas yr^{-1})
		Absolute	**Relative**	**Absolute**	**Relative**	
APM	Cambridge	0.30	0.08	0.3	0.07	~ 7
APS	Minnesota	0.50	0.20	0.2	—	~ 20
GAMMA	STScI	0.50	0.15	0.2	—	~ 4
PMM	Flagstaff	0.25	0.15	0.3	0.20	—
SuperCOSMOS	Edinburgh	0.20	0.03	0.3	0.07	~ 3

References:
McLean *et al.* (1997); Pennington *et al.* (1993); Hambly *et al.* (1998); Evans & Irwin (1995).

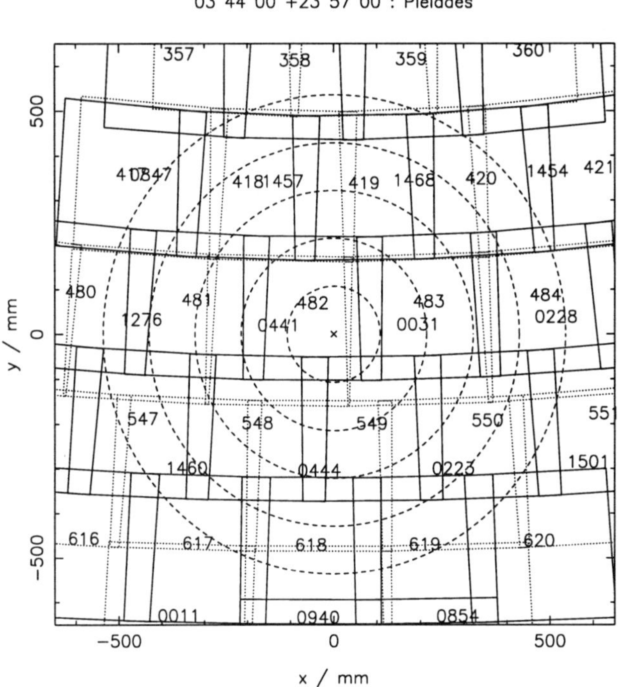

FIGURE 2. Mosaic of Palomar Sky Survey fields around the Pleiades cluster. Dashed circles have radii 2 to 10°.

references therein, is the latest in a line of microdensitometers designed at Edinburgh. The design philosophy of the machine was that the accuracy of the data produced should be limited, as far as possible, only by the plate being scanned. The machine is based around a Tesa–Leitz PMM 654 volumetric precision metrology unit. Motion in xy enables scanning of the plate while z motion allows for focussing on the plate over a 5 mm range (plates typically tilt and/or sag by up to $\sim$ 0.5 mm in the plate holder). Highly stable (both thermally and mechanically) granite blocks on air bearings provide for smooth motion in all co-ordinates. The performance specification includes a high scanning speed of 2.5 hr per plate; high scanning resolution of 10μm at 15–bit grey levels; high dynamic

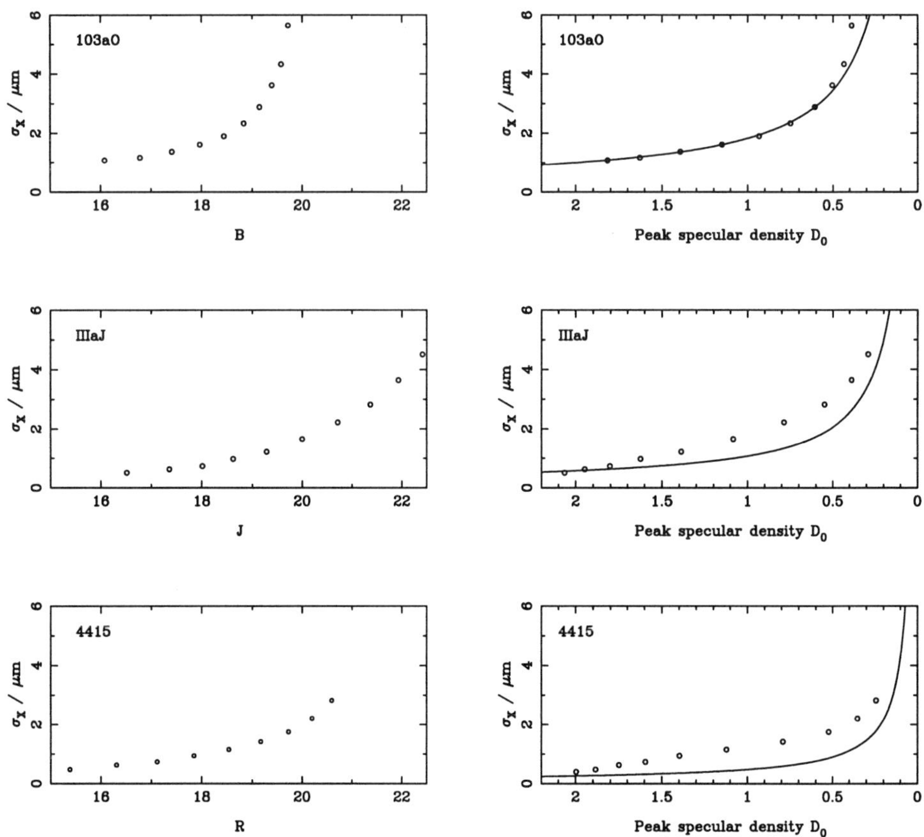

FIGURE 3. Plots of the RMS centroiding errors for plate pairs of various survey emulsions (see text for further explanation).

range which is diffraction limited in the imaging optics; and high positional accuracy with repeatability $< 0.1\mu$m in both xy co-ordinates. The principal components making this specification possible are, in addition to the granite air–bearing table, a linear CCD detector of 2048 pixels, and environmental isolation using a Class 100 clean room, temperature stable at $20.0 \pm 0.1°$C, the whole machine being supported on shock–absorbing feet.

Figure 3 shows the results of some astrometric tests with the system (for more details see Hambly *et al.* 1998). For the purposes of testing relative astrometry (fundamental to the measurement of proper motion) plate pairs of three emulsion types were chosen to elucidate the RMS error of a single measurement of stellar images as a function of magnitude. The left–hand panels of Figure 3 illustrate the measurement errors (open circles) as a function of magnitude (note that photographic 'J' is broadly equivalent to the Johnson B band), while the right–hand panels indicate the measurements as a function of photographic density to facilitate comparison with the theoretically optimum precision available for the particular emulsion using the analysis of Lee & van Altena (1983). These calculations are illustrated by the solid lines in the Figure, and the treatment is analogous to the generalised maximum information analysis of King (1983) and Irwin (1985). From the Figure, it can be seen that SuperCOSMOS can indeed centroid well exposed stars (i.e. those at 3 to 4 magnitudes above the plate limit) at the level of $\sim 0.5\mu$m for the modern, fine–grained emulsions.

4. A proper motion survey for Pleiades brown dwarfs

In Hambly *et al.* (1999) we report an I–band Schmidt plate proper motion survey for brown dwarfs, which extends the less sensitive R–band survey of HHJ. Several groups (eg. Zapatero–Osorio et al. 1997b; Cossburn et al. 1997; Stauffer et al. 1998) are using deep imaging with sensitive photoelectric detectors to probe the Pleiades membership at masses m< 0.05M$_\odot$, over areas of order one square degree. Here we took the different approach of employing less sensitive I band photographic Schmidt plates to survey the whole cluster, ie. several tens of square degrees. By pushing these plates to their limits, we have still reached masses just below the hydrogen burning threshold. We discovered a new sample of BD candidates having magnitudes in the range $17.8 < I < 18.3$, ie. as faint or fainter than the confirmed lithium BD member PPL 15. These stars all have proper motions, measured from positions on seven I–band plates spread in epoch over seven years, consistent with Pleiades membership and are extremely red, either from their measured (R–I) colours or by virute of the fact that they fail to appear in the R–band at all. The BD PPL 15, which previously had no proper motion measurement, appears in this list. Since the presence of lithium is the best indicator of substellar nature for such objects (eg. Rebolo et al. 1996), one ideally requires spectra of sufficient resolution and signal–to–noise ratio to measure the 6708Å lithium feature in order to confirm BD status and cluster membership for the candidates presented here. However, it has been shown in the past that infrared photometry is efficient at weeding out non–member contamination from candidate lists via the (I–K) temperature index (eg. Steele, Jameson & Hambly 1993; Zapatero–Osorio, Martín & Rebolo 1997a). Such data were obtained for these candidates, and we produced a mass function for the cluster that, for the first time, samples a mass bin within the substellar régime to an easily quantifiable degree of completeness.

4.1. *Results*

We selected all objects having $I > 12.5$ and $I < 2.875(R–I)+11.875$, ie. a conservative cut in the I, (R–I) colour–magnitude diagram to include all possible members, while at the same time minimising the non–member contamination. All images were required to be present on at least six I plates; images not present in the R stack (which reaches R ~ 21) were included as possible members with assumed large (R–I). Figure 1 illustrates the measured astrometric errors as a function of I magnitude for the best plate (I16448) and the worst plate (I12259). These curves show the characteristic form expected for centroiding accuracies of images in a photographic emulsion — see, for example, Hambly et al. (1998) and references therein. Because these errors increase dramatically towards the plate limit, the proper motion errors are correspondingly large in the magnitude range of most interest. We applied the Sanders (1971) membership probability technique, in an implementation described in Hambly et al. (1995a) and references therein, for three magnitude ranges: $12.5 < I_1 < 13.5 < I_2 < 14.5 < I_3 < 15.5$. Summing membership probabilities for all stars in these ranges enabled direct comparison with the survey of HHJ, which is known to be incomplete.

In Figure 4(a) we show the PMVPD for a sample of stars having $17.8 < I < 18.3$ (the bright limit is ~ 0.2 mag brighter than PPL 15 while the faint limit is the sensitivity limit of this survey). Typical astrometric error bars are shown — clearly, the errors are too large and the density of members too small for the Sanders fitting technique to be

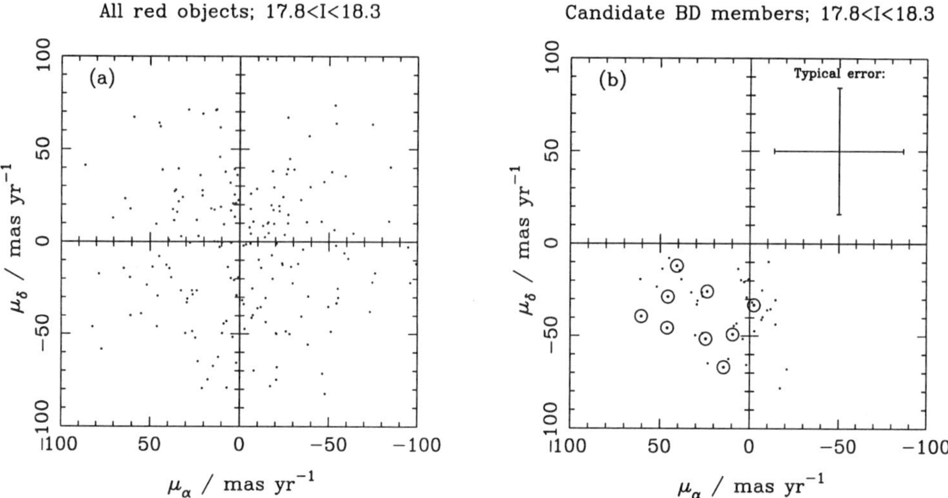

FIGURE 4. Proper motion vector–point diagrams, (a) for all red stars in the magnitude range $17.8 < \mathrm{I} < 18.3$, and (b) for all stars having pms within 1σ of the known cluster motion (a typical error bar is shown). In (b), the 9 candidate brown dwarfs finally selected via infrared photometry are circled.

stable. To make a selection of stars for photometric follow–up, we chose all objects having proper motions within 1σ of the known cluster motion $(\mu_\alpha, \mu_\delta) \equiv (+19, -43)$ mas yr^{-1} (eg. Jones 1973). This selection of 45 possible members is plotted in the PMVPD in Figure 4(b). In order to determine which objects in this list of 45 stars were good BD candidates, K band photometry was obtained at UKIRT. This yielded nine objects as high probability members from (I – K) colour; these 9 were subsequently measured for accurate photoelectric I photometry on the INT.

To provide a conversion of luminosity into mass, we used the data presented in Hambly et al. (1995b) to construct a relationship between the I–band bolometric correction, $\mathrm{BC_I}$, and $\log(T_{\mathrm{eff}})$. A cubic least–squares polynomial fit yields the following:

$$\mathrm{BC_I} = -93.8507 + 0.815193 \log_{10}(T_{\mathrm{eff}}) + 20.6998[\log(T_{\mathrm{eff}})]^2 - 3.78681[\log(T_{\mathrm{eff}})]^3.$$

Using the models of D'Antona & Mazzitelli (1994, Tables 3 and 5 – DM94) for masses $m > 0.8\mathrm{M_\odot}$ with very low mass stellar models from the Lyon group: Chabrier & Baraffe (1997) and an $\mathrm{M_I}$–mass relationship for BDs from Chabrier, Baraffe & Plez (1996) and Baraffe et al. (1998) – hereafter CBP – we constructed a mass–luminosity relationship. We found that our temperature scale agreed well with that resulting from the model atmosphere calculations of CBP and also that the higher mass ($m > 0.6\mathrm{M_\odot}$) models of DM94 agree well with those of Chabrier & Baraffe (1997); however the gray atmosphere models of DM94 were assumed to be out of date (eg. Stauffer et al. 1998) since a non–gray model atmosphere treatment of the boundary condition has been shown to be more accurate (eg. CBP) for very low mass stars and brown dwarfs. We therefore used the Lyon group models for masses below $0.8\mathrm{M_\odot}$, and DM94 with our own temperature scale above this value. We assumed an age for the Pleiades of 100 Myr (eg. Basri et al. 1996), a distance modulus of $(\mathrm{m}-\mathrm{M})_0=5.53$ (eg. Basri et al. 1996; Zapatero–Osorio et al. 1997c; Stauffer et al. 1998), an extinction $\mathrm{A_I}=0.07$ mag (Zapatero–Osorio et al. 1997c) and $\mathrm{M_{bol,\odot}} = 4.75$

(Allen 1973). Interpolating cubic polynomials were used to transform from given values of m_I to mass.

4.2. *Discussion*

In order to correct the number of stars in the lowest mass bin in this survey (ie. the mass bin containing the brown dwarf candidates) for completeness, two factors were employed. The first was derived from the proper motion selection. We chose objects having proper motions consistent with the cluster motion at the level of 1σ (ie. 68% confidence); assuming independent normally distributed errors in the two dimensions of the PMVPD we therefore multiplied the number of objects by a factor $1/0.68^2$. In addition, the survey is not 100% complete down to the magnitude limit I = 18.3. Comparing a number–magnitude histogram for all stars when paired with a catalogue of objects from a deep I stack of all available I–band plates, which is 100% complete down to $I > 19$ and fitting the log–linear part of the data were with a straight line yielded a total photometric completeness of 65% in the magnitude range $17.8 < I < 18.3$. The total correction factor for the lowest mass bin was therefore $1/(0.68^2 \times 0.65)$.

In Adams & Fatuzzo (1996), arguments were presented for the IMF approximating a normal distribution in the $\mathrm{d}N/\mathrm{d}m - \log m$ plane. This is equivalent to a quadratic form in the $\log \mathrm{d}N/\mathrm{d}m - \log m$ plane (the so called 'log–normal' form, eg. Miller & Scalo 1979). In Figure 5 we show our data in conjunction with higher mass points. The latter were derived using the Prosser & Stauffer compilation of Pleiades members (Prosser 1997), which lists objects from many sources and which we have therefore assumed to be largely complete. Mass–M_V relations as described in Pinfield (1998) were used to convert from V magnitudes to mass, again assuming $(m{-}M)_0{=}5.53$ and $A_V{=}0.12$ (eg. Stauffer et al. 1998). The data were fitted, using weighted linear least–squares, with a quadratic polynomial of the form

$$\log_{10} \xi\{m\} = A_0 + A_1(\log m) + A_2(\log m)^2$$

(eg. Miller & Scalo 1979). The data are clearly reasonably well represented, *in this mass range*, using this log–normal form. The coefficients of the fit are $A_0 = 2.3408$, $A_1 = -2.3134$ and $A_2 = -1.191$. For comparison with a power–law of the form $\xi\{m\} = \mathrm{d}N/\mathrm{d}m = km^{-\alpha}$, the gradient of the quadratic at any point is given by $-\alpha = A_1 + 2A_2 \log m$. So, at $m = 0.1\mathrm{M}_\odot$, $\alpha = +0.1$ and at $m = 1.0\mathrm{M}_\odot$, $\alpha = +2.3$, which compares favourably with the Salpeter (1955) value. At high masses this fit yields a gradient considerably steeper than Salpeter: at $m = 10\mathrm{M}_\odot$, $\alpha = +4.7$. It is possible that at such high masses the IMF is better represented as a single power–law having an index close to the Salpeter value, and the log–normal is only appropriate around the peak in the distribution (see, for example, the article by Adams elsewhere in these proceedings).

Needless to say, it is dangerous to extrapolate the symmetrical log–normal form below the lowest mass bin having data available here. There is some evidence that there is a local minimum in the IMF at $m \sim 0.1\mathrm{M}_\odot$, and that further into the BD régime the number of stars per unit mass begins to rise again (eg. Martín et al. 1998). However, such surveys only cover the cluster centre and are probably not representative of the cluster as a whole. Our data, by constrast, covers most of the effective area of the cluster. In addition, there is uncertainty in the distance and age of the cluster. For example, van Leeuwen & Hansen–Ruiz (1998) find a distance modulus of $(\mathrm{m} - \mathrm{M})_0 = 5.32$ from *Hipparcos* data, while Basri et al. (1996) discuss the age problem. Such uncertainties will have an effect on the shape of the derived mass–luminosity (ML) relationship, and for masses $m < 0.1\mathrm{M}_\odot$ the IMF is critically dependent on the gradient of the ML function; hence any turn–over seen at these masses should be treated with caution.

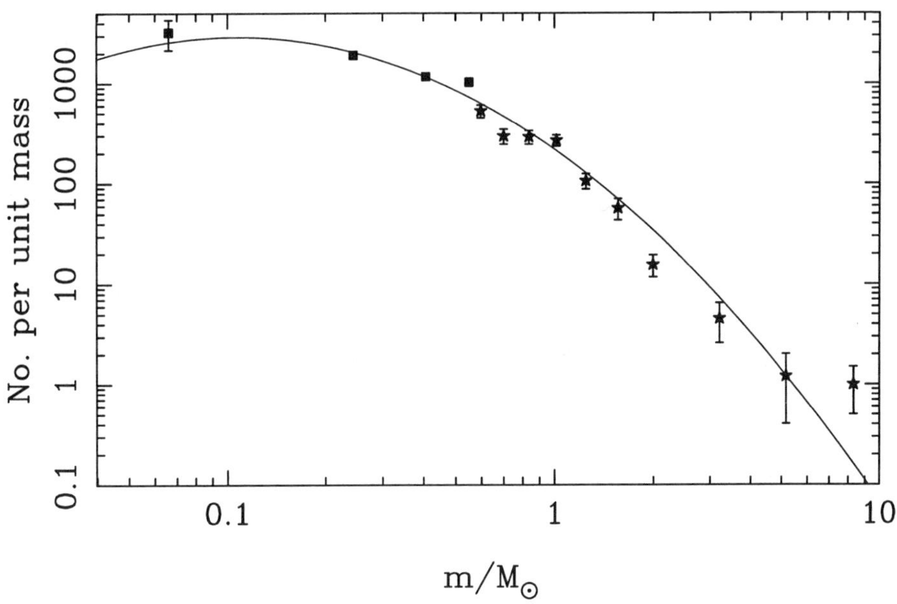

FIGURE 5. IMF for the Pleiades of the mass range $0.07 < m/\mathrm{M}_\odot < 10$. The solid squares are the data from this survey; the star symbols are from higher mass data as described in the text. The solid line is a quadratic fit to the data (see text).

5. Conclusions

The main conclusions of this presentation are as follows:

• the 2 epoch whole sky–survey Schmidt plate material provides proper motion information for all the nearby open clusters,

• a new proper motion survey of the Pleiades has identified nine candidate BDs, seven of which are new discoveries while one of the two previously discovered is the lithium BD PPL 15;

• the IMF of the Pleiades appears to be flat across the stellar/substellar boundary;

• the IMF in the mass range $0.07 < m/\mathrm{M}_\odot < 10$ is reasonably well represented by a log–normal function.

Finally, new digital sky surveys are now underway that will enable proper motion surveys for brown dwarfs in many open clusters (e.g. INT–WFC; SLOAN).

It is once again a pleasure to acknowledge the invitation of the SOC to present this review, and I am grateful to the LOC for organising a hugely enjoyable meeting at Los Cancajos, La Palma.

REFERENCES

Adams F.C., Fatuzzo M., 1996, ApJ, 464, 256

Allen C.W., 1973, Astrophysical Quantities, The Athlone Press, London

Baraffe I., Chabrier G., Allard F., Hauschildt P.H., 1998, A&A, in press

Basri G., Marcy G.W., Graham J.R., 1996, ApJ, 458, 600

Chabrier G., Baraffe I., Plez B., 1996, ApJ, 459, L91

Chabrier G., Baraffe I., 1997, A&A, 327, 1039

Chaffe F.H., Carbon D.F, Strom S.E., ApJ, 166, 593

Cossburn M.R., Hodgkin S.T., Jameson R.F., Pinfield D.J., 1997, MNRAS, 288, L23

D'Antona F., Mazzitelli I., 1994, ApJS, 90, 467 (DM94)

Evans D.W., Irwin, M.J., 1995, MNRAS, 277, 820

Hambly N.C., Hawkins M.R.S., Jameson R.F., 1993, A&AS, 100, 607 (HHJ)

Hambly N.C., Steele I.A., Hawkins M.R.S., Jameson R.F., 1995a, A&AS, 109, 29

Hambly N.C., Steele I.A., Hawkins M.R.S., Jameson R.F., 1995b, MNRAS, 273, 505

Hambly N.C., Miller L., MacGillivray H.T., Herd J.T., Cormack W.A., 1998, MNRAS, 298, 897

Hambly N.C., Hodgkin S.T., Cossburn M.R., Jameson R.F., 1999, MNRAS, in press

Irwin M.J., 1985, MNRAS, 214, 575

Jones B.F., 1973, A&AS, 9, 313

Jones B.F., 1997, In: Micela *et al.* (1997), 833

King I.R., 1983, PASP, 95, 163

Lee J.F., van Altena W., 1983, AJ, 88, 1683

Leggett S.K., Hawkins M.R.S., 1988, MNRAS, 234, 1065

Martín E.L., Zapatero–Osorio M.–R., Rebolo R., 1998, In: Rebolo, Martín & Zapatero–Osorio (1998), p. 507

McLean B.J., Hawkins G., Spagna A., Lattanzi M., Lasker B., Jenkner H., White R., In: New Horizons from Multi–Wavelength Sky Surveys, B.J. McLean *et al.* (eds.), IAU (Kluwer:Dordrecht), 431

Micela G., Pallavicini R., Sciortino S., 1997, Cool stars in CLusters and Associations, Journal of the Italian Astronomical Society, volume 68, number 4

Morgan D.H., Tritton S.B., Savage A., Hartley M., Cannon R.D., 1992, In: Digitised Optical Sky Surveys, H.T. MacGillivray & E.B. Thomson (eds.), Astrophysics and Space Science Library Proceedings volume 174 (Kluwer:Dordrecht), 11

Miller G.E., Scalo J.M., 1979, ApJS, 41, 513

Pallavicini R., Dupree A.K., 1996, Proceedings of the Ninth Cambridge Workshop on Cool Stars, Stellar Systems and the Sun, ASP Conf. Ser., volume 109

Pennington R.L., Humphreys R.M., Odewahn S.C., Zumach W., Thurmes P.M., 1993, PASP, 105, 521

Pinfield D.J., 1998, PhD Thesis, University of Leicester

Prosser C.F., 1997, private communication

Raboud D., Mermilliod J.C., 1998, A&A, 329, 101

Rebolo R., Martín E.L., Basri G., Marcy G.W., Zapatero–Osorio M.–R., 1996, ApJ, 469, L53

Rebolo R., Martín E.L., Zapatero–Osorio M.–R. (eds.), 1998, 'Brown Dwarfs and Extrasolar Planets', ASP Conf. Ser., volume 134

Salpeter E.E., 1955, ApJ, 121, 161

Sanders W.L., 1971, A&A, 14, 226

Staniero O., Chieffi A., Limongi M., In: Micela *et al.* (1997), 797

Stauffer J.R., 1984, ApJ, 280, 189

Stauffer J.R., Schild R., Barrado–y–Navascués D., Backman D.E., Angelova A., Kirkpatrick D., Hambly N.C., Vanzi L., 1998, ApJ, 504, 805

Steele I.A., Jameson R.F., Hambly N.C., 1993, MNRAS, 263, 647

van Leeuwen F., Hansen–Ruiz C.S., 1998, In: Proceedings of the Hipparcos Symposium, ESASP 402 (Garching:ESA), in press

Zapatero–Osorio M.–R., Martín E.L., Rebolo R., 1997a, A&A, 317, 164

Zapatero–Osorio M.–R., Martín E.L., Rebolo R., 1997b, A&A, 323, 105

Zapatero–Osorio M.–R., Rebolo R., Martín E.L., Basri G., Magazzù A., Hodgkin S.T., Jameson R.F., & Cossburn M.R., 1997c, ApJ, 491, L81

Parallaxes for Brown Dwarfs in Clusters

By C. G. TINNEY

Anglo-Australian Observatory, P.O.Box 296, Epping. 2121. Australia

The prospects for direct parallax measurements of brown dwarfs and very low-mass stars in stellar clusters are bleak indeed. However, significant progress in parallaxes for nearby faint dwarfs, and for brighter stars in clusters can be expected in the next few years. The current state of play for ground- and space-based parallaxes is reviewed, along with the prospects for the future and the scientific questions we hope such observations will address.

---

1. Introduction

There is a tendency in the wider astronomical community to view astrometrists in general, and those who measure parallaxes in particular, as musty old fuddy-duddies doing valuable and worthy, if dull, work. This has always seemed to me a strange attitude, given that such observations set the foundations for almost all areas of astronomy. Personally, I find measuring parallaxes to be just about the most rewarding type of observation I've ever performed. There's a certain satisfaction to be had in measuring a fundamental quantity whose only model dependence is on Euclidean geometry - there aren't many other areas in astronomy where *that's* possible.

This general view of astrometry is particularly surprising in view of the phenomenal demand amongst the astronomical community for the rapid release of HIPPARCOS results. To my mind the constant discussion to be heard over observatory dinner tables world-wide asking "just when would the HIPPARCOS data go public?" reinforces the fact that such data is of fundamental importance to every field of astronomy – from the study of the faintest "non-quite stars", to the study of galaxy formation and the early universe. I can only hope that this review will go some way towards highlighting the dramatic and exciting advances which have been made in recent years in the measurement of trigonometric distances, and to point out the areas where these advances have, and can be expected to, increase our understanding of the faintest members of stellar clusters.

2. Science goals

The first thing to point out, however, is that the prospects for actually measuring parallaxes for brown dwarfs or very low-mass stars (VLMs) in clusters are almost non-existent in the near future. The nearest clusters are all more than 100 pc away, which means that meaningful parallaxes must achieve 1-σ limits of $\lesssim 0.2$ mas† to obtain distances good to better than 5%. Moreover, VLMs and brown dwarfs in the nearest clusters are faint, with magnitudes of V $\gtrsim$ 23, I $\gtrsim$ 19 or K $\gtrsim$ 15. While there are future prospects for parallaxes at these precisions, *or* at these magnitudes, there is scant possibility of doing both within the next 10 years.

Having said that, there is still a lot we can learn from parallaxes which will inform our understanding of the bottom of the main sequence in clusters.

(*a*) Although considerable progress has been made in recent years in our understanding of very cool dwarf atmospheres, $T_{\rm eff}$ remains at best poorly measurable quantity for these objects, and at worst a downright misleading and systematically biased one. Luminosities, therefore, remain the one quantity we can precisely and empirically determine

† Throughout "mas" is used for milli-arcseconds, and "μas" for micro-arcseconds.

for cool dwarfs – in the first instance by obtaining wide wavelength coverage spectra and integrating to bolometric luminosities, and in the second instance by using this data to define precise bolometric corrections.

(*b*) It therefore follows that parallax studies on the brown dwarfs and VLMs in the solar neighbourhood will be needed to define the photometric and spectroscopic properties of VLMs and brown dwarfs *as a function of luminosity.* These results will be an essential input to the transformation of cluster luminosity functions into cluster mass functions. And cluster mass functions in turn offer the only hope of accurately determining the initial mass function in the brown dwarf regime.

(*c*) The determination of spectroscopic properties as a function of luminosity is particularly important for the "L" class of dwarfs. These objects in particular provide observational constraints on an entirely new area of photospheric "phase space", and precise luminosity determinations for them will be vital to our development of an understanding of their properties.

(*d*) Lastly, parallaxes of brighter cluster members are need to determine unbiased distances, and to understand the three dimensional structures of clusters as a function of age and dynamical evolution. It is hard to overstate the importance of cluster distances to astronomy in general, impacting as they do on stellar evolution, Galactic structure, and as the foundation for the cosmic distance ladder.

3. Current ground-based parallaxes

For long the mainstay of ground-based astrometry, photographic parallax programs have now been made essentially obsolete by the introduction of CCD programs. For example, the long running USNO program (Harrington *et al.* 1993) has now been terminated in favour of their CCD astrometric program (Dahn 1998). This revolution in ground-based astrometry has largely followed from Monet & Dahn's early experiments with CCD astrometry on the KPNO 4-m (Monet & Dahn 1983), which was then extended to a fully fledged program at the USNO's Flagstaff station, and then further implemented by numerous astronomers on telescopes world wide.

The major advantage which CCDs provide for astrometry is not so much that they have high quantum efficiency, but that their extremely regular matrix of pixels allows centroiding to be essentially photon-counting limited. This means that even in 1-2" seeing, precisions of 1-10 mas can be achieved simply by counting 10^6 photons in a given stellar image. On the ground, of course, these limits can not be pushed indefinitely. Because the typical field sizes required to do astrometry are larger than the projected size of an isoplanatic patch, the image motions produced by seeing in different objects in a given field will be somewhat independent. And the larger the separation between objects the more independent these motions will be. The precision with which a ground-based position can be measured, then, is not just determined by the number of photons counted, but also by the residual shift induced by seeing motions about each object's "true" mean position. This differential seeing produces the fundamental unavoidable limit to the precision of ground based astrometric programs.† The effects of differential seeing scale roughly linearly with seeing, and inversely with the square-root of telescope aperture, as the one-third power of the angle over which positions are being measured (Han & Gatewood 1995), and inversely with the square-root of exposure time (Han 1989).

† I say "unavoidable" to distinguish its effects from those of differential colour refraction, which are also significant, but which can be dealt with by careful and appropriate observing strategies (Monet *et al.* 1992; Tinney 1996).

Table 1 summarizes CCD parallax programs currently in operation (for more detailed information on each program I refer the reader to the 1998 review by Conard Dahn). Of the programs currently operating, only the USNO CCD program is operating essentially full time on a dedicated telescope. Between 1983 and 1995 this program used an 800×800 Texas Instruments CCD. In 1995 they switched to a 2K×2K detector, in order to obtain a larger field of view, as they were finding that precision was being limited not by the atmosphere, but by the lack of reference stars in their small field of view. This new device means that almost every target can be observed with an adequate reference frame, and has permitted a significant improvement in precision.

The USNO has also pioneered an important new development in their "ND9" program. "ND9" refers to the use of a 9 magnitude neutral density spot deposited on an optical flat placed 1 mm in front of their detector. This allows them to overcome the only serious astrometric deficiency of CCD's, which is that their limited dynamic range makes it impossible to use them to observe targets much brighter than 12th magnitude. It is impossible to detect lots of photons from both a bright target and a faint reference frame in a single exposure, unless a magnitude compensation scheme like this is used. However, the ND9 program has produced very promising results in its first years of operation, and there is every reason to expect that in the years to come it can produce the same sub-0.5 mas precisions obtained for fainter stars, but on much brighter targets which are of important astrophysical significance – in particular Cepheids, Miras, RR Lyraes and bright cluster members.

The remaining ground-based programs are all being operated on a somewhat smaller scale, and have generally been undertaken to address particular target samples of interest. In most cases, these samples have included solar neighbourhood VLMs and brown dwarf candidates, since in general these targets are ideally suited to CCD parallaxes, given that they nearby, faint and red. These programs have demonstrated that parallaxes need not be considered "something you wait for someone else to do", since with a little care, such programs can be carried out at almost any observatory with common-user instrumentation.

4. Current space-based parallaxes

4.1. *Hubble Space Telescope*

When it was first launched, HST was widely promoted as a facility for obtaining high precision astrometry. I think it has to be said, however, that the results actually obtained have failed to live up to expectation. This is largely so because, of the the two major instruments which were seen as powerful astrometric facilities, only the Fine Guidance Sensors (FGS) have produced significant results. These instruments (two of which are required to allow HST to track, which permits a third to be used for astrometry) have achieved 1-2 mas precisions for stars with V=10-12 (eg. van Altena *et al.* 1997), and have done useful work.

The Wide Field Planetary Cameras (WFPC1 & WFPC2), however, have proved disappointing. These cameras suffer from significant astrometric distortion, and heavily undersampled images. This, together with the fact that it has proved to be difficult to arrange for astrometric observations to be scheduled in any consistent way, has meant that no parallaxes have been obtained with WFPC1 or WFPC2. As an example, scheduling observations to have identical roll angles at multiple epochs has proved impossible, which means the highly differential nature of CCD ground-based astrometry has proved impossible to realise in space. This is particularly frustrating, as it is clear that in the

TABLE 1. Ground-based Astrometry Programs.

Program	Stars	Nights/yr	Precision	Duration
USNO 1.5m - TI800	175	240	0.7-2 mas	1983-1995
USNO 1.5m - TEK 2K	200	240	0.3-0.5 mas	1992-
USNO 1.5m - ND9	68	240	1-2 mas in 1yr	1996-
U.Virginia/SSO 40"	130	84	1-2 mas for R<13	1985-
			2-5 mas for I<17	
U.Chile/CTIO 1.5m	80	16	1-2 mas for R<19	1985-
Torino 1.05m	140	180 twilights	2-4 mas (expected)	1995-
ESO/Danish 1.5m				
Hawkins *et al.*	30	20×3^h twilights	4 mas	2yr
Forveille *et al.*	20	6	3-5 mas (expected)	few years
Palomar 1.5m	25	20	4-5 mas for I<17.5	1990-1995
ESO/MPIA 2.2m	12	8	3-7 mas for I<18	1992-1993

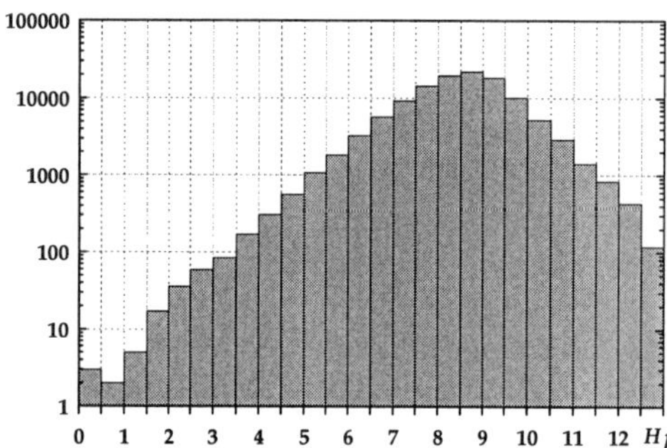

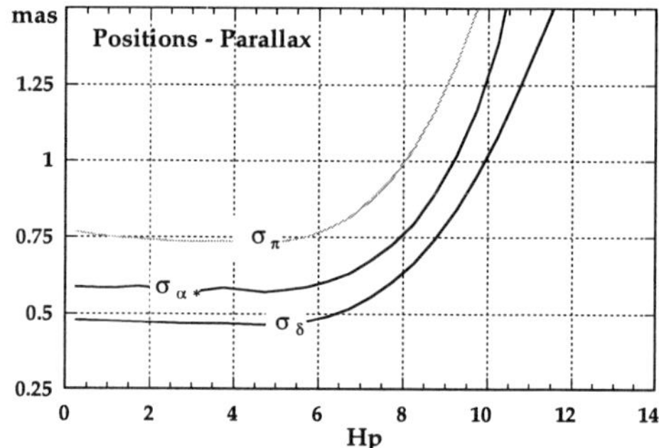

FIGURE 1. HIPPARCOS Main Mission magnitude distribution and precisions (Mignard 1997). H_p is the HIPPARCOS photometric magnitude, which is similar to V.

absence of the atmosphere, HST has the potential to achieve parallax precisions at level much below 0.1 mas.

4.2. *HIPPARCOS*

On a happier note, however, the HIPPARCOS mission has achieved stunning results in its lifetime, obtaining parallaxes as part of its "Main Mission" for 118218 stars at precisions of 1 mas for V<8, and 1.5 mas for V<10 (cf. Fig. 1). I'm sure that no-one who has seen the stunning colour-magnitude diagrams for the stars of the solar neighbourhood which this has produced, can be unaware of just what a phenomenal data base this represents for a variety of studies – and in particular for studies of the star formation history of the Galactic disk. As well as this database, the supplementary "Tycho" mission produced 25 mas astrometry for a further 1 million fainter stars.

5. Scientific results

While ground-based parallaxes have been ideal for studying faint, nearby targets, space-based parallaxes have had an important role in targeting bright, but more distant stars. As a result they have had an important impact on the question of the distances to star clusters.

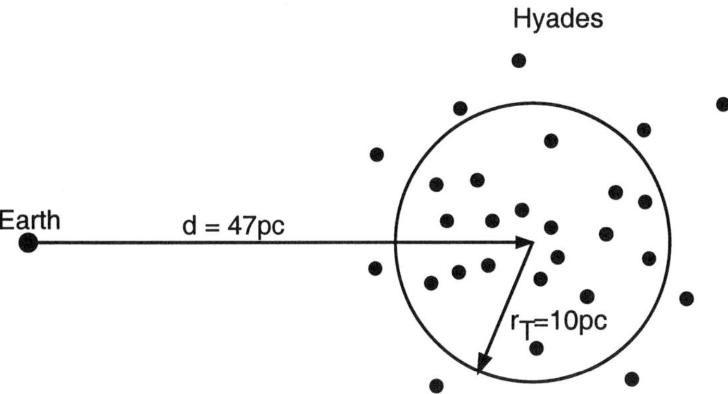

FIGURE 2. Schematic drawing of the Hyades.

5.1. *The Hyades*

The Hyades is an astrophysically important cluster, because its small distance means that actually measuring that distance is feasible. Unfortunately, this also means the radius of the cluster itself, is a significant fraction of distance to the cluster – as a result the distance to the Hyades is a somewhat fuzzy concept (cf. Fig. 2). The "mean" distance to the cluster that is measured, depends significantly on exactly which stars you choose. Any magnitude bias in your sample (say towards brighter magnitudes, which will preferentially pick objects closer to you) can heavily bias the mean distance you measure. In the case of the Hyades, therefore, distances are more useful for what they can tell you about individual stars and about the structure of the cluster, than what they tell us about the distance to the cluster as a whole (Perryman *et al.* 1998). Having said that, both HST/FGS (d=48.3±2.0 pc) and HIPPARCOS (46.3±0.3 pc) produce consistent Hyades distances (Perryman 1997; van Altena *et al.* 1997).

5.2. *The Pleiades*

The distance for the Pleiades is a better defined quantity, given that the ratio of the tidal radius to the distance is more like $\gtrsim 10$, instead of $\lesssim 5$ for the Hyades. Unfortunately, even with HIPPARCOS, measuring the distance to the Pleiades is a difficult proposition, since individual stars will have parallaxes of 7-9 mas, but measurement precisions of only 1-2 mas. In other words, each star gives only a 5-10-σ distance estimate.

The "accepted wisdom" pre-HIPPARCOS was that the Pleiades parallax was 7.7 mas, or 130 pc. However, the published distance to the Pleiades based on HIPPARCOS measures is 8.60±0.24 mas, or 116 pc. This places the Pleiades main sequence about 0.3 magnitudes too faint, and generally upsets the apple cart for models of stellar evolution within clusters, and relative cluster ages (cf. numerous papers in the HIPPARCOS "Venice 97" conference proceedings). There has been considerably discussion in the literature since the "Venice 97" meeting on the subject of this discrepancy.

One recurring theme is that the difference could be due to untreated biases of the Lutz-Kelker type in the HIPPARCOS data. Basically, these are caused by the fact that in a given absolute magnitude bin in sample of parallax targets, you'll tend to have more targets at greater true distances and smaller true absolute magnitudes. This produces a bias (which is a strong function of parallax uncertainty), which makes you under-estimate the mean distance. The detailed treatment of such a bias requires careful simulation for the particular sample in question. The "classical" Lutz-Kelker bias, for example, is appropriate for a uniform space distribution, and over-estimates the bias present in a

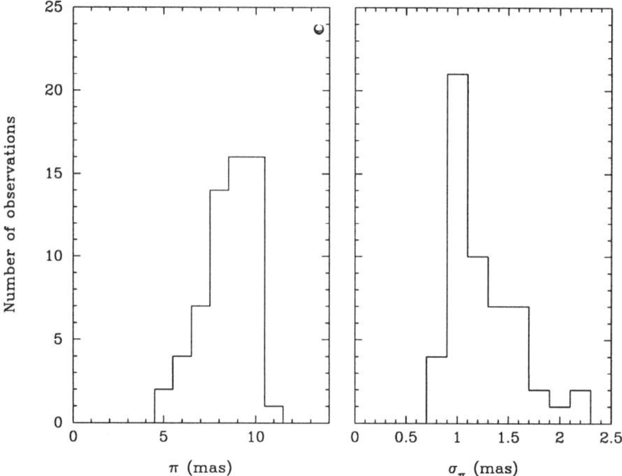

FIGURE 3. Parallax distributions for Pleiades stars (van Leeuwen & Hansen Ruiz 1997).

clustered sample. However, even this classical bias produces a correction of only 0.1 magnitudes to the Pleiades absolute magnitudes – not the 0.3 magnitude difference that stellar pundits would like to make go away.

It is worth bearing in mind that the very precise Pleiades distance quoted above, relies on determining a mean distance from a sample of much less precise measurements. In fact, it is known that because of systematic uncertainties these uncertainties do not beat down as $n^{-0.5}$, but more like $n^{-0.35}$ (Mermilliod *et al.* 1997). More worrying is the fact the magnitude distributions for these objects start out *extremely* non-Gaussian (cf. Fig. 3), which raises serious questions as to whether any averaging process can really produce a mean distance a factor of ten better than the individual parallaxes which make it up.

However, having said that it has to be emphasized that the HIPPARCOS parallaxes *are the best data we have to hand.* To argue that they must be in error because they don't fit with the pre-NIPPARCOS accepted wisdo' is truly putting the cart before the horse. The important lesson to take away here is that it is wise not to over-interpret the available data. It seems likely that the HIPPARCOS parallaxes for the Pleiades cannot produce a distance at the 0.2 mas level, simply because the raw parallaxes are not precise enough†. In the long run, this issue will be settled when 0.2 mas parallaxes for individual Pleiades cluster members are measured. In planning future parallax programs we should aim to measure parallaxes to the required precision for *each* cluster member, instead of relying on averaging to beat down the noise, if we want truly robust and unassailable distances. Thankfully, there is every prospect that the USNO ND9 program can perform just such measurements in the near future, and that future space-based astrometry missions will be able to do this for even more clusters.

6. Prospects for ground-based parallaxes

The prospects for ground-based astrometric programs to play a major role in cluster research for brown dwarfs and VLMs is particularly rosy. Over the next 2-5 years the

† Unfortunately, despite the incredible resource HIPPARCOS represents, this situation occurs in other areas of interest. To quote Trimble & McFadden (1998) "The measured values [for the proper motion of the LMC] differ from zero by only about one standard error, as turns out to be the case for Hipparcos numbers pertaining to all kinds of things you might want to study."

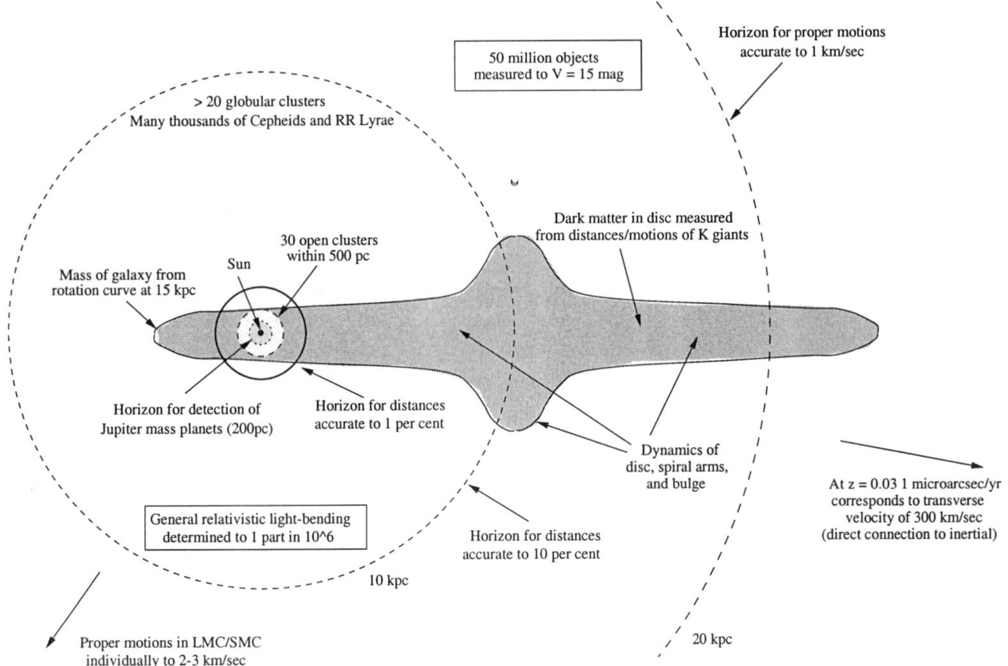

FIGURE 4. GAIA scientific goals (Perryman 1997).

2MASS (Skrutskie *et al.* 1998) and DENIS (Epchtein *et al.* 1998) infrared sky-surveys will produce large number of brown dwarfs, almost all of which will be readily accessible to ground-based astrometry. These observations will play an important role in defining the sequence of L-dwarf properties as a function of luminosity. The USNO's ND9 program should also produce particularly important results in the next few years, when it targets brighter members of nearby clusters.

7. Prospects for space-based parallaxes

Trying to review future space missions is an unenviable task – not only do their names and capabilities seem to change monthly, but trying to guess which proposed missions will actually get funded is about as straightforward as trying to predict stock markets. So I'll also restrict myself to just two missions – one of which is funded, and one of which *should* be funded. In particular, I'm going to neglect the interferometer missions like DARWIN and TPF, which are going to be primarily targetted at planet searches, and so which won't target particularly interesting stars from the point of view of cluster studies. Note that I've provide WWW URL's instead references to papers (which are outdated as soon as I've typed them, let alone by the time you read them) for these missions.

7.1. *The Space Interferometry Mission (SIM)*

SIM is a NASA mission planned for launch in 2005, which will provide an space interferometer with a 10 m baseline, operating in the 0.4-1 μm wavelength range (`http://sim.jpl.nasa.gov/sim/`). The interferometer itself will have a resolution of 10 mas, and will be capable obtaining 4 μas astrometry at V=20 in 15^h over a 15° field of view. This will give it the capability of obtaining 1% distances out to 2.5 kpc and 10% distances to 25 kpc. Since SIM will carry out a range of scheduled observations, rather than an all-inclusive survey, parallax programs will have to compete for observing time with other observations. However, it is to be hoped that observations of a suitable set of stars in a range of nearby clusters will be obtained in order to solve at least the Pleiades distance debate once and for all.

7.2. *GAIA*

GAIA (`http://astro.estec.esa.nl/SA-general/Projects/GAIA/gaia.html`) is a considerably more ambitious mission, which has been proposed to ESA as a successor to HIPPARCOS. The current proposal is to carry out an all-sky survey using an instrument at least conceptually similar to HIPPARCOS, which would observe 50 million stars down to V=15, obtaining 10 μas astrometry and multi-colour photometry with CCDs as detectors. This will provide 1% distances out to 1 kpc. and 10% distances to 10 kpc/ These are nearer distance limits than will be achieved by SIM, but they will be obtained for a *much* larger sample of objects. Fig. 4 (which is drawn from Perryman (1997)) highlights some of the scientific goals GAIA plans to address. I think the experience with HIPPARCOS cluster studies in the solar neighbourhood, however, reinforces the fact that the important radius in this diagram for cluster work is not that at 10 kpc, but the 1% radius at 1 kpc. Luckily this still includes an *awful* lot of interesting clusters!

8. Conclusions

Although parallax programs will not be able to target brown dwarf members of star clusters in the forseeable future, they will have an important role to play in the future of brown dwarf cluster research. Both space-based and ground-based programs will be required if we are to both understand the properties of brown dwarfs in clusters, and determine accurate distances and structures for clusters.

I would like to acknowledge the support provided by the conference organisers and the Director of the AAO, which permitted attendance at this meeting.

REFERENCES

DAHN, C. C. 1998 Review of CCD parallax measurements. In *on Fundamental Stellar Properties: The Interaction between Observation and Theory* Proceedings of IAU Symposium 189 (ed. T. R. Bedding, A. J. Booth and J. Davis), p. 19. Kluwer: Dordrecht.

EPCHTEIN, N. 1998 *The Impact of Large Scale Near-IR Sky Surveys* (eds Garzn, F. et al.) pp. 15-24, Kluwer, Dodrecht.

HARRINGTON, R.S. *et al.* 1993 U.S. Naval Observatory Photographic Parallaxes. List IX. *Astron.J.* **105**, 1571-1580.

HAN, I. 1989 The accuracy of differential astrometry limited by the atmospheric turbulence *Astron.J.* **97**, 607-610.

HAN, I. & GATEWOOD, G. 1995 A Study of the Accuracy of Narrow Field Astrometry using Star Trails taken with the CFHT *Publ.Astron.Soc.Pacif.* **107**, 399.

MERMILLIOD, J.-C. 1997 Distances of the Pleiades and Nearby Clusters In *Proceedings of the ESA Symposium "Hipparcos - Venice '97' "* ESA SP-402.

MIGNARD, F. 1997 Astrometric Properties of the HIPPARCOS Catalogue In *Proceedings of the ESA Symposium "Hipparcos - Venice '97' "* ESA SP-402.

MONET, D. G. & DAHN, C. C. 1983 CCD astrometry. I - Preliminary results from the KPNO 4-m/CCD parallax program *Astron.J.* **88**, 1489-1507.

MONET, D. G., DAHN, C. C., VRBA, F. J., HARRIS, H. C., PIER, J. R., LUGINBUHL, C. B. & ABLES, H. D. 1992 U.S. Naval Observatory CCD parallaxes of faint stars. I - Program description and first results *Astron.J.* **103**, 638-665.

PERRYMAN, M. A. C. 1997 The Scientific Goals of the GAIA Mission In *Proceedings of the ESA Symposium "Hipparcos - Venice '97' "* ESA SP-402.

PERRYMAN, M. A. C., BROWN, A. G. A., LEBRETON, Y., GOMEZ, A., TURON, C., DE STROBEL, G. CAYREL, MERMILLIOD, J. C., ROBICHON, N., KOVALEVSKY, J. & CRIFO, F. 1998 The Hyades: distance, structure, dynamics, and age *Astron.Astrop.* **331**, 81-120.

SKRUTSKIE, M.F. *et al.* 1998 In *The Impact of Large Scale Near-IR Sky Surveys* (eds Garzn, F. et al.) pp. 25-32, Kluwer, Dodrecht.

TINNEY, C.G. 1996 CCD astrometry of southern very low-mass stars *Mon.Not.R.Astron.Soc.* **281**, 644-65.

TRIMBLE, V. & MCFADDEN, L.-A. 1998 Astrophysics in 1997 *Publ.Astron.Soc.Pacif.* **110**, 223-267.

VAN ALTENA, W.F. *et al.* 1997 The Distance to the Hyades Cluster Based on Hubble Space Telescope Fine Guidance Sensor Parallaxes *Astrop.J.Lett.* **486**, 123.

VAN LEEUWEN, F. & HANSEN RUIZ, C.S. 1997 Distances of the Pleiades and Nearby Clusters In *Proceedings of the ESA Symposium "Hipparcos - Venice '97' "* ESA SP-402.

Very Low Mass Stars and Brown Dwarfs in the Belt of Orion

By S. J. WOLK[1] AND F. M. WALTER[2]

[1]Harvard–Smithsonian Center for Astrophysics Cambridge, MA 02138, USA

[2]SUNY Stony Brook, Stony Brook, New York 11794, USA

As part of our ongoing research into low–mass star formation in Orion, we have obtained deep photometric and spectroscopic observations of PMS objects in the σ Orionis cluster and near the other O stars in the belt of Orion. The photometry indicates the existence of objects with masses as low as 0.01 M$\odot$ (100 M_{jup}). Spectroscopic follow-up has confirmed the sub-stellar nature of the candidate object tested.

---

1. Introduction

The Orion OB associations are one of the richest star forming regions in the local galaxy. Recently, we have made a concentrated effort to study the stars near belt of Orion within the Orion OB1a and OB1b associations. ROSAT observations totaling 100 Ksec of this region have been supplemented with ground based spectroscopy and photometry (Wolk 1996). These data demonstrated the clear existence of a 1-5 Myr old pre–main sequence of stars with common space motion and density of sources reaching a maximum at the location of σ Orionis. The confluence of this data lead us to conclude that the stars near σ Orionis form a young stellar cluster with a central density of at least 50 stars/pc^3 (Walter et al. 1997, Walter et al. 1998). Similar clusters of older stars seem to exist near the other O stars in the belt of Orion.

These clusters represent one of the best opportunities to view very young PMS objects unobscured by large amounts of gas and dust. The faintest objects in the original photometric survey contained R–I colors consistent with the objects being near the mass limit for hydrogen burning on the main–sequence. We reobserved the σ cluster in January of 1998 to obtain sample spectroscopy of these candidate objects. We also took the opportunity to photometricly map the regions around the other belt stars. Here, we show the low–mass end of the color–magnitude diagrams of the clusters around each O star, and estimate the IMF of the σ Orionis cluster. We highlight the confirmation of a brown dwarf within this cluster and candidates discovered near the other belt stars.

2. Observations

2.1. *Photometry*

We photometricly observed four fields adjacent to σ Orionis and δ Orionis as well as two fields south of ϵ Orionis. Data were taken on 16-17 January 1998 UT using the 0.9 meter telescope at KPNO. We used the T2KA detector at f/13.5 which provided a 23$'$ field of view. The telescope was initially pointed at the O star and then off–pointed the array 1320$''$ to the northeast, northwest, southwest and southeast to avoid the glare of the central O and B stars. Observations were made using the V, R and I filters from the Harris set. Exposures were 15 minutes in V, and 10 minutes through the other two filters.

Data were debiased and flatfielded using zero exposures and skyflats taken that night.

We followed the usual IRAF processing prescription using the CCDPROC package. Photometric solutions for the two nights were made using the IRAF/DIGIPHOT package. Standards from Landolt (1993) were observed before and after each program field. Because we suspected the possibility of very thin cirrus clouds, we observed an average of 20 standard stars per hour during these observations. The overall errors induced by sky conditions proved to be less than 5%.

2.2. *Spectroscopy*

Spectra were taken 6 January 1998 of eight objects near σ Orionis, the colors of which indicated that they were of very low mass. The targets were chosen to test the reddening of these objects. Two objects were observed in each integral V magnitude between V=16 and V=19, inclusive. We used the RC spectrograph on the 4 meter Mayall telescope at Kitt Peak National Observatory. Grating KPC-22 was used and set to give us wavelength coverage from 6500Å to 9000Å, with about 2Å resolution.

We followed the usual IRAF prescription for reducing the data. This included the use of an optimal extraction algorithm (Horne 1986) to remove discrepant points. This was the only method available to removing cosmic rays from Star 5 which was observed at high airmass. Since the night was not photometric, the flux calibration is only taken as a guide for the relative flux across a given spectrum.

3. Results

3.1. *Photometry*

Colors were determined for about 10,000 stars between V=16 and V=24, complete to about V=23, with a range of V-I = from 0 to 5.5. This covers an effective temperature range (uncorrected for reddening) of 10,000K down to less than 2700K. Below 2700K optical colors are poorly mapped to effective temperature. The color-magnitude diagrams of the three regions are displayed below as Figures 1–3.

In Figure 3, objects marked with an 'X' were originally identified by their X–ray emission (see Wolk 1996 for more details). The PMS nature of most of these objects has been verified through spectroscopy (Walter et al. 1997, Walter et al. 1998). The two dashed lines indicate the region which would be occupied by young stellar objects coeval with the X-ray sources. The bulk of the region between these two lines runs parallel to the reddening vector so the reddening does not move a PMS object out of this region, nor does reddening move a star into this region. About 5% of the objects in this field lie along the locus defined by the PMS X-ray sources. In Figures 1–3 the vertical line indicates the boundary between stellar and sub-stellar objects assuming the 0.075 $M_\odot$ limit for Hydrogen burning, solar metallicity and an age of about 2 Million years (taken from Baraffe et al. 1998). A correction is made for the "missing opacity" discussed in this paper. Briefly, comparison of these theoretical tracks against field clusters indicates that below V−I$\sim$ 2 the colors predicted by the model are substantially too blue. The authors attribute this to an unknown source of opacity (possibly dust in the stellar photosphere). The authors provide a three–point calibration of the error which has been extrapolated here to provide a brown dwarf cut-off in color–magnitude space. To this offset, we added the mean reddening of the eight stars which we observed spectroscopically to determine a dividing line in V−I space between brown dwarfs and stars of V−I=4.03 for stars that are about 2 million years old.

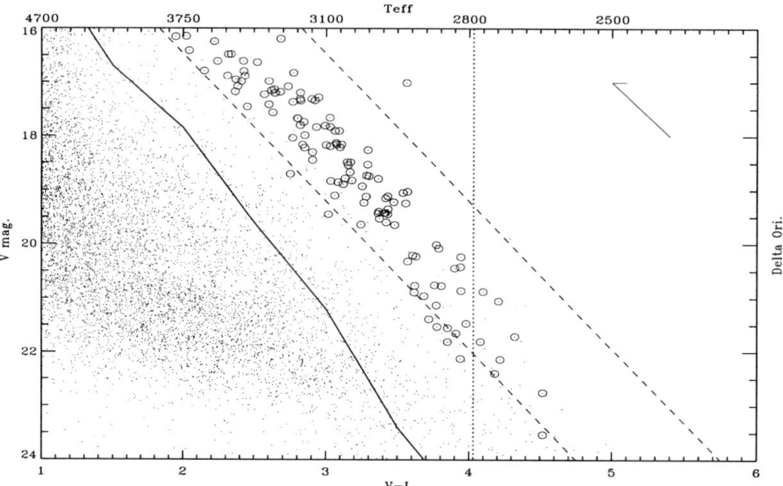

FIGURE 1. Color–magnitude diagram of stars in a 2200 square arcminute region centered on Delta Orionis. For this and the following figures, the effective temperature is derived from Bessell (1995). The solid line traces the pre–main sequence. The vertical indicates the brown dwarf cut–off derived in Baraffe et al. (1998) with corrections for "missing opacity" and mean reddening (see text). The $A_V = 1$ reddening vector is indicated in the right upper part of the figure.

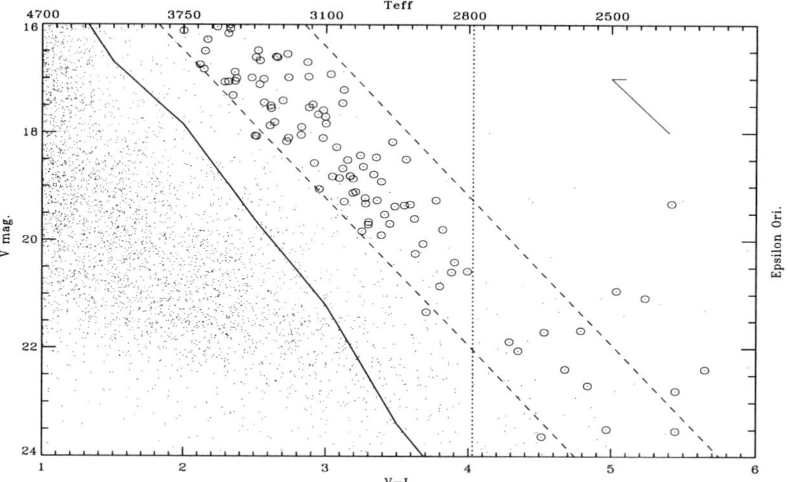

FIGURE 2. Color–magnitude diagram of stars for an 1100 square arcminute region south of epsilon Orionis. The circles in the figures indicate objects for which the V–R and R–I colors also matched those expected for 2–20 Myr old PMS stars. The $A_V = 1$ reddening vector is also indicated.

3.2. *Spectra*

The eight spectra taken are shown in Figure 4. We tabulate their spectral classification in Table 1. We calculated the spectral type in two ways. First we used the TiO5 decrement discussed by Reid et al. (1995). We also visually compared the spectra against standards from Kirkpatrick et al. (1993,1995) and used the pseudo–continuum ratios taken in 20Å bins centered at 6530Å, 7040Å, 7560Å, and 8130Å. We calculated four pseudo–

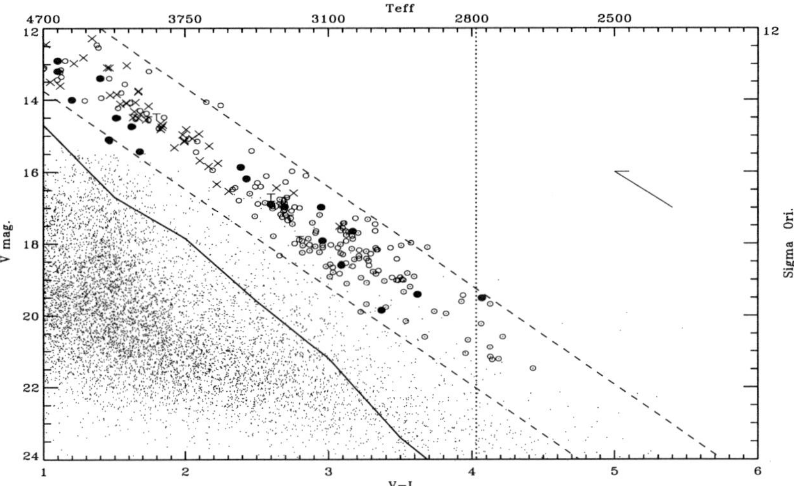

FIGURE 3. The color–magnitude diagram is for the 2200 square arcminute region centered on σ Ori. "X's" indicate X-Ray sources, filled circles indicate non–X–Ray detected PMS stars verified through spectra. "T's" indicate Classical T Tauri Stars. Open circles represent stars with colors consistent with PMS nature.

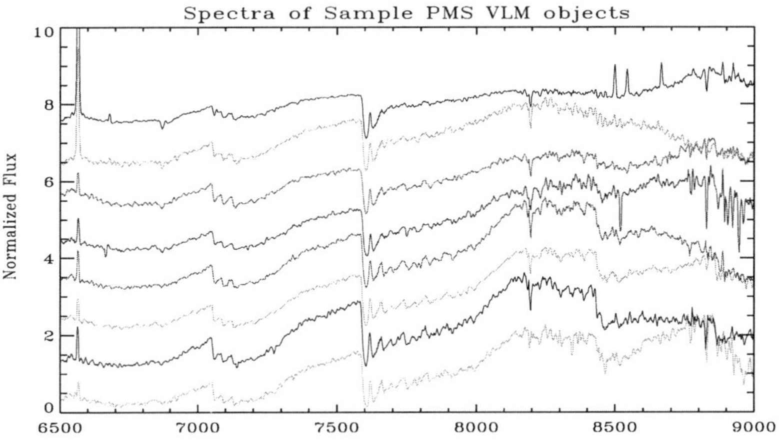

FIGURE 4. Spectra of eight late–type stars. The stars cool from top to bottom. The CTTS at the top are about 0.5 $M_\odot$. The large absorption near 8500 Å is indicative of M5 or later ($\approx$ 0.075 $M_\odot$). The spectra are vignetted beyond 8700Å. Stars are numbers from bottom to top for reference with the tables in this paper.

continuum ratios: (7040Å/6530Å), (7560Å/6530Å), (7560Å/7040Å) and (8130Å/7560Å). The results were compared against those of standards published by Montes & Martín (1998). The use of pseudo–continuum ratios was hampered by the existence of strong emission lines in many of the stars. Seven stars showed Hα in emission with equivalent width of more than 10Å in emission. In late M stars, the photometric flux is so weak at 6500Å that coronal emission alone can give rise to over 10Å EW emission at Hα. Such lines are not necessarily indicative a of Classical T Tauri disk unless other emission lines are present. Other emission lines *are* seen in Star 7 and Star 8. In our tabulation, we

TABLE 1. Spectral characteristics of sample stars.

Star	Pseudo	TiO5	Lum class	Hα	Comments
1	M5.5	M5.5	IV	−12.88	
2	M6	M5	IV	−16.78	
3	M5	M5	IV	−14.21	
4	M5	M5.5	IV	−12.42	
5	M4.5	M5	IV	−12.97	
6	M4	M4	IV	−5.23	
7	M4	M2.5	IV	−40.49	CTTS
8	M3	M1.5	IV	−63.85	CTTS
9	M0	M0.5	V	0.38	HD119850 – M1.5
10	K8	K8	V	0.61	GL546 – K8V
11	M3	M4.5	V	−−	GL447 – M4

include three spectral standards so that the accuracy of the two objective methodologies used could be independently verified. The differences are less than one subtype.

Detection of the Li I 6708Å is often used as a criterion for determining PMS or substellar nature. We did not observe Li I 6708Å at sufficient resolution to detect its presence in these stars. In any event, lithium could not have been used to distinguish a star from a brown dwarf as has been done for Calar 1 (Basri 1997) and others because at the estimated cluster age of 2 Myr, all stellar objects have lithium. Lithium was not a useful criterion for distinguishing cluster members from older background members of the Orion OB association either because all stellar objects are thought to possess lithium at the age of the oldest stars in the Orion OB association.

To determine the PMS nature of the objects we observed, we used spectral class diagnostics, specifically CaH, Na I, Ca II and Ti I line ratios as described in Kirkpatrick et al. (1993). In all cases, the classification indices indicated that star has surface gravity between that of giants and dwarfs. The sub-giant classification is expected for young stellar objects. All eight objects are extremely puffy, young and cool.

In Table 2 we list the colors of the stars for which we had also attained spectra. We determined a photometric spectral type by comparing the $V_J - I_C$ color index against calibrations presented by Leggett (1992) and Bessell (1995). In the cases of stars 1 through 6, the photometric and spectroscopic calibrations agree to within 1 spectral subtype. The standard deviation is about 0.25 of a spectral subtype. This is indicative of very little reddening along the line–of–sight to the cluster.

Stars 7 and 8 both show a larger discrepancy between the photometric and spectroscopic temperature determination. These stars also show the strongest Hα lines. Star 8 also shows strong Ca II emission. We take this as evidence that these stars are both Classical T-Tauri stars with circumstellar disks and masses about 1/2 that of the Sun. The additional reddening noted in the discrepancy between the photometric and spectroscopic temperature determinations is a result of highly localized reddening caused by the disk.

4. Discussion

4.1. *The Initial Mass Function*

Much can be learned directly from photometry alone. Direct comparison of the color–magnitude diagrams given in the first three figures reveals difference in the environs

TABLE 2. Photometric Characteristics of sample stars.

Star	V	V-R	R-I	SpT[l]	T_{eff}[l]	SpT[b]	T_{eff}[b]
1	19.51	2.01	2.06	M6	2600K	M6	2800K
2	19.41	1.71	1.91	M5.5	2800K	M5.5	2880K
3	19.85	1.63	1.74	M5	2880K	M5	2950K
4	18.60	1.32	1.77	M5	3020K	M4.5	3050K
5	17.66	1.35	1.82	M5	3000K	M4.5	3025K
6	16.98	1.33	1.62	M4	3080K	M4	3090K
7	17.89	1.41	1.39	M3.5	3130K	M4	3160K
8	17.30	1.26	1.47	M3.5	3150K	M4	3180K

[b]–Derived based on Bessell (1995).
[l]–Derived based on Leggett (1992).

around the three O stars. First, we can identify the PMS candidate objects. We do this using objects which have V−R and R−I colors which are constant with PMS nature. These regions are roughly indicated by the dashed lines in Figures 1–3. The σ Ori cluster contains 231 PMS candidates between V=16 and V=24 in a 2200 square arcminute region centered on σ Ori. In the same volume of space around δ Ori 188 PMS candidates have been identified. There are 136 PMS star candidates in the 1100 square arcminutes south of ϵ Orionis. The ϵ Ori region, seems to be the most fertile, but this region has the most work to be done in terms of data collection.

The relative ages of the three groups of stars can also be directly compared from the first three figures. In the case of the σ Orionis cluster, the bulk of the stars lie directly between the the two dashed guide lines. The one σ deviation covers about 40% of the width between the guide lines. In the case of δ Ori, the PMS candidates are all toward the main sequence side of the guide lines. This indicates a somewhat older age, maybe about 10 million years, with almost no objects remaining near the youngest portions of the diagram. The ϵ Orionis cluster seems intermediate in age with a few stars within V−I=0.2 of the high guide line. The stars near ϵ Orionis, also seem to have a much wider age gradient than the other two clusters. Epsilon Orionis straddles the border between Orion OB1a and OB1b, so we my be seeing stars from both groups.

Each cluster shows several brown dwarf candidates. Being somewhat younger, the stars near σ Ori tend to be brighter and this made them the obvious first targets for spectroscopic follow-up. The first result of the spectroscopy was that the reddening was quite low and fairly uniform with the exception of the classical T Tauri stars. The spectra allow us the make a rough cut at the initial mass function. To do this, each of the 88 PMS candidates near σ Ori was dereddened by $A_V = 0.5$ this color was compared with the evolutionary models compiled in Baraffe et al. (1998) for two million year old stars. The models were adjusted for missing opacity using the three point fit given in the text of their paper. The resulting IMF is given in Figure 5. We find a peak in the mass function at $0.45M_\odot \pm 0.1M_\odot$. This is similar to results for the Pleiades (Bouvier et al. 1998).

4.2. *Brown dwarfs*

It is not possible to assign an object the status of a brown dwarf purely on the basis of optical photometry. The fact that the reddening vector runs parallel to the pre–main sequence makes reddening difficult to detect in optical color–magnitude or two–color diagrams. Thus, it is impossible to distinguish an object with R−I of 2 from an object with R−I of 1 with $A_V = 2.5$. This is what prompted us to take the spectra

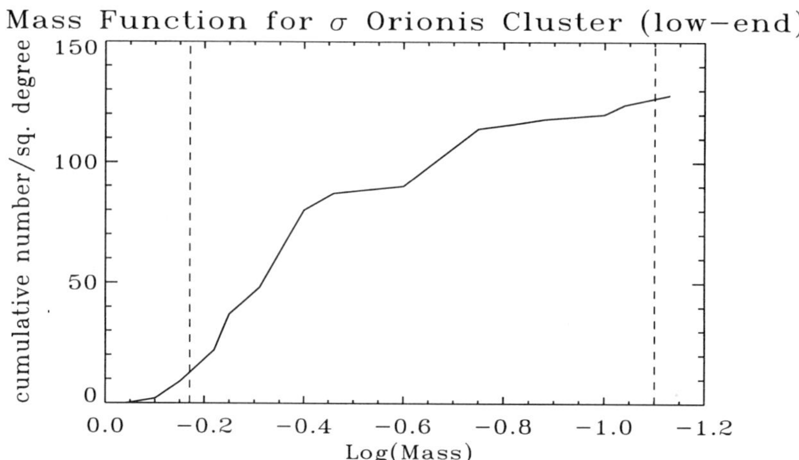

FIGURE 5. A preliminary mass function for the σ Orionis cluster. The dashed vertical lines indicate completeness limits. The total area studied is about 2200 square arcminutes.

discussed. The results of the spectra indicate that reddening was very low to all eight objects observed ($< A_V > \ < 0.5$). Further, that average is dominated by the classical T Tauri stars which make up about 1/3 of the PMS stars in the σ Orionis cluster (see Walter et al. 1997, Walter et al. 1998) and less in the other two regions.

While the reddening is small, the effect on the perceived temperature is about 100K which is substantial when trying to estimate the mass of very low mass objects. To determine stellar mass and hence sub-stellar status, we place objects on the temperature–luminosity diagram and compare with theoretical tracks. One of the advantages to comparing the stars to the work of Baraffe et al. (1998) it that the theoretical calculations are transformed to the observer frame with final values of M_{filter} tabulated for easy comparison to observation after corrections are made.

We can define the brown dwarf cut–off using the tracks of Baraffe et al. (1998) with corrections for "missing opacity" and field reddening. We find that at an age of about 2 million years the cut–off is V−I=4.03, this is what is used in Figures 1–3. We use this criterion and exclude stars outside of the guide lines. In the three regions, covering 5500 arcminutes, we detected 25 objects with V magnitudes between 20 and 24 which had V, R and I colors consistent with their being brown dwarfs. In addition, there are 71 other objects with V−I consistent with being sub–stellar, but measurement errors are considered too large to categorize them as candidates. However, using the same tracks at an age of 10 million years, a more likely age for the stars near δ and ϵ Orionis, the cut–off moves blueward to V−I$\approx$ 3.8. This admits a dozen additional candidates from these two regions.

Of course, there are many groups working on modeling the evolution of very low mass, young stellar objects. Recently the work of Burrows et al. (1997) has also explored the evolution of very low mass objects using a different set of non-gray models. This work focused exclusively on the sub-stellar domain and shows that at an age of 2 million years, objects of $0.07M_\odot$ have and effective temperature of about 2900K (objects with $M= 0.08M_\odot$ are about 2950K). From the temperatures given in Table 1 we can see that the first three stars all lie within the sub-stellar domain. The results are not highly dependent on our age estimate for the σ Orionis cluster since, based on these models, temperatures for objects above $0.04M_\odot$ are fairly static between 1 and 20 million years.

We have a total of about 25 objects which qualify as brown dwarf candidates based on comparison with both models. We have obtained one spectra so far; Star 1 on our tables. It has a spectral type of M5.5–M6. Even by the most conservative (warmest) temperature estimates this object is less massive than the hydrogen burning limits based on the mass tracks of Baraffe et al. (1998), Burrows et al. (1997) and D'Antona and Mazettelli (1994).

This work has been funded by the AXAF Science Center through grant number NAS8-39073.

REFERENCES

ALLARD, F., 1990. Ph.D. Thesis, Ruprecht Karls Universität Heidelberg.

BARAFFE, I., CHABRIER, G., ALLARD, F., HAUSCHILDT, P.H. 1998. Evolutionary Models for Solar Metallicity Low-Mass Stars Mass-Magnitude Relationships and Color-Magnitude Diagrams. *Astron. Astrophys.* 337: 403–412.

BASRI, G. 1997 Lithium Near the Substellar Boundary: a New Age Diagnostic In *Memorie, S.A.It.* **68** N.4, 917–924.

BESSELL, M.S., 1995 The Temperature Scale for Cool Dwarfs in *The Bottom of the Main Sequence-and Beyond* (ed. G.G. Tinney) pp. 123–131. Springer.

BOUVIER, J., STAUFFER, J., MARTÍN, E. 1998 Results of a Deeper, Wider Area Brown Dwarf Search in the Pleiades In *Cool Stars Stellar Systems and the Sun 10* (ed. R. Donahue and J. Bookbinder) pp. 1793–1799. A.S.P.

BURROWS, A., MARLEY, M., HUBBARD, W.B., LUNINE, J.I., GUILLOT, T., SAUMON, D., FREEDMAN, R., SUDARSKY, D. SHARP, C. 1997 A Nongray Theory of Extrasolar Giant Planets and Brown Dwarfs *Astrophys. Jour.* **491**, 856–875.

D'ANTONA, F., MAZZITELLI, I. 1994 New Pre-Main-Sequence Tracks for M Less Than or Equal to 2.5 Solar Mass as Tests of Opacities and Convection Model. *Astrophy. Jour. Supp.* **90**, 467–500.

HORNE, K. 1986 An Optimal Extraction Algorithm for CCD Spectroscopy *Pub. Astr. Soci. Paci.* **98**, 609–617.

KIRKPATRICK, J.D., 1995 Classification Spectroscopy of M Dwarfs from 0.6 to 2.4 Microns in *The Bottom of the Main Sequence -and Beyond* (ed. G.G. Tinney) pp. 140–150. Springer.

KIRKPATRICK, J.D., KELLY, D. M., RIEKE, G.H., LIEBERT, J., ALLARD, F., WEHRSE, R. 1993 M Dwarf Spectra from 0.6 to 1.5 Micron - A Spectral Sequence, Model Atmosphere Fitting, and the Temperature Scale *Astrophys. Jour.* **402**, 643–654.

KIRKPATRICK, J.D., HENRY, T.J., MCCARTHY, D.W., JR. 1991 A Standard Stellar Spectral Sequence in the Red/Near-infrared - Classes K5 to M9 *Astrophys. Jour. Supp.* **77**, 417–440.

LEGGET, S.K. 1992 Infrared Colors of Low-Mass Stars *Astrophys. Jour. Supp.* **82**, 351–394.

MONTES, D., MARTÍN, E. L. 1998 Library of High-Resolution UES Echelle Spectra of F, G, K and M Field Dwarf Stars. *Astron. & Astrophys Supp.* **128**, 485–495.

REID, I.N., HAWLEY, S.L., GIZIS, J.E. 1995 The Palomar/MSU Nearby-Star Spectroscopic Survey. I. The Northern M Dwarfs-Bandstrengths and Kinematics *Astron. Jour.* **110**, 1838–1859.

WALTER, F.M., WOLK, S.J., SHERRY, W, SCHMITT, J.H.M.M 1998 The σ Orionis Cluster. In *Cool Stars Stellar Systems and the Sun 10* (ed. R. Donahue and J. Bookbinder) pp. 1793–1799. A.S.P.

WALTER, F.M., WOLK, S.J., FREYBERG, M, SCHMITT, J.H.M.M 1997 The Discovery of the σ Orionis Cluster. In *Memorie, S.A.It.* **68** N.4, 1081–1088.

WOLK, S.J. 1996 Watching the Stars go 'Round and 'Round PhD thesis, State University of New York, Stony Brook.

Photometric Surveys in Open Clusters

By M. R. ZAPATERO OSORIO

Instituto de Astrofísica de Canarias, E–38200 La Laguna, Tenerife, Spain

Photometric surveys in nearby, young open clusters have provided a large amount of very low-mass stars and brown dwarfs over the last two decades. These clusters offer a number of advantages like known distance, metallicity and age, which make feasible the identification of such objects. Furthermore, deep searches do constitute one of the most direct means for measuring the mass function through the whole stellar (and brown dwarf) mass range. In this paper it will be reviewed the progress of recent work on several young open clusters leading to the findings of unambiguous brown dwarfs and very low mass stars approaching the substellar mass limit. These discoveries, particularly in the Pleiades, imply a rising mass function ($\alpha = 0.75 \pm 0.25$, $\mathrm{d}N/\mathrm{d}M \sim M^{-\alpha}$) in the very low mass stellar and substellar domains down to $0.04\,M_\odot$. The detection of reliable *free-floating* candidate members with estimated masses of only 0.04–0.015 $M_\odot$ does provide substantial evidence on the formation of such low mass objects and thus, on the extension of the initial mass function down to the deuterium burning mass limit.

---

1. Introduction

Our knowledge of the low mass stellar content in open clusters has increased considerably during the last decade. For a relatively large amount of nearby young clusters, like α Per, Pleiades, Praesepe and Hyades, membership lists extending down to the hydrogen-burning limit ($\sim 0.08\,M_\odot$) are now available (see the reviews by Stauffer 1996, and Hambly 1998). Although depending on the age of the stellar members, these stars present spectra classified within the K- and M-type dwarf regime approaching the main sequence.

Regarding brown dwarfs ($M \leq 0.075\,M_\odot$), the search of these elusive objects was initially carried out in the Pleiades and Hyades, due in part to the fact that the stellar end of the main sequence of these clusters was rather well defined, making possible a reasonable extrapolation towards fainter magnitudes. The search for substellar objects has been recently extended to several other nearby, young clusters, like α Persei, Praesepe, IC 348, and stellar associations, like Orion, Taurus and ρ Ophiuchi (see the review by Hambly 1998). Nowadays, in a baseline of only 4–5 years since the discovery of the first free-floating brown dwarf in the Pleiades, around 40 substellar Pleiads are already catalogued, and various good candidates do exist in many other young clusters. The spectral energy distribution of the young brown dwarfs resembles that of late M-type dwarfs, and the cooler ones do belong to the new defined L-class (Martín et al. 1999; Kirkpatrick et al. 1999).

Three major techniques of finding young cluster members have been used: proper motion surveys, deep photometric imaging and X-ray emission; the two later approaches make use of the properties of youth, which well differenciate young members (overluminous photometric sequence, active chromospheres and coronae) from older, field contaminants. X-ray imaging is useful for young compact clusters, and will obviously produce a membership list biased to active binaries. Nevertheless, this method if applied in star forming regions is successfull in identifying the presence of clusters/associations which can be further investigated. So far the technique that facilitates large numbers of member candidates is proper motion measurements, providing an unbiased sample down to the magnitude limit of the study. However, proper motions in combination with photometry and spectroscopy constitute the most powerful tool for selecting true cluster members, and this approach has been widely used by different authors in the study of several clusters. This paper summarizes some of these surveys in the following sections. All these techniques suffer from severe limitations as they are clearly biased towards binaries near their limiting magnitude; therefore, none of them should be considered complete. Additionally, because of mass segregation spatial limits in the surveys are also a source of incompleteness. Regardless all difficulties, recent studies on young clusters of different ages reveal that the very low mass stellar and substellar domains are quite well populated, and that objects in this mass domain may largely populate the Galaxy.

2. Pleiades

Table 1 summarizes the recent surveys carried out in several young clusters with the main goal of identifying their very low mass stellar and substellar populations. In the case of the Pleiades a list of photometric/proper motion searches of the last ten years is provided. The completeness I magnitude, the area covered and the estimated mass of the faintest, confirmed candidates arising in each survey are indicated in the third, fifth and fourth columns, respectively.

2.1. *Pioneering surveys*

One of the pioneering CCD-based photometric searches aimed at discovering brown dwarfs (or at least one brown dwarf) in the Pleiades was performed by Jameson & Skillen (1989). These authors obtained RI photometry of an area of 125 arcmin^2 in the central region of the cluster, and found 9 candidates to very low mass stars and brown dwarfs. These objects were selected based on their location in the I *versus* $R-I$ colour-magnitude diagram because they appeared as the reddest objects per interval of magnitude. Follow-up proper motion and optical and IR observations (Zapatero Osorio *et al.* 1997a, 1997b, and references therein) of the JS candidates yield that none of them are cluster members.

Almost simultaneously, Stauffer *et al.* (1989) presented a larger CCD-based survey using VI filters which covered 900 arcmin^2. These authors argued that the choice of these broader based colour filters helps to identify very red objects. However, because brown dwarfs emit most of their flux at red wavelengths, this survey is limited by the sensitivity reached in the V-band. For some of their candidates Stauffer *et al.* obtained IR photometry and/or optical low-resolution spectroscopy, showing that they were cool objects indeed. Some years later, Stauffer *et al.* (1994) extended their original search to 1440 arcmin^2 reaching similar limiting magnitudes. The 16 objects selected from these surveys, named with the acronym "PPl" (which stands for "Palomar Pleiades"), could be cluster low mass stars, cluster brown dwarfs or field contaminants. The mass of the candidates is estimated by direct comparison of their photometry with available theoretical isochrones (Burrows *et al.* 1993; D'Antona & Mazzitelli 1994; Baraffe *et al.* 1995), but this procedure did not disentangle the real stellar/substellar nature of the candidates (if they are true members) given the uncertainties in coversion from the colour-magnitude observational plane to the theoretical plane, and viceversa. More recently, one of the faintest candidates, PPl 15, with I around 18 magnitude was observed at high-resolution looking for lithium in its atmosphere (Basri *et al.* 1996), a signature which can define brown dwarfs at the age of the Pleiades (the "lithium test" was proposed by Rebolo *et al.* 1992, and Magazzù *et al.* 1993; see Basri 1998 for a detailed discussion). The lithium feature at 607.8 nm was detected implying that the internal temperature of this object is not high enough yet for depleting this light element. Another implication is that PPl 15 is a true Pleiad. According to models, its mass should be around $0.07\,M_{\odot}$, i.e.

TABLE 1. Recent photometric surveys in some young clusters.

Cluster	Age (Myr)	I_{comp}	Mass ($M_{\odot}$)	Area ($arcmin^2$)	Filters	Source
ρ Ophiuchi	≤ 1	–	0.030	200	HK	Comerón *et al.* 93
		–	0.100	–	K^a	Greene & Meyer 95
		–	0.025	–	opt^b	Luhman *et al.* 99
		–	0.020	–	K^a	Wilking *et al.* 99
IC 348	3	19.0	0.100	380	VRI	Herbig 98
		–	0.040	25	K^a	Luhman *et al.* 98
σ Orionis	3	20.0	0.025	870	IZ	Béjar *et al.* 99
		–	0.015	–	opt^b	Zapatero Osorio *et al.* 99b
IC 2391	50	17.0	0.090	216	$BVRI$	Rolleston & Byrne 97
		20.5	0.065	7200	RI	Barrado *et al.* 98,99
		16.5	0.090	2880	VRI	Patten & Pavlovsky 99
α Persei	80	18.0	0.080	108000	VI	Prosser 92, 94
		–	0.090	–	opt^b	Zapatero Osorio *et al.* 96
		19.5	0.063	21600	RI	Stauffer *et al.* 99
Pleiades	120	19.5	–	125	RI	Jameson & Skillen 89
		19.0	0.070	1440	VI	Stauffer *et al.* 89, 94
		17.5	0.080	57600	RI	Hambly *et al.* 91, 93
		20.5	–	200	IK	Simons & Becklin 92
		19.5	0.075	400	VIK	Williams *et al.* 96
		21.0	0.040	1044	IJK	Festin 97, 98
		20.5	0.045	100	IZ	Cossburn *et al.* 97
		20.0	0.050	575	RI	Zapatero Osorio *et al.* 97a
		21.0	0.060	3600	VI	Stauffer *et al.* 98
		22.5	0.045	9000	RI	Bouvier *et al.* 98
		20.0	0.055	180000	IZ	Pinfield *et al.* 98
		21.5	0.035	3600	IZ	Zapatero Osorio *et al.* 99
		18.5	0.070	129600	RI	Hambly *et al.* 99
Praesepe	600	18.0	0.100	68400	RI	Hambly *et al.* 95
		21.5	0.070	3600	RIZ	Pinfield *et al.* 97
		21.5	0.070	790	IZ	Magazzù *et al.* 98
Hyades	600	18.5	0.090	90000	VI	Leggett *et al.* 94
		17.5	0.100	136360	phot	Bryja *et al.* 94
		–	–	100800	JHK	Gizis *et al.* 99
		–	–	–	opt^b	Reid & Hawley *et al.* 99

[a] These are K-band spectroscopic surveys. Candidates are selected from previous photometric searches.
[b] Based on optical spectroscopy. The candidates are selected from previous photometric surveys.

close to the substellar limit. Based on the IR $(I-K)$ colour excess observed in PPl 15, Zapatero Osorio *et al.* (1997b) pointed out that this object was a likely binary system. High-resolution spectroscopy taken at the Keck telescope by Basri & Martín (1999) has proved that PPl 15 is indeed a spectroscopic binary with a measured orbital period of

5.8 days and a mass ratio near 1. Thus, PPl 15 has become the first identification of a binary brown dwarf.

Simons & Becklin (1992) performed a photometric search using a slightly different approach. They surveyed an area of 200 arcmin^2 in IK filters taking advantage of the fact that cool objects are brighter at these wavelenghts. In addition, the $I-K$ colour is very sensitive to temperature which makes the use of these two filters a very powerful tool for the detection of very low mass stars and brown dwarfs. These authors identified more than 30 objects in their survey whose estimated masses could be between 4 and 10% solar, and concluded that the Pleiades mass function in the regime of M stars and brown dwarfs showed a slope similar to that of Salpeter. But candidates have to be confirmed as cluster members before a definitive conclusion can be achieved. Particularly, at the time this work was published, it provided a relatively large number of candidates with the faintest magnitudes ($I > 19$). Thus, aimed at identifying the least massive members of the Pleiades, it became important to establish their membership in the cluster. Zapatero Osorio (1997c) re-observed the Simons & Becklin's fields at RI bands down to $I = 20$ mag. Apart from 3 of the SB candidates, all of them now have revised I photometry. Compared to Simons & Becklin's data this new photometry is brighter by about half magnitude in most of the cases, making the SB objects blue in $R-I$. Only SB 27 shows photometry marginally consistent with the Pleiades sequence. It is also surprising the fact that some known proper motion cluster members present in the SB fields were not found by the authors. Follow-up low-resolution spectroscopy (Martín 1993) of 4 SB candidates yielded spectral types (early M dwarfs) that are not consistent with membership. Therefore, the substellar Pleiades mass function derived by Simons & Becklin (1992) must be taken with extreme caution.

Hambly *et al.* (1991, 1993) complemented the investigation of the low mass population of the Pleiades with an astrometric survey covering 20 deg^2 in the photographic RI filters. The authors compared plates separated in time by about 50 years. Fortunately, this cluster has a quite distinct proper motion, so that it is rather easy to discriminate members from field objects. Hambly *et al.*'s survey, whose candidates are known as the "HHJ" objects, is so far the most complete source of low mass and very low mass Pleiades stars. The lithium test applied on the faintest HHJ objects with I magnitudes in the range 16.5–17.5 (Martín *et al.* 1994; Marcy *et al.* 1994) yields that none have preserved their initial lithium content, and thus, their masses must be greater than 0.08 $M_\odot$, i.e. slightly above the hydrogen-burning mass limit. Hambly *et al.* (1991) give an estimation of the Pleiades star mass function down to 0.1 $M_\odot$ with an index slope, α, around unity ($\mathrm{d}N/\mathrm{d}M \sim M^{-\alpha}$), implying that the low mass stellar population of the cluster is very numerous (more than 400 early- and mid-M stars).

2.2. *Recent work: the substellar mass domain*

The discovery of the unambiguous brown dwarfs Teide 1 (M8, Rebolo *et al.* 1995), Calar 3 (M8, Martín *et al.* 1996) and PPl 15 (M6.5) conformed the first step in the study of the Pleiades substellar domain. The description of the photometric survey where the two former brown dwarfs were found is provided in Zapatero Osorio *et al.* (1997a), and the detection of lithium in their atmospheres confirming their substellar nature is presented in Rebolo *et al.* (1996). Both Teide 1 and PPl 15 show proper motions consistent with being members of the Pleiades. These three substellar Pleiads with masses of 0.055 $M_\odot$ (Teide 1 and Calar 3) and around 0.07 $M_\odot$ (PPl 15) have been extensively used as a reference in all the later photometric and spectroscopic surveys carried out in the Pleiades and in many other young clusters. Several goals are after these deep searches: to identify less massive brown dwarfs, to study the properties of young brown dwarfs of different ages,

TABLE 2. Estimated magnitudes and spectral types for the substellar mass limit (0.075 $M_\odot$).

Cluster	**Age** (Myr)	d (pc)	R	I	J	K	I–K	**SpT**
IC 2391	30	140	18.4	16.5	14.6	13.6	2.9	M5–M6
IC 4665	50	450	21.5	19.5	17.5	16.5	3.0	M5–M6
α Persei	80	165	19.8	17.8	15.7	14.7	3.1	M5.5–M6
Pleiades*	120	125	20.0	17.8	15.6	14.5	3.3	M6–M6.5*
Hyades	600	45	20.7	18.0	14.7	13.7	4.3	M8.5–M9
Praesepe	600	175	23.7	21.0	17.7	16.7	4.3	M8.5–M9
	900	175	24.3	21.5	18.0	17.0	4.5	M9–L0

* The substellar limit in the Pleiades has been established from lithium observations of cluster members. The estimations for other clusters are based on these results and on the NextGen theoretical evolutionary tracks by Baraffe *et al.* (1995).

to deep on the knowledge of the cluster substellar population as this can have meaningful implications on the initial mass function and on the origin, formation and evolution of very low mass objects.

The lithium test has been applied so far to about 20 Pleiades members with I magnitudes in the range 16.5 down to 19 (Marcy *et al.* 1994; Basri *et al.* 1996; Rebolo *et al.* 1996; Stauffer *et al.* 1998; Martín *et al.* 1998a). According to models, the substellar mass borderline and the lithium depletion mass boundary do coincide at the age of the Pleiades. Thus, the positive/negative detections of lithium have provided for the first time an accurate measurement of the location of the substellar mass limit in this cluster (Martín *et al.* 1998a; Stauffer *et al.* 1998). The $RIJK$ magnitudes and the spectral type of the Pleiades stellar/brown dwarf frontier are given in Table 2. Using theoretical evolutionary models it is possible to predict the magnitudes, colours and spectral types for the location of such borderline in other young clusters. Table 2 displays the results (magnitudes are corrected for the distance to the cluster and its age). Some of these predictions will be proved within the following two–three years. So far, the clusters α Persei (~80 Myr) and IC 2391 (~50 Myr) have been deeply investigated by Stauffer *et al.* (1999) and Barrado y Navascués *et al.* (1999), respectively. These authors have lithium detections in some very low mass stellar and substellar cluster members. Very low mass stars of clusters younger than the Pleiades are expected to preserve their atmospheric lithium; thus, the lithium boundary here takes place at greater luminosities than the substellar mass limit. The results by Stauffer *et al.* (1999) and Barrado y Navascués *et al.* (1999) fully agree with the numbers of Table 2. This is a consistency test for models.

The most recent surveys in the Pleiades (see Table 1) were looking for brown dwarfs fainter and less massive than Teide 1 and Calar 3. Because the $(R - I)$ colour saturates at around M8 spectral type ($R - I = 2.6$) in the Pleiades (it does at M7 type in the field), and because cooler objects become brighter in the IR, only those searches using filters redder than the R-band have succeeded. Various groups have covered large areas in the cluster with the IZJ filters (Cossburn *et al.* 1997; Zapatero Osorio *et al.* 1997d, 1999a; Festin 1997, 1998; Pinfield *et al.* 1998). The least massive Pleiades brown dwarf confirmed so far is Roque 25 ($I = 21.2$, Martín *et al.* 1998b; this is the abridged name, the full name is Roque Pleiades 25). Its mass is estimated at 0.035 $M_\odot$ for an age of 120 Myr and solar metallicity. This young brown dwarf has a cool $T_{\rm eff}$ around 2000 K,

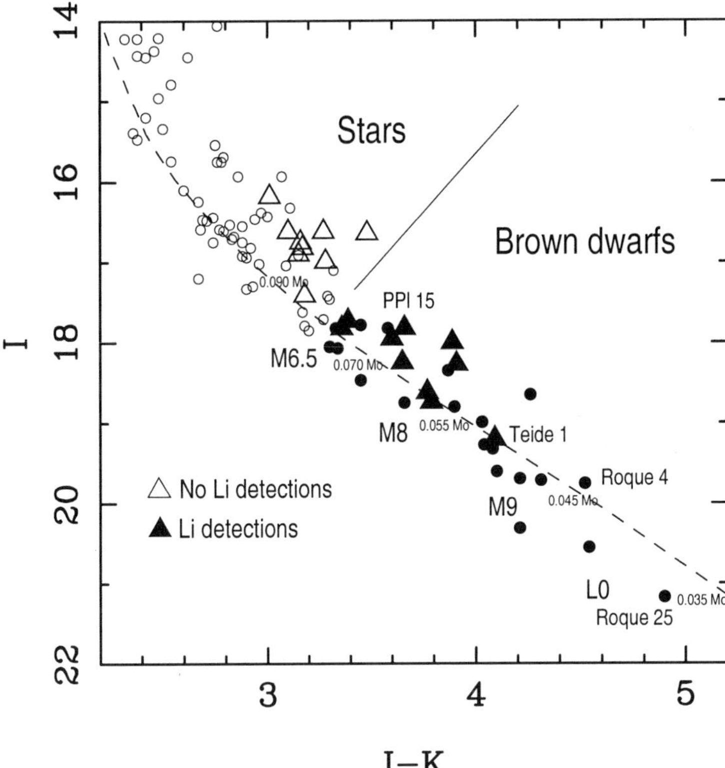

FIGURE 1. Optical-IR colour-magnitude diagram for Pleiades low mass stars (open symbols) and brown dwarfs (filled symbols). The substellar limit at the age of the Pleiades is given by the reappearance of the lithium feature in the spectra of cluster members (denoted with triangle-shaped symbols in this figure). Spectral types are provided on the left side of the cluster sequence. Overplotted to the data is the theoretical dust-free, NextGen 120 My-isochrone (dashed line) from Baraffe *et al.* (1995) shifted to the distance and reddening of the Pleiades. Masses in solar units are indicated.

therefore with a spectral energy distribution beyond the M-class. Classifed as an L0.5 object (Martín *et al.* 1999), Roque 25 represents the transition from M type to L type in the Pleiades as well as it is a benchmark brown dwarf that may serve as a guide for future deep searches in young clusters and the field.

The substellar photometric and spectroscopic sequences in the Pleiades are becoming quite well defined as it is inferred from the numerous candidates being confirmed in the last years. Brown dwarfs in this cluster have spectral types later than M6.5 and their spectroscopic properties are described elsewhere in these proceedings (see papers by Martín, Basri and Pavlenko). As expected for such cool objects, they show red colours particularly in the IR. Future searches are recommended to be done using filters centered at effective wavelengths larger than 800 nm since colours like $I - Z$, $I - J$ keep increasing down to temperatures as cool as 1000 K. Figure 1 shows the I *versus* $I - K$ diagram for very low mass stars and brown dwarfs in the Pleiades as it is known to date (unfortunately not all confirmed members have K photometry available, thus they are not plotted in this figure). Some objects mentioned in this paper are labelled and different symbols are used in order to summarize several items described previously in the text.

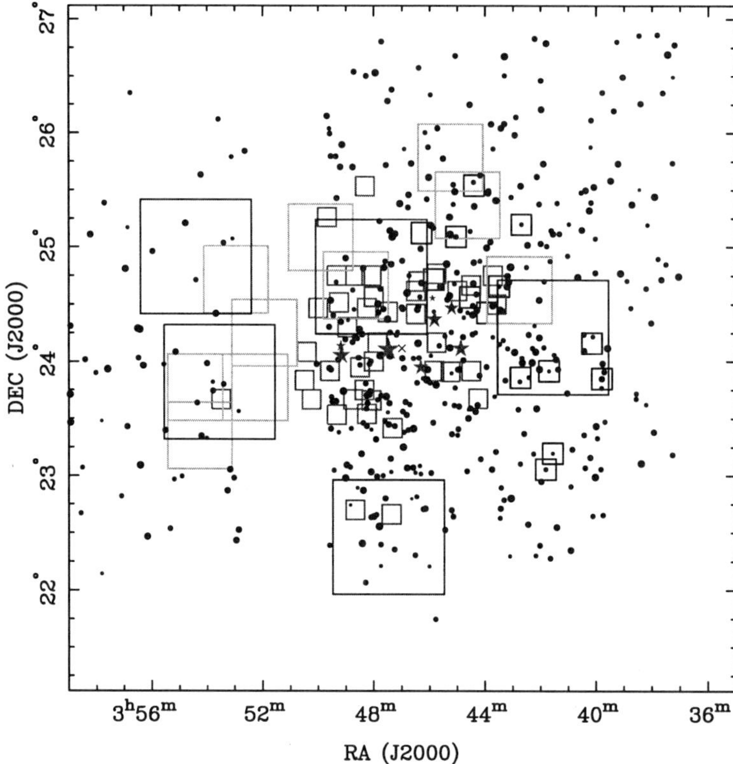

FIGURE 2. The Pleiades region (6° × 6°). The area covered by several recent CCD-based surveys is indicated with squares of different sizes (Zapatero Osorio *et al.* 1999a; Bouvier *et al.* 1998; Pinfield *et al.* 1998; Stauffer *et al.* 1998). HHJ stars are plotted, and the "Seven Sisters" are denoted with the star symbol. Recent proper motion studies in the cluster (Hambly *et al.* 1999) yield that low mass members are found at radii from the centre of up to ~3°.

2.3. *The low mass stellar and substellar mass function*

Membership confirmations for complete lists of Pleiades candidates have been achieved by proper motions, optical/IR photometry and spectroscopy. This procedure is crucial for a reliable study of the cluster mass function. Several authors (Hambly *et al.* 1991, 1999; Zapatero Osorio *et al.* 1997d; Bouvier *et al.* 1998; Festin 1998; Pinfield *et al.* 1998; see Jameson *et al.* these proceedings) have addressed this topic based on their own surveys. Among these searches only those using photographic plates (proper motion analysis) cover the whole cluster area (extending over 30 deg^2) with completeness down to magnitudes nearby the substellar limit (Hambly *et al.* 1991, 1999). Thus, they provide accurate luminosity and mass functions for the Pleiades low mass stellar regime. Deeper surveys are then expected to facilitate data complete enough for deriving the mass function in the brown dwarf domain. Figure 2 provides an overview of the most recent, largest CCD-imaged areas in the Pleiades region; these surveys have sampled a significant fraction of the cluster.

Based on the photometric work by Zapatero Osorio *et al.* (1999a), one of the deepest searches performed on the central 1 deg^2 of the Pleiades (see Table 1), we have derived the cluster substellar mass function illustrated in Fig. 3. Zapatero Osorio *et al.*'s survey is complete in the mass range 0.07–0.04 $M_\odot$, and all their candidates have been investigated for membership (optical/IR photometry and optical spectroscopy are available). The

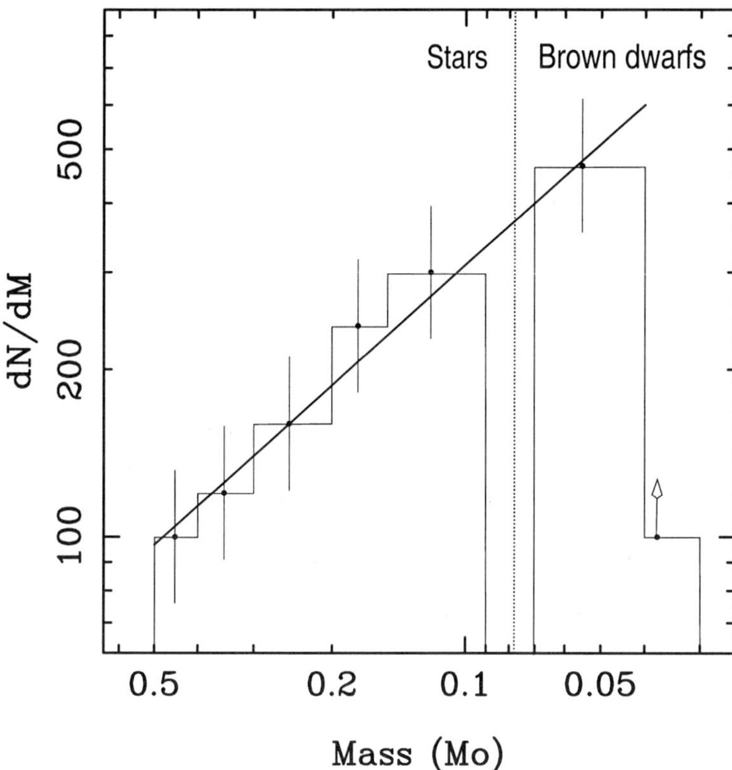

FIGURE 3. The Pleiades mass function for very low mass stars and brown dwarfs down to $0.04\,M_{\odot}$ ($M_I = 15.1$ at the age of 120 Myr) within the central 1 deg^{2}region of the cluster. Overplotted to the data is the power law fit with $\alpha = 0.75 \pm 0.25$ (dN/d$M \sim M^{-\alpha}$). The mass bin at $0.035\,M_{\odot}$ represents a lower limit as it corresponds to uncompleteness in Zapatero Osorio *et al.*'s (1999a) survey. Poissonian error bars are plotted.

function displayed in Fig. 3 only considers those star and brown dwarf candidates that have turned out to be real Pleiads. In this sense, our mass function is the first one to include bona-fide substellar examples with masses as low as 4% solar. We have adopted an age of 120 Myr and the mass–luminosity relationship given by the NextGen isochrone from Baraffe *et al.* (1995). The mass function for cluster stars has been obtained counting the HHJ proper-motion members that overlap with the surveyed area. This stellar part of the function shows the same slope and structure than that from Hambly *et al.* (1991, 1999), indicating that our brown dwarf mass function must also be representative of the whole Pleiades.

It can be seen from the diagram of Fig. 3 that the mass function (in a log-normal representation) is increasing through both the stellar and substellar regimes, and that a single power law fit ($\alpha = 0.75 \pm 0.25$, dN/d$M \sim M^{-\alpha}$) facilitates a reasonable representation of the Pleiades data in the mass range 0.5–0.04 $M_{\odot}$. This result is fully consistent with the values of α (0.6–1.3) found in the literature for this cluster. Brown dwarfs down to $0.04\,M_{\odot}$ have been originated in the Pleiades quite copiously and a few hundreds of them are expected to populate the cluster. The mass function for lower masses is still uncertain due to the uncompleteness of all surveys at such faint magnitudes. The extrapolation of our mass function in the mass range 0.04–0.03 $M_{\odot}$ yields that about 5 brown dwarfs should have been found in the 1 deg^2-survey by Zapatero Osorio *et al.* in the case of completeness. So far only Roque 25 ($0.035\,M_\odot$) has been confirmed as a likely cluster member. Other similar deep surveys (Festin 1997, 1998) provide with fainter candidates for which follow-up spectroscopic observations are important in order to assess their membership status. Nevertheless, the number of photometric candidates is rather small; this might be a hint for the Pleiades substellar mass function flattening beyond $0.04\,M_\odot$ at the central region of the cluster. This must be addressed using very deep photometric and spectroscopic surveys and covering a significant area of the Pleiades. Our mass function imply that brown dwarfs with masses 0.075–$0.04\,M_\odot$ contribute only ~1–2% to the total mass of the cluster, and a limit of 10% can be imposed to the total mass contribution of objects down to the deuterium mass limit.

3. Other young clusters

3.1. *Hyades and Praesepe*

It is of great observational and theoretical interest to identify very low mass stars and brown dwarfs with different ages. This will help to characterize and understand their evolution. Therefore, photometric and spectroscopic surveys have to be extended to other galactic clusters. Large proper motion and broad-band photometric studies have been recently conducted in Praesepe (Hambly *et al.* 1995; Pinfield *et al.* 1997; Magazzù *et al.* 1998) and the Hyades (Leggett *et al.* 1994; Bryja *et al.* 1994; Reid & Hawley 1999), allowing the determination of the photometric and spectroscopic sequences of the low mass star members down to the hydrogen burning limit. These clusters (600–900 Myr) are rich in their content and show a stellar mass function that resembles that of the Pleiades, which may indicate that the formation of stars in these three galactic clusters was not distinct. Because of the large distance to Praesepe brown dwars in this cluster are quite faint ($I, K > 21.5$, 17, see Table 2), and their discovery requires large-size telescopes. The Hyades is much closer, but no substellar members have been unambiguously identified so far. Brown dwarfs at the age of 600–900 Myr have cooled down considerably and therefore, they have very low luminosities and must display spectral energy distributions within the L-class. Current surveys in the Hyades are not sufficiently sensitive to detect isolated substellar-mass dwarfs. The recent search based on 2MASS JHK data (Gizis *et al.* 1999) is so far the deepest IR survey in the Hyades and its detectability for brown dwarfs is only marginal: the limiting magnitudes of the 2MASS survey are very close to those expected for the substellar limit at the age and distance of the cluster (see Table 2).

3.2. *Younger clusters*

The search for substellar objects in clusters younger than the Pleiades and in young stellar associations sheds new light on the knowledge of the initial mass function and on the theories of substellar formation. Intermediate-age clusters like α Persei and IC 2391 have been extensively surveyed in the very recent years (Prosser 1992, 1994; Stauffer *et al.* 1999; Barrado y Navascués *et al.* 1998, 1999) resulting in new findings of massive brown dwarf candidates. The lithium test has been applied to them and because of their high lithium abundance and their location in the luminosity-temperature diagram they must be substellar.

Age is a crucial parameter for discovering low mass and very low mass brown dwarfs ($M < 0.04\,M_\odot$). The youth of clusters like IC 348, and star-formation regions like Orion, ρ Ophiuchi and Chamaleon more than compensates for their large distances. Substellar-mass dwarfs of 0.04–$0.025\,M_\odot$ are identified in ρ Ophiuchi (Luhman *et al.* 1999; Wilking *et al.* 1999), IC 348 (Luhman *et al.* 1998), and σ Orionis (Béjar *et al.* 1999). In particular, the σ Orionis cluster (around the multiple O9.5V-type σ star of the constellation of Orion) is an ideal target where the identification of objects with all possible masses (stellar, brown dwarf and planetary masses) can be feasible. With an estimated age smaller than 5 Myr no strong nor variable reddening is known to affect the cluster area. The least massive brown dwarf known so far, S Ori 47 (Zapatero Osorio *et al.* 1999b), has been found in σ Orionis with a mass estimated at the deuterium burning limit ($\sim$0.015 $M_\odot$). This object is the second example of an L-type young brown dwarf in a cluster. Classified as an early-L class substellar low-gravity dwarf, S Ori 47 represents for the σ Orionis cluster the frontier between brown dwarfs and planets (between substellar objects that will deplete deuterium at any time in their lifes from those that will never). It will be very interesting to see the results of deeper surveys being performed on this cluster. The initial low mass stellar and substellar mass function (in the range 0.5–0.04 $M_\odot$) inferred for these very young clusters (see references above) behaves similarly (α spanning from 0.7 up to 1.8) to that presented in Fig. 3 for the Pleiades. This may be indicative of the fact that related mechanisms are taking place in the origin of stars and brown dwarfs in the Galaxy. The existence of substellar members around the deuterium burning mass limit populating very young clusters is an evidence for the initial mass function extending down to such low masses.

4. Conclusions

Solar metallicity, free-floating brown dwarfs have been identified as members of young open clusters (Pleiades, α Persei, IC 2391, σ Orionis, IC 348, ρ Ophiuchi) with masses spanning from the substellar boundary (0.075 $M_\odot$) down to the deuterium burning mass limit ($\sim$0.015 $M_\odot$, S Ori 47 in σ Orionis). These findings have provided an accurate determination of the substellar mass limit in the Pleiades cluster. It is located at M6.5 spectral type, and absolute magnitudes $M_I = 12.3$, $M_J = 10.1$ and $M_K = 9.0$. The identification of the first examples of L-type brown dwarfs in young clusters allow us to determine that the transition between M- and L-class ($T_{\rm eff}$ $\sim$ 2200 K) happens at a mass of 0.04 $M_\odot$ at the age of the Pleiades (120 Myr) and at 0.02 $M_\odot$ at the age of the σ Orionis cluster ($\sim$5 Myr). So far the number of accepted brown dwarfs populating clusters younger than 150 Myr outnumbers 50, and a similar quantity of photometric candidates still remain confirmation.

Young open clusters provide the most direct means of measuring the initial mass function through the whole mass range. The Pleiades low mass stellar and substellar mass function in the interval 0.5–0.04 $M_\odot$ can be fitted by a single power law with $\alpha = 0.75 \pm 0.25$ ($\mathrm{d}N/\mathrm{d}M \sim M^{-\alpha}$); other young clusters significantly present similar initial mass functions. This function for lower masses is quite uncertain in the Pleiades due to the incompleteness of the surveys. The detection of reliable candidate members with masses in the interval 0.04–0.015 $M_\odot$ in younger clusters does provide substantial evidence for the initial mass function extending down to the deuterium burning limit. Future deep surveys in compact clusters with ages smaller than 50 Myr are encouraged as they can provide important data for the knowledge of the initial mass function at very low masses, and they will contribute to the understanding of the origin and formation processes of these objects.

REFERENCES

BARAFFE, I., CHABRIER, G., ALLARD, F., HAUSCHILDT, P. 1995 New Evolutionary Tracks for Very Low Mass Stars *ApJ* **466**, L35

Barrado y Navascués, D., Stauffer, J. R., Briceño, C. 1998 Deep photometric survey of IC 2391. Poster contribution of this meeting

Barrado y Navascués, D., Stauffer, J. R., Patten, B. M. 1999 The lithium-depletion boundary and the age of the young open cluster IC 2391 *ApJ* **522**, L53

Basri, G. 1998 The lithium test for young brown dwarfs *ASP Conf. Series* **134**, 394

Basri, G., Marcy, G. W., Graham, J. R. 1996 Lithium in Brown Dwarf Candidates: The Mass and Age of the Faintest Pleiades Stars *ApJ* **458**, 600

Basri, G., Martín, E. L. 1999 PPL 15: The First Brown Dwarf Spectroscopic Binary *AJ* **118**, 2460

Béjar, V. J. S., Zapatero Osorio, M. R., Rebolo, R. 1999 A search for very low mass stars and brown dwarfs in the young σ Orionis Cluster *ApJ* **521**, 671

Bryja, C., Humphreys, R. M., Jones, T. J. 1994 The lowest mass stars in the Hyades *AJ* **107**, 246

Bouvier, J., Stauffer, J. R., Martín, E. L., *et al.* 1998 Brown dwarfs and very low-mass stars in the Pleiades cluster: a deep wide-field imaging survey *A&A* **336**, 490

Burrows, A., Hubbard, W. B., Saumon, D., Lunine, J. I. 1993 An expanded set of brown dwarf and very low mass star models *ApJ* **406**, 158

Cossburn, M. R., Hodgkin, S. T., Jameson, R. F., Pinfield, D. J. 1997 Discovery of the lowest mass brown dwarf in the Pleiades *MNRAS* **288**, 23

D'Antona, F., Mazzitelli, I. 1994 New pre-main-sequence tracks for M less than or equal to 2.5 solar mass as tests of opacities and convection model *ApJS* **90**, 467

Festin, L. 1997 Brown dwarfs in the Pleiades. A deep IJK survey *A&A* **322**, 455

Festin, L. 1998 Brown dwarfs in the Pleiades. II. A deep optical and near infrared survey *A&A* **333**, 497

Gizis, J. E., Reid, I. N., Monet, D. G. 1999 A 2MASS survey for brown dwarfs toward the Hyades *AJ* **118**, 997

Greene, T. P., Meyer, M. R. 1995 An infrared spectroscopic survey of the ρ Ophiuchi young stellar cluster: masses and ages from the H-R diagram *ASP Conf. Series* **134**, 11

Hambly, N. C. 1998 Deep searches in open clusters *ASP Conf. Series* **134**, 11

Hambly, N. C., Hodgkin, S. T., Cossburn, M. R., Jameson, R. F. 1999 Brown dwarfs in the Pleiades and the initial mass function across the stellar/substellar boundary *MNRAS* **303**, 835

Hambly, N. C., Hawkins, M. R. S., Jameson, R. F. 1993 Very low mass proper motion members in the Pleiades *A&AS* **100**, 607

Hambly, N. C., Jameson, R. F., Hawkins, M. R. S. 1991 A deep proper motion survey of the Pleiades for low-mass stars and brown dwarfs *MNRAS* **253**, 1

Hambly, N. C., Steele, I. A., Hawkins, M. R. S., Jameson, R. F. 1995 The very low-mass main sequence in the Galactic cluster Praesepe *MNRAS* **273**, 505

Herbig, G. H. 1998 The young cluster IC 348 *ApJ* **497**, 736

Jameson, R. F., Skillen, I. 1989 A search for low-mass stars and brown dwarfs in the Pleiades *MNRAS* **239**, 247

Kirkpatrick, J. D., Reid, I. N., Liebert, J., et al. 1999 Dwarfs Cooler than "M": The Definition of Spectral Type "L" Using Discoveries from the 2 Micron All-Sky Survey (2MASS) *ApJ* **519**, 802

Leggett, S., K., Harris, H. C., Dahn, C. C. 1994 Low-mass stars in the central region of the Hyades cluster *AJ* **108**, 944

Luhman, K. L., Liebert, J., Rieke, G. H. 1999 Spectroscopy of a young brown dwarf in the ρ Ophiuchi cluster 1 *ApJ* **489**, L165

Luhman, K. L., Rieke, G. H., Lada, C. J., Lada, E. A. 1998 Low-mass star formation and the initial mass function in IC 348 *ApJ* **508**, 347

Magazzù, A., Martín , E. L., Rebolo, R. 1993 A spectroscopic test for substellar objects *ApJ* **404**, L17

MAGAZZÙ, A., REBOLO, R., ZAPATERO OSORIO, M. R., MARTÍN, E. L., HODGKIN, S. T. 1998 A Brown Dwarf Candidate in the Praesepe Open Cluster *ApJ* **497**, L47

MARCY, G. W., BASRI, G., GRAHAM. J. R. 1994 A search for lithium in Pleiades brown dwarf candidates using the Keck HIRES echelle *ApJ* **428**, L57

MARTÍN, E. L. 1993 T Tauri stars and brown dwarfs. PhD. Thesis, Univ. of La Laguna, Tenerife, Spain

MARTÍN, E. L., BASRI, G., GALLEGOS, J. E., REBOLO, R., ZAPATERO OSORIO, M. R., BÉJAR, V. J. S. 1998a A New Pleiades Member at the Lithium Substellar Boundary *ApJ* **499**, L61

MARTÍN, E. L., BASRI, G., ZAPATERO OSORIO, M. R., REBOLO, R., GARCÍA LÓPEZ, R. J. 1998b The First L-Type Brown Dwarf in the Pleiades *ApJ* **507**, L41

MARTÍN, E. L., DELFOSSE, X., BASRI, G., GOLDMAN, B., FORVEILLE, T., ZAPATERO OSORIO, M. R. 1999 Spectroscopic classification of late-M and L field dwarfs *AJ* **118**, 2466

MARTÍN, E. L., REBOLO, R., MAGAZZÙ, A. 1994 Constraints to the masses of brown dwarf candidates form the lithium test *ApJ* **436**, 262

MARTÍN, E. L., REBOLO, R., ZAPATERO OSORIO, M. R. 1996 Spectroscopy of New Substellar Candidates in the Pleiades: Toward a Spectral Sequence for Young Brown Dwarfs *ApJ* **469**, 706

PATTEN, B. M., PAVLOVSKY, C. M. 1999 A Photometric Survey for Low-Mass Members of IC 2391 *PASP* **111**, 210

PINFIELD, D. J., COSSBURN, M. R., HODGKIN, S. T., JAMESON, R. F. 1998 A deep 5 square degree Pleiades brown dwarf survey. Poster contribution at this meeting.

PINFIELD, D. J., HODGKIN, S. T., JAMESON, R. F., COSSBURN, M. R., VON HIPPEL, T. 1997 Brown dwarf candidates in Praesepe *MNRAS* **287**, 180

PROSSER, C. F. 1992 Membership of low-mass stars in the open cluster Alpha Persei *AJ* **103**, 488

PROSSER, C. F. 1994 Photometry and spectroscopy in the open cluster alpha Persei, 2 *AJ* **107**, 625

REBOLO, R., MARTÍN, E. L., BASRI, G., MARCY, G. W., ZAPATERO OSORIO, M. R. 1996 Brown Dwarfs in the Pleiades Cluster Confirmed by the Lithium Test *ApJ* **469**, L53

REBOLO, R., MARTÍN, E. L., MAGAZZÙ, A. 1992 Spectroscopy of a brown dwarf candidate in the Alpha Persei open cluster *ApJ* **389**, L83

REBOLO, R., ZAPATERO OSORIO, M. R., MARTÍN, E. L. 1995 Discovery of a Brown Dwarf in the Pleiades Star Cluster *Nature* **377**, 129

REID, I. N., HAWLEY, L. S. 1999 Brown dwarfs in the Hyades and beyond? *AJ* **117**, 343

ROLLESTON, W. R., J., BYRNE, P. B. 1997 Photometrically determined membership of the young, open cluster IC 2391 *A&AS* **126**, 357

SIMONS, D. A., BECKLIN, E. E. 1992 A near-infrared search for brown dwarfs in the Pleiades *ApJ* **390**, 431

STAUFFER, J. R. 1996 Open cluster membership, photometry and rotation *ASP Conf. Series* **109**, 305

STAUFFER, J. R., BARRADO Y NAVASCUÉS, D., *et al.* 1999 Keck spectra of brown dwarf candidates and a precise determination of the lithium depletion boundary in the α Persei open cluster *AJ* **527**, 219

STAUFFER, J. R., HAMILTON, D., PROBST, R. 1994 A CCD-based search for very low mass members of the Pleiades cluster *AJ* **108**, 155

STAUFFER, J. R., HAMILTON, D., PROBST, R., RIEKE, G., MATEO, M. 1989 Possible Pleiades members with M of about 0.07 solar mass - Identification of brown dwarf candidates of known age, distance, and metallicity *ApJ* **344**, L21

STAUFFER, J. R., SCHULTZ, G., KIRKPATRICK, J. D. 1998 Keck spectra of Pleiades brown dwarf candidates and a precise determination of the lithium depletion edge in the Pleiades *ApJ* **499**, L199

WILKING, B. A., GREENE, T. P., MEYER, M. R. 1999 Spectroscopy of brown dwarf candidates in the ρ Ophiuchi molecular core *AJ* **117**, 469

WILLIAMS, D. M., BOYLE, R. P., MORGAN, W. T., *et al.* 1996 Very low mass stars and substellar objects in the Pleiades *ApJ* **464**, 238

ZAPATERO OSORIO, M. R. 1997c Brown Dwarfs in the Pleiades. PhD. Thesis, Univ. of La Laguna, Tenerife, Spain

ZAPATERO OSORIO, M. R., BÉJAR, V. J. S., REBOLO, R., MARTÍN, E. L., BASRI, G. 1999b An L-type substellar object in Orion: reaching the mass boundary between brown dwarfs and giant planets *ApJ* **524**, L115

ZAPATERO OSORIO, M. R., MARTÍN, E. L., M. R., REBOLO, R. 1997b Brown dwarfs in the Pleiades cluster. II. J, H and K photometry *A&A* **323**, 105

ZAPATERO OSORIO, M. R., REBOLO, R., MARTÍN, E. L. 1997a Brown dwarfs in the Pleiades cluster: a CCD-based R, I survey *A&A* **317**, 164

ZAPATERO OSORIO, M. R., REBOLO, R., MARTÍN, E. L., *et al.* 1997d New brown dwarfs in the Pleiades cluster *ApJ* **491**, L81

ZAPATERO OSORIO, M. R., REBOLO, R., MARTÍN, E. L., *et al.* 1999a Brown dwarfs in the Pleiades cluster. III. A deep IZ survey *A&AS* **134**, 537

ZAPATERO OSORIO, M. R., REBOLO, R., MARTÍN, E. L., GARCÍA LÓPEZ, R. J. 1996 Stars approaching the substellar limit in the alpha Persei open cluster *A&A* **305**, 519

The Mass Function of the Pleiades

By R. F. JAMESON[1], S. T. HODGKIN[1], D. PINFIELD[2], AND M. R. COSSBURN[1]

[1]Department of Physics and Astronomy, University of Leicester, University Road, Leicester, LE1 7RH, UK

[2]Department of Pure and Applied Physics, The Queen's University of Belfast, Belfast, BT7 1NN, Northern Ireland

We combine the results from two CCD surveys covering a large area of the custer at I and Z wavebands. We have obtained follow-up K photometry for many of the numerous brown dwarf candidates discovered in these surveys which we employ as a test for cluster membership. From these data we derive the mass function of the whole Pleiades cluster down to 0.04 $M_{\odot}$. We emphasise the importance of a careful consideration of the spatial distribution within the cluster and find the core radius for brown dwarfs to be 2±1 parsecs. The contribution of brown dwarfs to the total mass of the cluster is about 1%.

---

1. Introduction

The Pleiades has long been recognised as one of the best places to search for brown dwarfs, e.g. Jameson & Skillen (1989), Stauffer et al. (1989, 1994), Simons & Becklin (1992), Rebolo et al. (1995), Cossburn et al. (1997), Zapatero Osorio et al. (1997), Bouvier et al. (1998), Festin (1998).

The cluster is both reasonably close (but not so close as to cover too large an area of the sky) and young, so that brown dwarfs are not too faint. Controversy still rages over the precise distance to the Pleiades, which Hipparcos places significantly closer (at 118 parsecs) than ground based measurements (at typically 133 parsecs). The Hipparcos results have been published by Van Leeuwen & Hansen Ruiz (1997) and Mermilliod et al. (1997) and critically discussed by Pinnsoneault et al. (1998). In this paper we adopt a distance modulus of 5.65 (135 parsecs) as we feel that the Hipparcos distance has not yet been proved correct.

We take the age of the cluster to be $\sim$ 125 million years, after Stauffer et al (1998) extended the Lithium dating technique first employed by Basri et al. (1996) for the brown dwarf PPL 15 (now shown to be a binary, see Basri & Martin this volume).

One of the original reasons for searching for and studying brown dwarfs was to see if they might contribute significantly to local baryonic dark matter. If the Pleiades is assumed to be typical of Population I stars, then the cluster currently offers the best chance of answering this question, at least for Population I material.

2. The surveys

The candidate lists for this paper come from two recent surveys. The first was carried out using the prime focus CCD camera on the 2.5m Isaac Newton Telescope (INT) at the Observatorio del Roque de los Muchachos on La Palma. Following Cossburn et al. (1997), we used I and Z filters (with a few fields also measured at R) to survey one square degree in the central region of the cluster down to a completeness limit of I=21, Z=20.5. This survey, named the ITP survey, is described in Zapatero Osorio et al. (1997, 1998).

The second survey, also using I and Z, was made with the Burrell Schmidt telescope (0.6m) at Kitt Peak. It covered a massive 5 square degrees of the cluster, but suffered a much larger pixel size (2 arcseconds) and was less sensitive, complete to I=19.5 (see Pinfield 1997).

These two surveys generated a long list of possible candidates, selected from the I, I–Z colour magnitude diagrams, see Zapatero Osorio *et al.* (1998), Pinfield (1997).

3. Brown dwarf selection

Before constructing the luminosity and mass functions it is crucial to define a sample which is as unbiased and complete as possible. We have restricted ourselves to a magnitude range (and therefore mass range) over which both surveys are complete. We set the upper limit to be at I=17.6 which approximately marks the substellar limit in the Pleiades. The lower bound for the Kitt Peak survey, which is complete to I_C=19.5, corresponds to a mass of 0.05 $M_\odot$ according to Baraffe et al. (1998). For the ITP data, the lower bound, at I=21.0, is equivalent to 0.04 $M_\odot$ from this same model.

We then observed as many of these Pleiades brown dwarfs as possible using IRCAM3 at the United Kingdom Infra Red Telescope (UKIRT) in Hawaii. We observed targets at K so that we could place them on the I, I–K diagram, which gives a much longer baseline colour and improves separation between the cluster sequence and the field. There is no obvious clear gap between where we believe the cluster sequence lies (the upper isochrone) and the field. Remaining contaminants are likely to be late M dwarfs which are roughly at the same distance as the cluster, or reddened background stars. We expect no contamination from heavily reddened background giants (see e.g. Festin 1998). To aid candidate selection we therefore argue that a Pleiades cluster member must be brighter then the field star sequence at the distance of the Pleiades. This approach should not lose any Pleiades members from the sample, though would not reject a very cool field star at the front of the cluster. However, the space density of such objects is so low that we expect negligable contribution to our luminosity and mass functions (see Festin 1998). In Table 1 we list candidates from the two surveys, not yet ruled out on the basis of infrared photometry, i.e. they satisfy the above criteria (plotted in Figure 3). There are still a handful of candidates with no infrared photometry. We have decided to include these in our sample and they count towards the luminosity and mass function at present.

4. Spatial distribution

In order to derive the luminosty and mass functions for the whole cluster we need to understand how both the stars and the brown dwarfs are distributed so that we can correctly compute the number of objects in unsurveyed areas. To improve counting statistics we combine objects from both surveys (and many are common to both surveys, see Table 1) where the two datasets are overlapping, i.e. for the range I=17.6–19.5.

Pinfield et al. (1998) have shown that the spatial distribution of Pleiades stars, divided into mass bins, are well fitted by King (1962) profiles. They derive a tidal radius of 13.1 parsecs for the cluster, and show that the core radius (r_c) of these profiles increases as $m^{-\frac{1}{2}}$, where m is the average stellar mass, as predicted by theory for a relaxed cluster. There is a suggestion in Pinfield et al. (1998) that the lowest mass bin may not be following the $m^{-\frac{1}{2}}$ relation, but is flattening off at about $r_c \sim 3$ parsecs. One explanation is that the lowest stars are not fully relaxed. To follow this up, we have subdivided the lowest mass bin from Pinfield et al. (1998) into several bins and redetermined the core radii (see Figure 4). It is now rather clear that the core radius flattens off at the lowest masses.

It is crucial, therefore, to determine the core radius for brown dwarfs. Surveys for

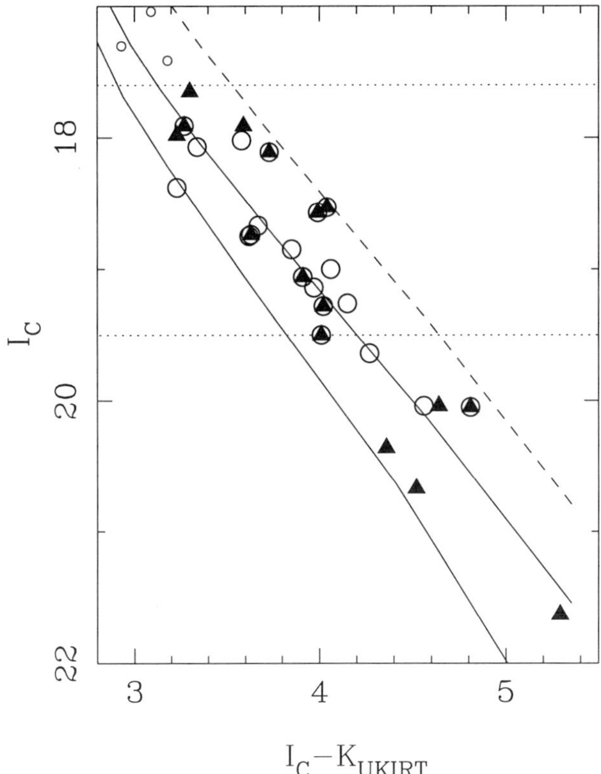

FIGURE 1. The I, I-K diagram for Pleiades brown dwarf candidates from the two surveys considered in this paper. Cluster non-members are not shown. Objects from the ITP survey are shown with a solid triangle, and Kitt Peak objects are shown with a large open circle (the smaller circles at the top of the figure are three of the faintest HHJ stars). The leftmost curve is the 5 Gyr isochrone (for solar metallicity) from Baraffe et al. (1998) offset to the distance of the Pleiades (m-M=5.65). The single star (solid) and binary (dashed) sequences for 120 Myr are also shown, again from Baraffe et al. (1998). The horizontal dotted lines represent the upper and lower bounds for inclusion in the spatial analysis.

brown dwarfs have naturally concentrated on the central region of the cluster in order to maximise the chance of finding brown dwarfs. Clearly, if we combine the central surface density with a brown dwarf core radius of say 6 parsecs, as might be expected for a fully relaxed cluster, we will obtain an erroneous mass function.

In Figure 4 we show the spatial distribution of brown dwarfs discovered by the two surveys (in the magnitude range I=17.6–19.5). We indicate the surveyed areas with rectangles; the FOV of the Kitt Peak survey was one square degree, while the ITP surveyed covered 100 square arcminutes in each image. The concentric circles divide the cluster into annuli, enabling us to compute the surface density of brown dwarfs as a function of radius from the cluster centre. The King function for surface density is given by

$$f = k\left[\frac{1}{\sqrt{1+x}} - \frac{1}{\sqrt{1+x_t}}\right]^2 \tag{4.1}$$

where $x = (r/r_c)^2$ and $x_t = (r_t/r_c)^2$, with r_c the core radius, r_t the tidal radius. We take $r_t = 13.1$ parsecs from Pinfield et al. (1998) and can thus in principle fit the observed

TABLE 1. RIZK photometry for Pleiades brown dwarfs

Name						I	I–Z	R–I	K
Burrell	**ITP**	**CFHT**	**NOT**	**Proper Motion**	**Other**				
	Roque 15				PPL 1	17.65	1.02	2.46	14.35
IZpl43		CFHT-PL-9				17.71	1.00	2.18	
				IPMBD23	PPL 15	17.91	1.13	2.28	14.32
IZpl32	Roque 16	CFHT-PL-11				17.91	0.96	2.21	14.64
IZpl36						17.91	0.96		
	Roque 47			IPMBD20		17.98	0.86		14.75
IZpl74		CFHT-PL-12				18.00	1.12	2.18	
IZpl59		CFHT-PL-13			Teide 2	18.02	0.97	2.21	14.44
IZpl85				IPMBD11		18.07	1.08		14.73
IZpl2						18.09	1.02		
IZpl29	Roque 17					18.11	1.14		14.38
IZpl80						18.38	1.07		15.15
IZpl21	Roque 14					18.53	1.11		14.49
IZpl14	Roque 13					18.57	1.10		14.58
IZpl5						18.67	1.11		15.0
IZpl39	Roque 12		NPL 36			18.74	0.96		15.11
IZpl55						18.75	1.15		15.13
IZpl15						18.85	1.37		15.00
IZpl52		CFHT-PL-21			Calar 3	19.00	1.20	2.50	14.94
IZpl26	Roque 11		NPL 37			19.06	1.30		15.15
IZpl79						19.14	1.13		15.17
IZpl27			NPL 39		Teide 1	19.26	1.28	2.74	15.11
IZpl18	Roque 9					19.28	1.02		15.26
IZpl6	Roque 7	CFHT-PL-24				19.50	1.11	2.61	15.49
IZpl42						19.64	1.34		15.37
IZpl77		CFHT-PL-25				19.69	1.26	2.57	
	Roque 5					20.04	1.14		15.40
IZpl58						20.04	1.75		15.48
IZpl8	Roque 4					20.05	1.13		15.24
	Roque 33		NPL 40			20.36	1.25		16.00
	Roque 30					20.67	1.22		16.15
	Roque 25					21.63	1.49		16.34

We list objects from the Kitt Peak survey (Pinfield 1997) and the ITP survey (Zapatero Osorio et al. 1998) with cross-identifications from Bouvier et al. (1998), Festin (1998), Hambly et al. (1998), Stauffer et al. (1998), Martin et al. (1998), Rebolo et al. (1996). Optical photometry is compiled from these references (with the least weight given to the Kitt Peak survey). Infrared photometry is predominantly measured by ourselves with a few K magnitudes derived from the literature: PPL 15 from Basri et al. (1996), Teide 2 from Martin et al. (1998), Rebolo et al. (1996).

surface density and determine r_c and k. The number of brown dwarfs in the survey is small however, despite having surveyed a relatively large fraction of the central annuli (around 43% out to 5.17 parsecs) leading to large uncertainties in k and r_c. However, we can also make use of the cumulative distribution, obtained by integrating Equation 1 with respect to $2\pi r dr$, which gives the total number of stars in projection within a distance r (or $x = r/r_c$) of the centre of the cluster.

$$n(x) = \pi r_c^2 k \left[\ln(1+x) - 4\frac{\sqrt{1+x}-1}{\sqrt{1+x_t}} + \frac{x}{1+x_t} \right] \tag{4.2}$$

We can therefore measure the ratio $f/n(x)$ and from Equations 1 and 2 eliminate k and solve numerically to constrain r_c. Using 1σ counting statistics we find $r_c = 2.0 \pm 1.0$ parsecs. A core radius of 2 parsecs, means that the core radius has turned down at the lowest masses. In Figure 4 we show the spatial density as a function of radius compared with the King profile (Equation 1) for a core radius of 2 parsecs. Clearly we need to observe a larger fraction of the cluster much deeper to make any significant improvement

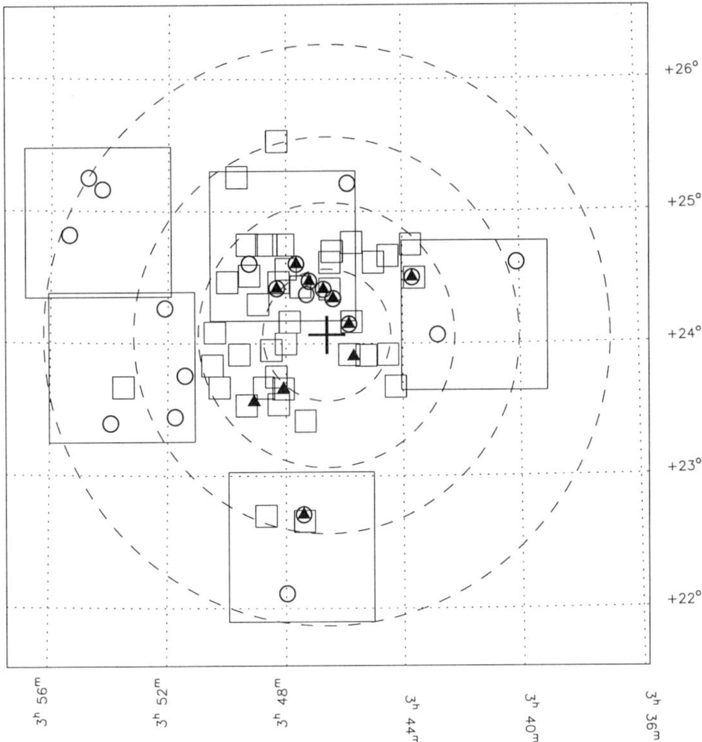

FIGURE 2. The spatial distribution of brown dwarfs in the Pleiades, within the range I=17.6–19.5, from the ITP and Kitt Peak surveys. Symbols as in Figure 1. Annuli are drawn at radii of 0.5 ,1.0, 1.5 and 2.2 degrees (1.17, 2.35, 3.52 and 5.17 parsecs).

to the counting statistics and therefore confirm this result. At the present age of the Pleiades significant numbers of brown dwarfs will not have escaped the cluster and the lowest mass objects are not relaxed. A turnover in the core radius would suggest that brown dwarfs are formed preferentially near the cluster centre.

5. The luminosity and mass functions

To determine the luminosity function for the whole cluster we have assumed $r_c = 3$ parsecs. This we regard as the most 'conservative' assumption, i.e. brown dwarfs are like the low mass stars and are not relaxed. Actually this assumption makes very little difference to the luminosity function, since k trades with r_c to give $\pi r_c^2 k$ about the same value. This is only true when r_c has been determined as above; using $r_c = 6$ parsecs would of course, as remarked earlier, given an incorrect luminosity function.

The luminosity function for the whole cluster is simply determined by calculating the cumulative distribution out to the tidal radius. We have assumed that a core radius of 3 parsecs applies equally to all brown dwarfs including the lowest mass brown dwarfs, i.e. beyond I=19.5. We can only use the ITP survey for these lowest bins, from I=19.5–21.0, as the Kitt Peak data is incomplete here. For the ITP survey, we calculate the cumulative distribution out to $r = 2.35$ parsecs (1 degree) before applying the correction to obtain the total number at r_t, while for the Kitt Peak data we make the correction

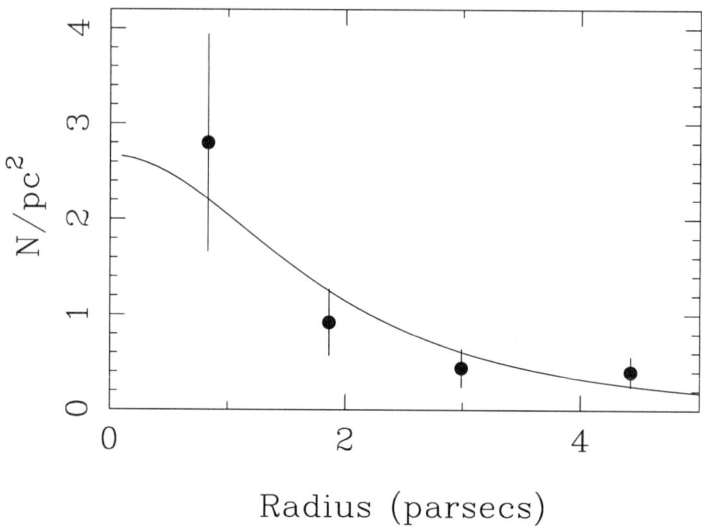

FIGURE 3. Spatial density profile for brown dwarfs with $17.6 \leq I_C \leq 19.5$ with the best fit King profile (Equation 2) for a core radius of 2 parsecs.

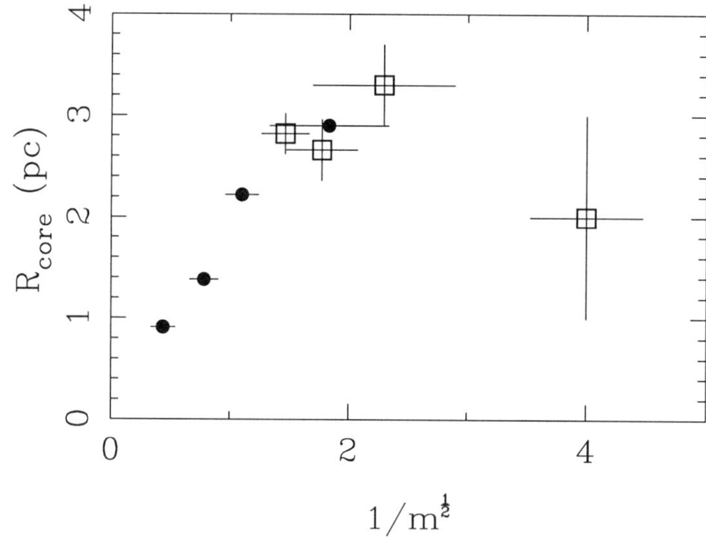

FIGURE 4. Core radii for low mass stars and brown dwarfs derived from King fits to spatial distributions. Solid circles are from Pinfield et al. (1998), open squares are our analysis.

from $r = 5.17$ parsecs (2.2 degrees). The luminosity function for the whole cluster is shown in Figure 5, including stellar data from Hambly et al. (1998).

The mass function for the whole cluster is then calculated by multiplying the luminosity function by the slope of the I-band magnitude-mass relation, again returning to the models of Baraffe et al. (1998). The stellar data is from Hambly et al. (1998) and Van Leeuwen (1980) and is plotted in Figure 6. The mass function is nearly flat across the substellar boundary and into the brown dwarf regime with no evidence for a turnover by 0.04 $M_\odot$.

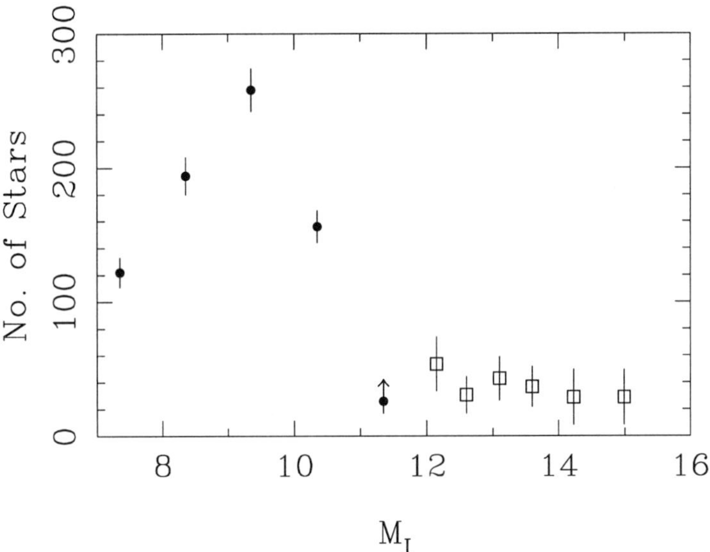

FIGURE 5. Luminosity function for the Pleiades, i.e. the number of stars per unit magnitude in the whole cluster versus absolute I magnitude. Open squares from this analysis. Solid circles from Hambly et al. (1998).

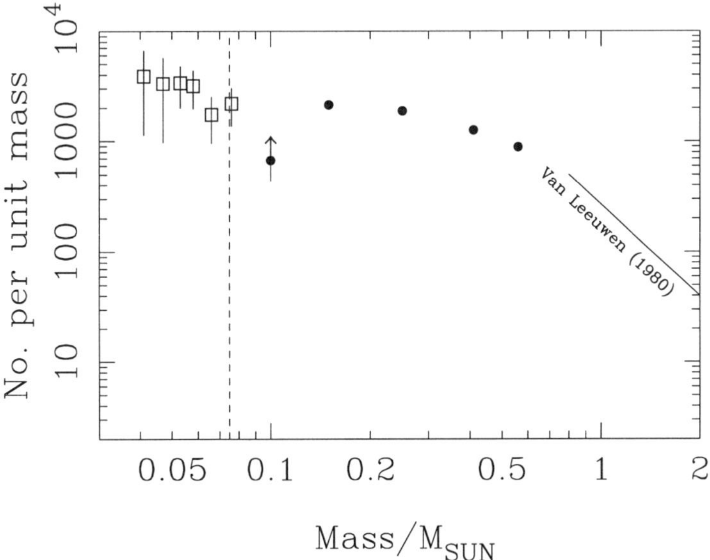

FIGURE 6. Mass function for the Pleiades, i.e. the number of stars per unit mass in the whole cluster as a function of mass. The slope for high mass stars is from Van Leeuwen (1980), symbols as in Figure 5. The substellar limit is shown with a dashed line (at 0.075 $M_\odot$).

6. Conclusions

We have studied the spatial distribution of both stars and brown dwarfs in the Pleiades and find that they can be fitted by King profiles. The core radius for brown dwarfs appears to be less than for the lowest mass stars, but this is only a tentative result, due to the poor statistics.

We have determined the luminosity and mass functions for the whole cluster and find a mass function that is essentially flat for brown dwarfs with no evidence for a turnover by 0.04 $M_{\odot}$. The total mass of brown dwarfs in the cluster is estimated to be about 8 $M_{\odot}$ or about 1% of the total mass of the cluster (750 $M_{\odot}$ according to Pinfield et al. 1998). Thus if the Pleiades are regarded as typical of Population I stars, brown dwarfs cannot make a significant contribution to local dark matter.

We gratefully thank the LOC for providing the invitation and means to attend this conference. Thanks also to Isabelle Baraffe for sending electronic versions of her models. STH acknowledges funding from PPARC.

REFERENCES

BARAFFE, I., CHABRIER, G., ALLARD, F., HAUSCHILDT, P. H. 1998 Evolutionary models for solar metallicity low-mass stars: mass-magnitude relationships and color-magnitude diagrams *A&A* **337**, 403–412.

BASRI, G., MARCY, G. W., GRAHAM, J. R. 1996 Lithium in brown dwarf candidates: the mass and age of the faintest Pleiades stars *ApJ* **458**, 600–609.

BOUVIER, J., STAUFFER, J. R., MARTIN, E. L., BARRADO Y NAVASCUES, D., WALLACE, B., BEJAR, V. J. S. 1998 Brown dwarfs and very low-mass stars in the Pleiades cluster: a deep wide-field imaging survey *A&A* **336**, 490–502.

COSSBURN, M. R., HODGKIN, S. T., JAMESON, R. F., PINFIELD, D. J. 1997 Discovery of the lowest mass brown dwarf in the Pleiades *MNRAS* **288**, L23–L27.

FESTIN, L. 1998 Brown dwarfs in the Pleiades. II. A deep optical and near infrared survey *A&A* **333**, 497–504.

HAMBLY, N. C., HODGKIN, S. T., COSSBURN, M. R., JAMESON, R. F. 1998 Brown dwarfs in the Pleiades and the initial mass function across the stellar/substellar boundary *MNRAS* in press.

JAMESON, R. F. & SKILLEN I. 1989 A search for low-mass stars and brown dwarfs in the Pleiades *MNRAS* **239**, 247–253.

KING, I. 1962 The structure of star clusters. I. An empirical density law *AJ* **67**, 471–485.

MERMILLIOD, J.-C., TURON, C., ROBICHON, N., ARENOU, F., LEBRETON, Y. 1997 The distance of the Pleiades and nearby clusters. In *Proceedings of the ESA Symposium 'Hipparcos – Venice '97'*. ESA SP-402, pp. 643–650.

MARTIN, E. L., BASRI, G., GALLEGOS, J. E., REBOLO, R., ZAPATERO OSORIO, M. R., BEJAR, V. J. S. 1998 A new Pleiades member at the lithium substellar boundary *ApJ* **499**, L61–L64

PINFIELD, D. J. 1997 Low mass stars and brown dwarfs in open clusters. PhD thesis, University of Leicester, UK.

PINFIELD, D. J., JAMESON, R. F., HODGKIN, S. T. 1998 The mass of the Pleiades *MNRAS* **299**, 955–964.

PINSONNEAULT, M. H., STAUFFER, J. R., SODERBLOM, D. R., KING, J. R., HANSON, R. B. 1998 The problem of Hipparcos distances to open clusters. I. Constraints from multicolour main-sequence fitting. *ApJ* **504**, 170–191.

REBOLO, R., ZAPATERO OSORIO, M. R., MARTIN, E. L. 1995 Discovery of a brown dwarf in the Pleiades star cluster *Nature* **377**, 129.

REBOLO, R., MARTIN, E. L., BASRI, G., MARCY, G. W., ZAPATERO OSORIO, M. R. 1996 Brown dwarfs in the Pleiades cluster confirmed by the lithium test *ApJ* **469**, L53–L56.

SIMONS, D. A., & BECKLIN, E. E. 1992 A near-infrared search for brown dwarfs in the Pleiades *ApJ* **390**, 431–439.

STAUFFER, J. R., HAMILTON, D., PROBST, R. G., RIEKE, G., MATEO, M. 1989 Possible

Pleiades members with M of about 0.07 solar mass - Identification of brown dwarf candidates of known age, distance, and metallicity *ApJ* **344**, L21–L24.

STAUFFER, J. R., HAMILTON, D., PROBST, R. G. 1994 A CCD-based search for very low mass members of the Pleiades cluster *AJ* **108**, 155–159.

STAUFFER, J. R., SCHULTZ, G., KIRKPATRICK, J. D. 1998 Keck Spectra of Pleiades Brown Dwarf Candidates and a Precise Determination of the Lithium Depletion Edge in the Pleiades *ApJ* **499**, L199-L203.

VAN LEEUWEN, F. 1980 Mass and luminosity functions of the Pleiades. In *Star clusters; proceedings of the symposium.* IAUS, 85, pp157–162. Dordrecht. pp. 689–692.

VAN LEEUWEN, F. & HANSEN RUIZ, C. S. 1997 The parallax of the Pleiades cluster. In *Proceedings of the ESA Symposium 'Hipparcos – Venice '97'.* ESA SP-402, pp. 689–692.

ZAPATERO OSORIO, M. R., REBOLO, R., MARTIN, E. L., BASRI, G., MAGAZZÙ, A., HODGKIN, S. T., JAMESON, R. F., COSSBURN, M. R. 1997 New brown dwarfs in the Pleiades cluster *ApJ* **491**, L81–L84.

ZAPATERO OSORIO, M. R., REBOLO, R., MARTIN, E. L., HODGKIN, S. T., COSSBURN, M. R., MAGAZZÙ, A., STEELE, I. A., JAMESON, R. F. 1998 Brown dwarfs in the Pleiades cluster. III. A deep IZ survey *A&AS* **134**, 537

Brown Dwarfs and the Low-Mass Initial Mass Function in Young Clusters

By KEVIN L. LUHMAN

Harvard-Smithsonian Center for Astrophysics, 60 Garden Street, Cambridge, MA 02140, USA

We have conducted an extensive program of optical and IR imaging and spectroscopy targeted at the low-mass populations of nearby ($\leq$ 300 pc) young ($\sim$ 1-10 Myr old) clusters: L1495E, IC 348, and ρ Oph. By combining the spectroscopic data with IR luminosity function modeling, we arrive at mass functions which are roughly flat or slowly declining in logarithmic mass units below $\sim$0.4 $M_\odot$ into the substellar regime. With the discovery of several likely brown dwarfs, we demonstrate the potential of young clusters in studying the formation and mass functions of substellar objects.

---

1. Introduction

Young, nearby (< 500 pc) clusters offer unique advantages in the search for brown dwarfs and the study of the low-mass initial mass function (IMF). Young (< 10 Myr) low-mass stars and brown dwarfs are quite luminous relative to evolved (> 1 Gyr) objects found in the field. Because young clusters often occupy small regions on the sky (D $\sim$ 10′), many low-mass candidates can be identified in only a limited amount of imaging. In addition, the mass function can be studied in the context of a compact, well-defined region of star formation where the stars have a common history and origin. Compared to open cluster studies, contamination by background stars is reduced significantly by extinction of the natal molecular cloud and the compact nature of the cluster. These factors also facilitate completeness estimates, which can be highly problematic in studies of low-mass objects in the field. Although proper motions are difficult to obtain for these embedded sources, several properties can indicate cluster membership in very young populations, such as X-ray and Hα emission, Li absorption, IR excess emission, giant-like spectral features (e.g., Luhman, Liebert, & Rieke 1997; Martín, Rebolo, & Zapatero Osorio 1996) However, there are a few disadvantages in using young clusters for measuring the low-mass IMF. Many of the low-mass stars and brown dwarf candidates in these clusters are heavily embedded and cannot be studied in the optical. Only recently have high-performance IR cameras and spectrometers been available. In addition, masses derived from H-R diagrams of young low-mass sources are sensitive to uncertainties in 1) the conversion of pre-main-sequence late-M spectral types to effective temperatures and 2) the evolutionary tracks at extremely low masses and young ages. In the following discussion, we briefly review our observations of low-mass candidates in three young clusters and the resulting mass functions.

2. Identifying brown dwarfs and measuring the Mass Function

2.1. *Imaging*

Studies of both open and embedded clusters require deep imaging to identify low-mass candidates, but the approaches differ due to the large amounts of extinction in the latter. In open clusters, large format CCDs can survey several square degrees at R, I, and Z in a reasonable amount of observing time. In the resulting color-magnitude diagrams, most background stars fall below the main sequence and thus are easily identified and

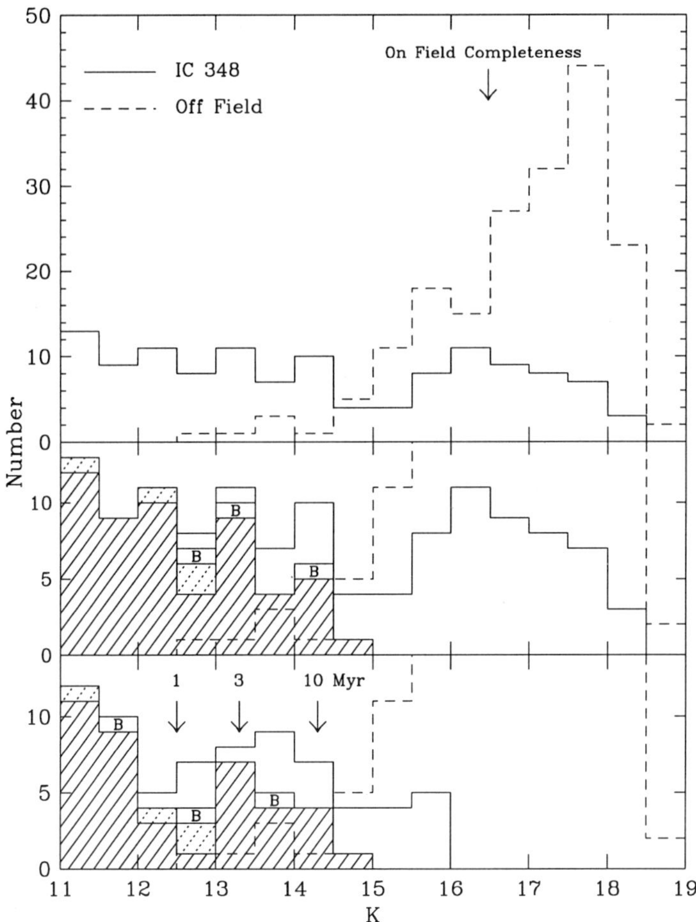

FIGURE 1. The K-band luminosity functions towards the $5' \times 5'$ core of IC 348 and a nearby off-field location of the same size are given in the top panel. The next panel shows magnified versions of these distributions, where solid stripes indicate stars which are plotted in the H-R diagram of IC 348 in Figure 3. Sources with uncertain late-type spectral types (>K5) are represented by dotted stripes. In the lower panel, all sources brighter than $K = 16$ in IC 348 have been dereddened. The resulting KLF can be directly compared to the off-field KLF to estimate the number of background stars contaminating the IC 348 KLF at faint magnitudes. Sources identified as background stars in the spectroscopic sample are represented by "B". The arrows indicate the K-band magnitudes of DM94 corresponding to 0.08 $M_\odot$ at 1, 3, and 10 Myr. If all stars are younger than 10 Myr, then the addition of sources without spectroscopy above $K = 14.5$ (open boxes) to the IMF in Figure 4 results in a mass completeness limit of 0.08 $M_\odot$.

rejected. The remaining faint, red candidates are then examined individually through spectroscopy or proper motions. In young (< 10 Myr), embedded ($A_V > 5$) clusters, this technique reaches only a fraction of the cluster members which happen to have little obscuration. Instead, near-IR arrays must be used to obtain a complete and unbiased census. Although these devices have smaller fields of view than CCDs, young clusters are much more compact than open clusters.

The three clusters examined in this study are L1495E in Taurus (150 pc), IC 348

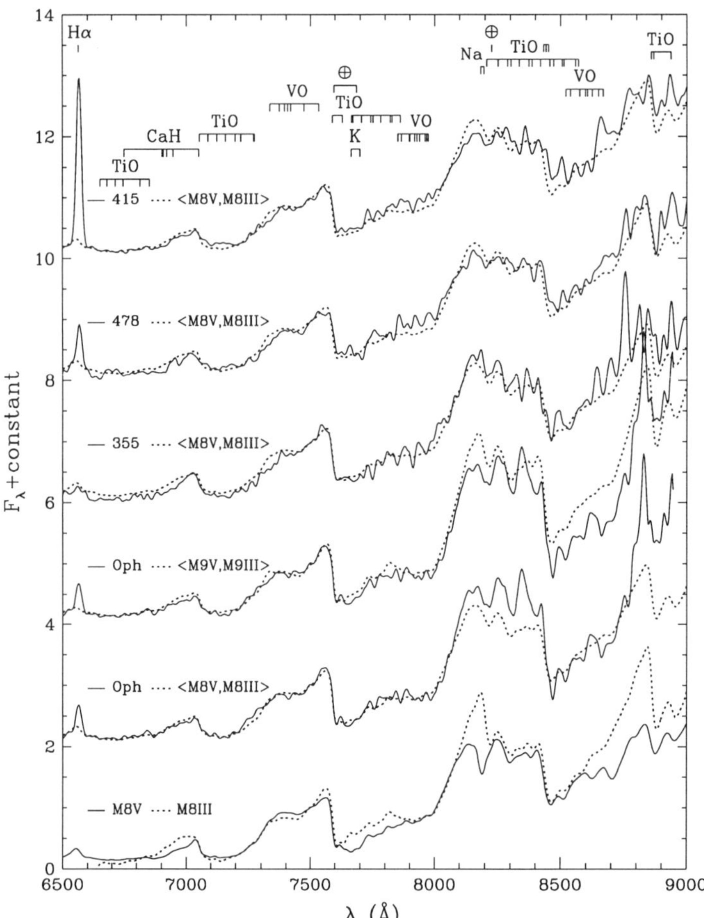

FIGURE 2. The three latest sources observed in IC 348 (415=M7.5, 478=M7.5, 355=M8) and ρ Oph 162349.8-242601 (M8.5) (solid lines) are plotted with averages of standard M8 and M9 dwarfs and giants (dotted lines). Features which are sensitive to surface gravity are apparent in the comparison of the M8 V and M8 III spectra. All spectra are normalized at 7500 Å.

(300 pc), and ρ Oph (150 pc). These regions were selected since they are all nearby, have fairly rich populations (> 50 stars), and are compact ($\sim 10' \times 10'$). In L1495E, Strom & Strom (1994) obtained IR images containing several low-mass candidates. More recently, Luhman & Rieke (1998) have performed deeper J and K imaging, revealing additional candidates. A comparison of the cluster and off-field K-band luminosity functions (KLFs) reveals the magnitude at which background stars are like to appear in cluster photometry, as illustrated in the data for IC 348 in Figure 1. In ρ Oph, a different technique is used to identify likely brown dwarfs. Substellar cluster members should have less reddening than stars which are behind the entire thickness of the cloud core ($A_V > 50$), where sources with $H - K < 2$ are possible cluster members. With this criterion, we have used deep H and K images to identify several extremely low-mass (5-20 M_J) candidates (Luhman & Rieke in prep).

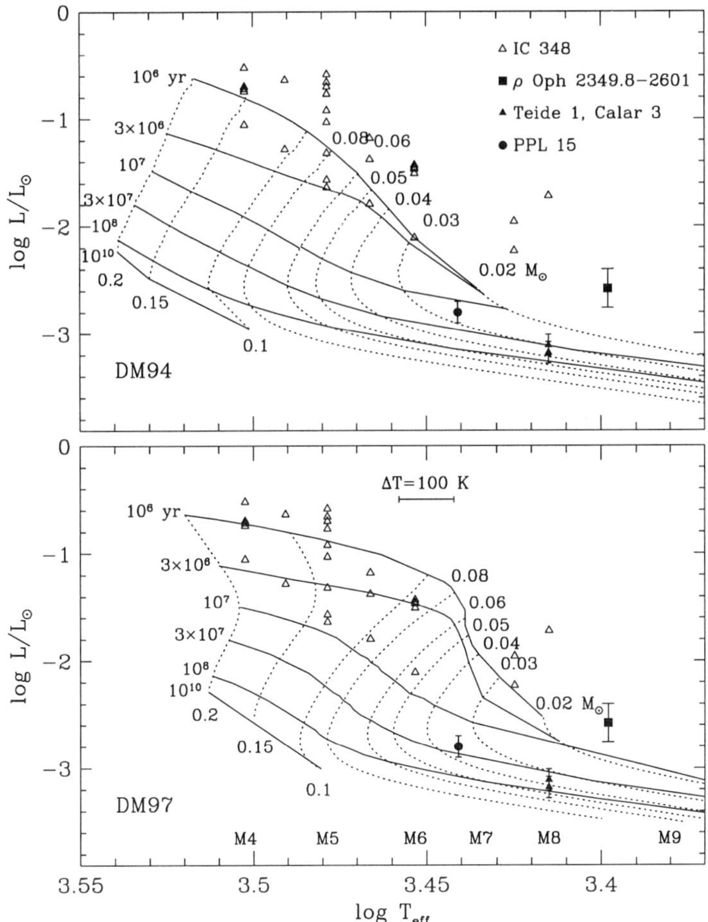

FIGURE 3. The H-R diagram for all late-type sources observed in IC 348 with evolutionary tracks of DM94 and DM97. The brown dwarf ρ Oph 162349.8-242601 and Pleiades brown dwarfs PPL 15, Teide 1, and Calar 3 are shown for reference. Uncertainties in spectral types and temperature scale are typically ± 0.5 subclass and ± 100 K, respectively.

2.2. *Spectroscopy*

Objects at an age of < 10 Myr with spectral types later than M6 are likely to be substellar (Luhman et al. 1998a). Previous studies of young clusters have frequently failed to locate any objects with spectral types later than M5 to M6. However, photometric studies have inferred the existence of numerous brown dwarfs (e.g., Comerón et al. 1993). Spectroscopy can help determine the cluster memberships and spectral type (and hence mass) of the candidates in L1495E, IC 348, and ρ Oph.

In a study of the small stellar aggregate in L1495E, Luhman & Rieke (1998) developed a technique of K-band spectral classification ($R \sim 1000$) to derive the spectral types and continuum veilings of young, late-type stars ($\sim$ 1 Myr, >G0). This method was then used to classify $\sim$ 100 stars in IC 348, where the optical and IR spectral types were in reasonable agreement. The subtleties of K-band classification of young stars are discussed by Luhman & Rieke (1998) and Luhman et al. (1998b). We have also

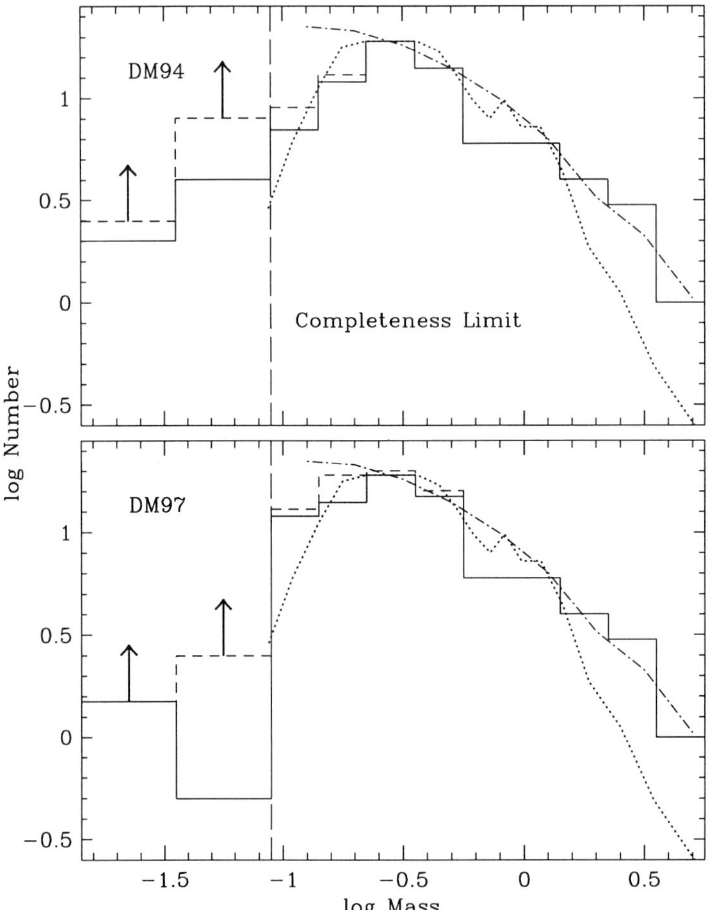

FIGURE 4. The solid histograms are the IMFs derived from the spectroscopic sample in IC 348 with the tracks of DM94 and DM97. The sources without spectroscopy in the lower panel of Figure 1 are added to the IMF as a completeness correction, represented by the dashed histogram. After such a correction, the completeness is indicated by the vertical dashed line and a lower limit to the substellar IMF is provided below this boundary. For reference, the field IMFs of Miller & Scalo (1979) (dot-dashed) and Scalo (1986) (dotted) are given.

performed IR spectroscopy complete to 0.1 $M_\odot$ within the cloud core of ρ Oph. Although most brown dwarf candidates in these regions are beyond the reach ($K > 12$) of our IR spectral classification technique even on a 4 meter telescope, several sources in IC 348 are sufficiently lightly obscured that we have been able to obtain spectra in the far red (see Figure 2). One unobscured brown dwarf candidate ($I = 18.5$) in ρ Oph has been confirmed through optical spectral classification (M8.5) (Luhman et al. 1997). Similar spectral types have been measured for several faint (K = 12-14), embedded sources through K-band steam features (Wilking, Greene, & Meyer 1998).

2.3. *H-R Diagram*

To interpret the H-R diagram of ρ Oph, several sets of low-mass evolutionary tracks are available, where those of D'Antona & Mazzitelli (1994) (hereafter DM94) have been used predominantly for young clusters. Calculations of both DM94 and DM97 were used by Luhman et al. (1998b) in the analysis of data for IC 348 to provide continuity with previous studies. Virtually all the faint objects with spectroscopy that would be assigned brown dwarf candidacy from our photometry are indeed very cool and appear as low-mass stars or brown dwarfs when placed on the H-R diagram in Figure 3.

2.4. *Mass estimates and the IMF*

Accurate masses for young, late-M objects await observational tests, preferably in the form of spectroscopic binaries, of both the tracks and temperature scales at very young ages and cool temperatures. Without such data, the masses derived for these objects depend on the adopted set of evolutionary tracks and the choice of temperature scale. In the H-R diagram shown here, we use the dwarf conversion of spectral types to effective temperatures described by Luhman & Rieke (1998). The adoption of a temperature scale for giants, which is warmer by 100-200 K and may be partially applicable to pre-main-sequence stars, moves the late-M stars to higher masses on the evolutionary tracks. However, regardless of the choice of temperature scale or evolutionary tracks, at least four sources in Figure 3 appear substellar.

In conjunction with the evolutionary tracks of DM97, we can use $T_{\rm eff}$ and $L_{\rm bol}$ to estimate masses for individual sources and construct cluster IMFs. After constructing this spectroscopic IMF for each cluster, we have applied a completeness correction by 1) developing a model for the background stars shining through each molecular cloud at K, 2) identifying likely cluster members at this wavelength which lack spectroscopy, and 3) estimating their masses by combining canonical cluster ages from the H-R diagrams with photometry and evolutionary tracks. The resulting IMF for IC 348 before and after this completeness correction is given in Figure 4. At masses higher than $\sim 0.4\ M_\odot$, the DM97 IMF matches that of Miller & Scalo (1979), while falling less steeply than that of Scalo (1986). After the peak at $\sim 0.4\ M_\odot$, the IMF slowly declines to the hydrogen burning limit with a slope of ~ -0.4 in logarithmic units (where Salpeter is 1.35). Similar mass functions have been measured for L1495E and ρ Oph (Luhman & Rieke 1998; Luhman & Rieke in prep) and the Pleiades open cluster (Bouvier et al. 1998).

3. Conclusions

We have obtained deep near-IR imaging of nearby young stellar clusters and identified likely low-mass members. After showing IR spectral classification to be robust, we have used it in conjunction with optical spectroscopy to observe large numbers of these candidates. Spectral classification and KLF modeling have been combined to derive the IMF down to the hydrogen burning limit (0.08 $M_\odot$) and lower limits to the mass function in the substellar regime. The IMF in young clusters is roughly flat or slowly declining (in logarithmic units), with no substantial variation in the slope as a function of stellar density. Several young brown dwarfs have been discovered through this spectroscopic survey. In addition, the photometric techniques used in identifying young brown dwarfs have been refined considerably, resulting in a large number of new, extremely low-mass candidates (5-30 M_J) for future followup spectroscopy.

We are grateful to F. Allard, I. Baraffe, A. Burrows, and F. D'Antona for providing their most recent calculations and useful advice. This work was supported by NASA grant NAGW-4083 under the Origins of Solar Systems program.

REFERENCES

BOUVIER, J., STAUFFER, J. R., MARTÍN, E. L., BARRADO Y NAVASCUÉS, D., WALLACE, B., & BEJAR, V. J. S. 1998 Brown Dwarfs and Very Low-Mass Stars in the Pleiades Cluster: A Deep Wide-Feild Imaging Survey *Astronomy & Astrophysics* **336**, 490-502.

COMERÓN, F., RIEKE, G. H., BURROWS, A., & RIEKE, M. J. 1993 The Stellar Population in the ρ Ophiuchi Cluster *Astrophysical J.* **416**, 185-203.

D'ANTONA, F., & MAZZITELLI, I. 1994 New Pre-Main-Sequence Tracks for $M \leq 2.5\ M_\odot$ as Tests of Opacities and Convection Model *Astrophysical J. Supplement Series* **90**, 467-500.

D'ANTONA, F., & MAZZITELLI, I. 1997 Evolution of Low-Mass Stars *Cool stars in Clusters and Associations* **68**, 807-822.

LUHMAN, K. L., BRICEÑO, C., RIEKE, G. H., & HARTMANN, L. W. 1998 A Young Star Near the Hydrogen Burning Limit *Astrophysical J.* **493**, 909-913.

LUHMAN, K. L., LIEBERT, J., & RIEKE, G. H. 1997 Spectroscopy of a Young Brown Dwarf in the ρ Ophiuchi Cluster *Astrophysical J.* **489**, L165-L168.

LUHMAN, K. L., & RIEKE, G. H. 1998 The Low-Mass Initial Mass Function in Young Clusters: L1495E *Astrophysical J.* **497**, 354-369.

LUHMAN, K. L., RIEKE, G. H., LADA, C. J., & LADA, E. A. 1998 Low-Mass Star Formation and the Initial Mass Function in IC 348 *Astrophysical J.* **508**, in press.

MARTÍN, E. L., REBOLO, R., & ZAPATERO OSORIO, M. R. 1996 Spectroscopy of New Substellar Candidates in the Pleiades: Towards a Spectral Sequence for Young Brown Dwarfs *Astrophysical J.* **469**, 706-714.

MILLER, G. E., & SCALO, J. M. 1979 The Initial Mass Function and Stellar Birthrate in the Solar Neighborhood *Astrophysical J. Supplement Series* **41**, 513-547.

SCALO, J. 1986 The Stellar Initial Mass Function *Fund. Cosmic Phys.* **11**, 1-278.

STROM, K. M., & STROM, S. E. 1994 A Multiwavelength Study of Star Formation in the L1495E Cloud in Taurus *Astrophysical J.* **424**, 237-256.

WILKING, B. A., GREENE, T. P., & MEYER, M. R. 1998 Spectroscopy of Brown Dwarf Candidates in the ρ Ophiuchi Molecular Core *Astronomical J.*, in press.

Very-Low-Mass Stars in Globular Clusters

By IVAN R. KING[1], AND GIAMPAOLO PIOTTO[2]

[1]Astronomy Dept., University of California, Berkeley, CA 92720-3411, USA

[2]Dipartimento di Astronomia, Università di Padova, Vicolo dell' Osservatorio 5, I-35122 Padova, Italy

We discuss the low-mass ends of mass functions in globular clusters, and extrapolate them to estimate the number of brown dwarfs. Although the brown dwarfs can be quite numerous, they probably contain only a small fraction of the mass of a cluster. We show how the mass function can be pursued observationally down close to the hydrogen-burning limit, and how these observations can be used to derive an empirical mass–luminosity relation for this region. We mention briefly a projected microlensing observation that may actually reveal the presence of brown dwarfs in one cluster.

1. Introduction

This paper has three parts. First will be an estimate of how many brown dwarfs there ought to be in globular clusters, by following their observed mass functions as close as possible to the hydrogen-burning limit, and then naïvely extrapolating the mass function beyond that. Next will be a discussion of the H-burning limit and how we can try to locate it observationally, by pushing luminosity functions as faint as possible. This part will conclude with a demonstration of how the observations can guide the theoreticians toward more accurate models in that region, by telling us something about how the MLR *must* go. And finally we will give a brief description of a microlensing experiment that some one else has underway, that may actually tell us how many brown dwarfs one particular globular cluster contains.

2. Mass functions

First let us look at the mass functions (MFs) of globular clusters. We begin by summarizing how we derive them. What we observe, of course, is the luminosity function (LF). To convert it into an MF we need a mass–luminosity relation (MLR). Since almost nothing is known empirically about the masses of metal-poor stars, we are forced to rely on theory for the MLR. But the theory has its difficulties, and the theoreticians do not agree very well with each other. Moreover, the transformation is particularly sensitive, because to transform a LF into an MF we use not the MLR itself, but its *slope*. This is clear from the equations that relate the MF and the LF. These two functions are actually distribution functions that group the same stars in different ways:

$$f(m)\,dm = L(M_V)\,dM_V. \tag{1}$$

Thus the transformation is

$$f(m) = L(M_V)\,\frac{dM_V}{dm}, \tag{2}$$

where not only is there a transformation from luminosity to mass; there is also a multiplication by the slope of the MLR.

One more problem is that the distance moduli of clusters are often not well-enough known. We have to know the distance modulus, of course, because the LF is in apparent magnitudes while the use of the MLR requires absolute magnitudes.

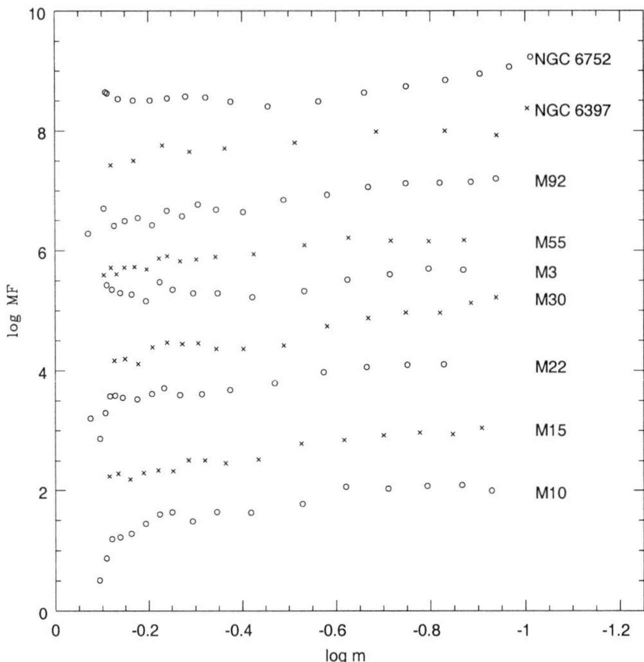

FIGURE 1. Mass functions of 9 globular clusters. In each case, the LF was transformed to an MF using an MLR from Baraffe et al. (1997), and a multi-mass King model was then used to transform the local MF to a global one.

Still another difficulty is that the LF is usually determined only in a small part of the cluster, which will not in general be typical of the cluster as a whole, because stars of different mass tend to be distributed differently. So one needs to have a dynamical model of a cluster, so as to be able to go from a local LF to the global LF.

In Figure 1 are 9 MFs (global, of course) that we have assembled from various data that we had on hand. Although they are plotted against $\log m$, the mass functions that are shown are actually the number per unit interval of m, not $\log m$.

A small digression is appropriate here: It is customary to plot an MF against $\log m$ rather than against m, because stellar masses go through such a large range of sizes. For the distribution function itself, however, two choices are almost equally common: to define the MF as the number of stars per unit m, or else per unit $\log m$. For brevity, the two choices are often referred to as dN/dm and $dN/d\log m$, respectively. Since $d\log m$ is proportional to dm/m, the two functions clearly differ in shape by one power of m.

Astronomers have a tendency, sometimes excessive, to force straight lines through their data. In a log-log plot such as Fig. 1, a straight line corresponds to a power-law relationship between f and m. Thus an MF is often approximated by

$$f(m) \propto m^{-\alpha}. \tag{3}$$

Similarly, MFs that are expressed with respect to $\log m$ are often written as

$$\phi(m) \propto m^{-x}. \tag{4}$$

From the relationship between dm and $d\log m$, it is easy to see that

$$\alpha = 1 + x. \tag{5}$$

TABLE 1. Number fraction and mass fraction between lower limit indicated and $0.09\,m_\odot$.

α	$N_{0.01}$	$N_{0.02}$	$m_{0.01}$	$m_{0.02}$
0.0	0.1013	0.0897	0.0125	0.0120
0.1	0.1230	0.1077	0.0155	0.0148
0.2	0.1486	0.1286	0.0192	0.0183
0.3	0.1784	0.1526	0.0238	0.0225
0.4	0.2128	0.1799	0.0295	0.0277
0.5	0.2518	0.2107	0.0364	0.0339
0.6	0.2952	0.2446	0.0449	0.0414
0.7	0.3427	0.2817	0.0552	0.0505
0.8	0.3935	0.3216	0.0678	0.0614
0.9	0.4468	0.3638	0.0830	0.0744
1.0	0.5014	0.4077	0.1013	0.0897

(For orientation, the traditional slope of the Salpeter mass function is $x = 1.35$, $\alpha = 2.35$.)

This dual terminology has caused an inordinate amount of confusion; one always has to be sure in which dialect a purveyor of MFs is speaking.

The mass functions in Fig. 1 have values of α, near the low-mass end that we shall make use of later in this paper, that range from about 0 to 1 (positive numbers for an upward slope of $m^{-\alpha}$ because mass increases to the left). At higher masses, larger values of α would apply—about 0.5 to 1.2.

We note in passing that this range in MF slopes does not necessarily indicate that globular clusters were born with such a large range of MFs. The clusters do in fact evolve dynamically (Gnedin & Ostriker 1997, Vesperini & Heggie 1997), and some of the shallowest slopes in Fig. 1 do indeed belong to clusters that are expected to have evolved rapidly, in a direction that depletes their low-mass stars. (See, e.g., the discussion in Piotto, Cool, & King 1997.)

There is some tendency among astronomers to use the Salpeter value $\alpha = 2.35$ blindly over the entire range of stellar mass. The above discussion shows how wrong such a value is for the low-mass end of globular clusters. The same stricture applies also to the solar-metal-abundance stars of the Galactic disk; Gould, Bahcall, and Flynn (1997) find that for these stars the slope of the MF below $0.5\,m_\odot$ has $\alpha \sim 0.5$. (Unfortunately their paper itself uses the symbol α for what we would here call $-x$.)

3. Estimating the number of brown dwarfs

Since these observed MFs approach reasonably close to the limit of hydrogen burning (which is somewhere near $\log m \sim -1.05$, depending on the metallicity— and on which theoretical group you believe) and are generally rather smooth and not far from a power law, a reasonable way to estimate the number of brown dwarfs (BD) is by extrapolating the MF into the BD region. What we do, therefore, is to assume that the MF follows a power law all the way, and integrate to get the number of brown dwarfs and the total mass that they contribute, using power laws with $0 \leq \alpha \leq 1$. The results obviously depend on how far into the BD region we carry the integration; we have arbitrarily chosen two limits for the least massive BD: $0.01\,m_\odot$ and $0.02\,m_\odot$. They are given in Table 1. The integrations began at $0.8\,m_\odot$, approximately the main-sequence turnoff mass in a globular cluster, and ended at the lower limits indicated.

TABLE 2. As in Table 1, but for improbably high values of α.

α	$N_{0.01}$	$N_{0.02}$	$m_{0.01}$	$m_{0.02}$
1.1	0.5560	0.4527	0.1230	0.1077
1.2	0.6092	0.4978	0.1486	0.1286
1.3	0.6600	0.5426	0.1784	0.1526
1.4	0.7073	0.5861	0.2128	0.1799
1.5	0.7506	0.6279	0.2518	0.2107
2.35	0.9511	0.8748	0.6841	0.5645

It is clear that we can expect that there are a considerable number of brown dwarfs in a globular cluster, depending of course on the slope of the MF. In terms of mass, however, they do not amount to nearly as much. Given the uncertainties in the observations and in model-fitting, it seems unlikely that their dynamical effect could be noticed.

That situation would change somewhat if the MFs were steeper. Table 2 gives a continuation to steeper MFs. We are not aware of any globular clusters that have MFs in this range of steepness near the low-mass end, although mass functions as steep as this have indeed been given in the literature. Based on our experience here, we believe such mass functions (some of which have been attributed to clusters that appear in our Figure 1) to be totally unreliable.

To emphasize a point made earlier, we have added a final line to Table 2, with the Salpeter value 2.35; it is clear what ridiculous distortions are created by misuse of a slope that is appropriate only for the MF of high-mass stars. If the low-mass end of the MF were so steep, the brown dwarfs would play an important role in the dynamics of the cluster; with more realistic MFs, however, their role is quite minor.

We emphasize that we have given here only a crude estimate of the number of brown dwarfs to be expected in globular clusters. It is hard to see how to do better at the present time—although in our final section we will mention a way in which brown dwarfs may actually be detected in one globular.

4. Faint LFs in globular clusters

We move on now to the study of the faint stars in the neighborhood of the H-burning limit. In globular clusters these stars are *very* faint, since the smallest distance modulus among globular clusters is more than 12 magnitudes. So if we want to go to $M_V \sim 14$ or 15, this means we are aiming at $V \sim 27$. Ground-based telescopes are limited to about 24th or 25th magnitude. (The 10-meter Keck telescopes, and the 8-meter VLT, should theoretically be able to reach $V \sim 27$ or fainter in reasonable exposure times, but they have not achieved that yet.)

The two closest globulars are NGC 6397 $[(m-M)_I \simeq 12.05]$ and M4 $[(m-M)_I \simeq 12.4]$; they obviously offer the chance of going farthest down the main sequence. Unfortunately, however, both these clusters lie in rich star fields, and the faint ends of their CMDs look hopelessly swamped with field stars. With the excellent resolution of HST, however, it is possible over a rather short time interval to measure proper motions, and to use them to separate the faint cluster stars from those of the field. We have succeeded in doing this in NGC 6397; Figure 2 (from King et al. 1998) shows how good the separation is.

It looks as if the MS of the cluster is petering out at the faint end, because the field stars continue in considerable numbers at a magnitude where the number of cluster stars

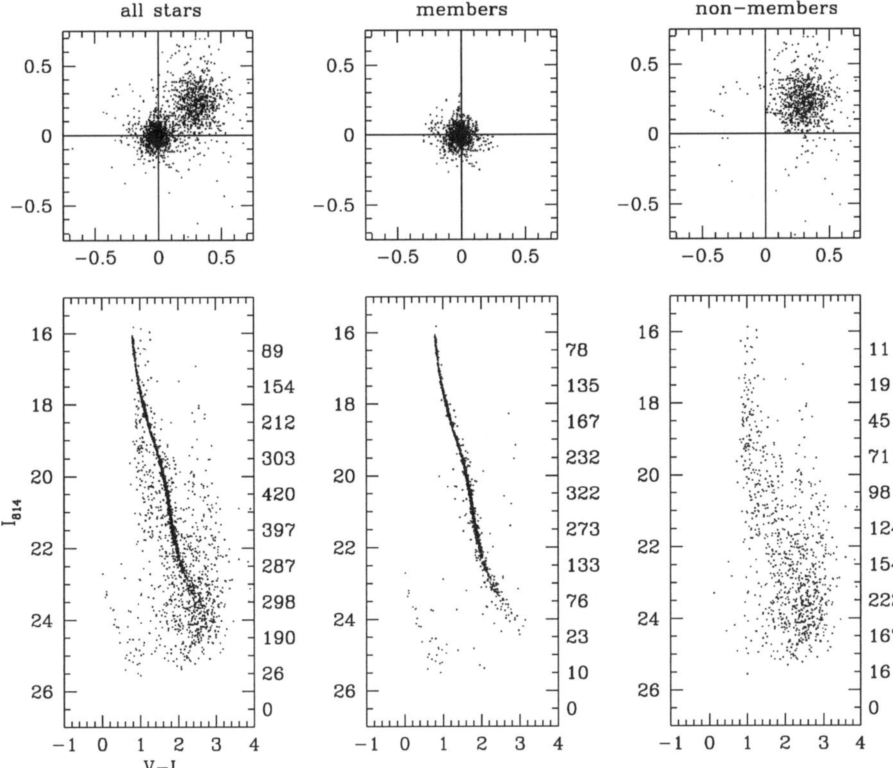

FIGURE 2. Proper-motion distributions, above, and color–magnitude diagrams, below. The scale of the proper motions is displacement in WFC pixels over the 32-month time baseline; a full WFC pixel of displacement would correspond to 37.5 mas/yr. Since all reference stars were cluster members, the zero point of motion is the mean motion of cluster stars. Left: the entire sample; center: stars within a proper-motion region chosen to isolate the cluster stars; right: stars outside this region. Numbers at right are stars per unit-magnitude bin.

decreases sharply. Incompleteness at the faint end is an important problem, however. Observers of faint CMDs routinely test this by throwing artificial stars (constructed from the actual PSF) into an image and seeing what fraction of the stars are recovered, as a function of magnitude. We did this, of course, and allowed for the completeness factors in plotting our LF, which is shown in the upper half of Figure 3. The figure also shows our earlier result, before we had proper motions. It does not go as faint, and the last point was clearly not corrected enough for field contamination.

Our new LF drops steeply at its faint end, so much so that our faintest half-magnitude bin has no cluster stars at all. Over the magnitude range that this bin covers, our completeness figures ran from 0.78 to 0.35, so the absence of stars is certainly significant. It is a problem to deal with a zero count statistically, however. What we did was to ask, for various putative values of the LF at $I = 24.75$, what the probability would be of our observing a count of zero in this bin. The points plotted are such that if the true value of the LF were at that point, the probability of observing zero stars in that bin would be, respectively, 0.1, 0.25, and 0.5, for the three points plotted. The results suggest that this bin very probably continues the steep plunge of the LF.

We had two different MLRs available, one published by Baraffe et al. (1997) and the

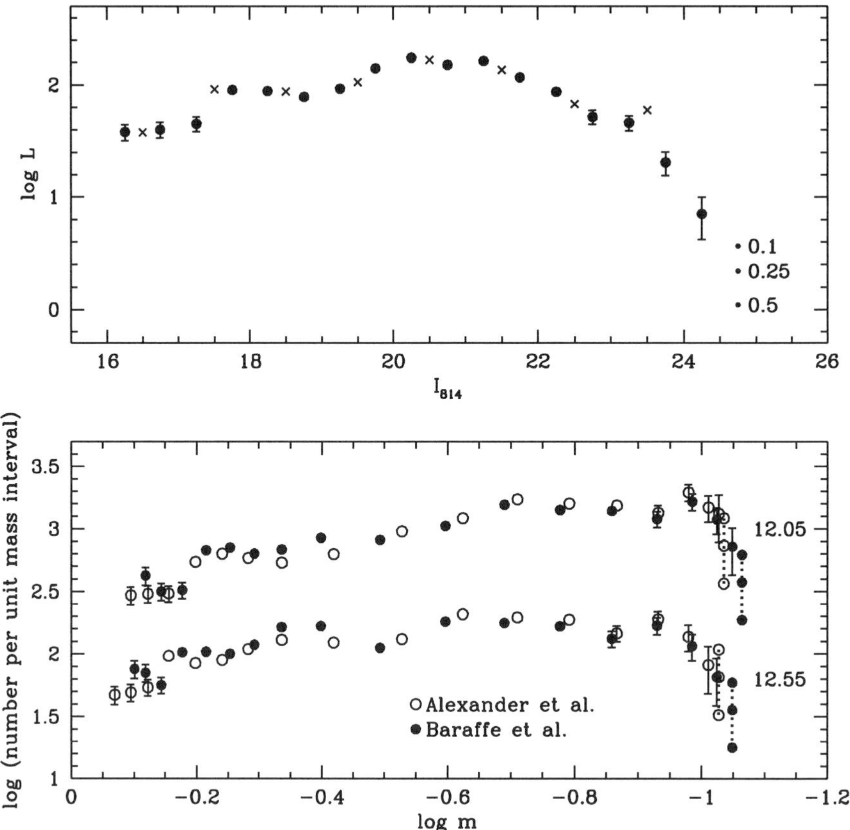

FIGURE 3. (top) Our new luminosity function of NGC 6397, with Poisson error bars (plotted only when they are larger than the sizes of the symbols). The vertical array of three small dots is explained in the text. The crosses are the LF given in an earlier paper (Cool, Piotto, & King 1996), converted to the present field size. (bottom) Mass functions, as derived from each of the two MLRs indicated. For clarity the sets of 3 points representing the empty bin have been connected with lines. The error bars arise from those of the log LF points. The MFs are shown for two different assumed distance moduli, as labeled.

other supplied to us by the authors of Alexander et al. (1996). We used each of these, and to allow for uncertainties in the distance modulus we tried two values of the latter. The MFs derived in this way are shown in the lower panel of Figure 3.

The rather sudden drop at the low-mass end of the MF is highly suspicious. It seems rather implausible that the mass spectrum of pre-stellar fragmentation would have a sudden bend at this point; much more likely is the hypothesis that the MLRs are not reliable in this range. We shall explore this possibility in the following section.

5. Extending the mass–luminosity relation empirically

The boundary in mass between luminous stars and brown dwarfs is very sharp. Pre-stellar configurations just above the H-burning limit have been able to sustain H-burning and maintain a luminosity in the stellar range, while those that were below the limit have by now faded into the darkness of the brown-dwarf condition. Theory predicts that after more than 10 Gyr the boundary between these two extreme states will span only a miniscule range of mass. In any reasonable MF, very few of the stars in a cluster will fall in this tiny range of mass, which by now must stretch over a luminosity range of many magnitudes. Thus just above the critical mass, the luminosity function (LF) must begin a precipitous drop.

Theory predicts the H-burning limit to occur at about $0.075\,m_\odot$ for solar metallicity, and at 0.08–$0.095\,m_\odot$ for the metallicity range that characterizes globular clusters. Just above the limit, a star has a luminosity appropriate to the main sequence, but stars whose mass is just a little lower have a luminosity that is hardly above zero in the optical wavelength range. Thus, as just indicated, the few stars in this tiny interval of mass are spread over a very large number of magnitudes, and the LF plunges abruptly. In the globular cluster of smallest distance modulus, NGC 6397 $[(m-M)_I \simeq 12.05]$, the precipitous drop in the luminosity function takes place around $I \sim 24$, and in M4 $[(m-M)_I \simeq 12.4]$ it is predicted to be seen only a few tenths of a magnitude fainter.

The abruptness of the plunge in the LF allows us to use it in a new way, which actually constructs a part of the MLR from the shape of the LF. What allows us to do this is precisely the fact that the region spans such a small mass range. This means that the MF in this mass range can be estimated quite reliably by extrapolation of the values of the MF at slightly higher masses; then from this extended MF we can infer the masses of the individual stars in this range, as follows: The mass function, by definition, tells us how many stars there are per unit interval of mass, centered on a given mass. If we order the stars by decreasing mass (which we can do by ordering them in decreasing luminosity), the MF will tell us explicitly the mass interval Δm between successive stars. To restate this in different words, the MF can be thought of as dN/dm, where N is the cumulative number of stars, and this Δm is then just the dm that corresponds to $dN = 1$.

What we do then is to (1) arrange the stars in order of faintness, (2) use the MLR to calculate the mass of some star that is near the onset of the plunge, and (3) use the Δm's described above to find the mass of each star that is fainter than that starting point. Note that the masses were inferred from the MF, without using the luminosities, except to set the order in which the stars were arranged. We can therefore plot luminosity against mass for these stars, and thus get an empirical MLR for them. The result depends on the MLR that we used in getting the MF from the LF; thus it tells how *that* MLR *ought to* behave just above the H-burning limit. The location and character of the turndown in the LF therefore provide an important constraint for theories of the structure of low-mass stars.

Without belaboring the point, we also note that the above procedure is easily modified to allow for the increasing incompleteness of our data as they approach our magnitude limit.

One aspect of this process deserves special note: in using an MLR to convert an LF to an MF, we must know the distance modulus, because the MLR is specified in *absolute* magnitude. Thus the procedure, and its results, depend on the value we have chosen for the distance modulus of the cluster. We intend to make a thorough study of the distance modulus of each cluster that we observe, along with its uncertainty—but we note that once our LFs become available, the above procedure can be carried out by anyone who prefers a different distance modulus (or MLR).

The faint LF recently derived for NGC 6397 by King et al. (1998), which features a dramatic turndown at $I \sim 24$, makes it possible to illustrate this procedure of extracting the MLR from the LF. In Figure 4 we demonstrate it for two different theoretical MLRs. The detailed steps of the procedure are described in the figure caption.

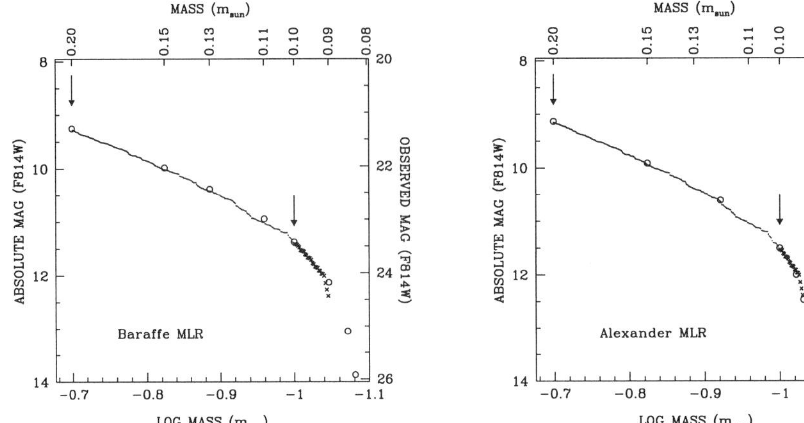

FIGURE 4. Demonstration of use of a faint LF to deduce the lower end of an MLR. The stellar masses were calculated, in the way described in the text, for two different MLRs: one given by Baraffe et al. (1997) and another supplied to us by Alexander et al. (1996). The procedure that we used was as follows: (1) Using the given MLR, we converted each observed M_I to a mass, and we used these mass values to get an MF. (2) Noting that the MF was well fitted by a power law in the range $0.2 \geq m \geq 0.1\, m_{\odot}$ (between the two arrows), we extrapolated the MF downward and (3) used this MF to derive the Δm values described in the text and assign a mass to each star fainter than the stars at $0.1\, m_{\odot}$. To test the power law, masses for the brighter stars were also generated from Δm's derived from the power law; these are plotted as small dots (which blend into a continuous line), and their agreement with the open circles shows how well the power law fits. The masses derived from the extrapolated MF are plotted as crosses. They constitute an empirical extension of the MLR below $0.1\, m_{\odot}$, and they serve as a test of the lowest part of the MLR, whose points are shown as open circles. The fit, which also depends on the distance modulus assumed, is better for one theoretical MLR than for the other; but it would obviously be desirable to have a larger number of stars.

We thus use our faintward extension of the CMD as a check not only on the mass–radius relation (which determines the shape of the main sequence in the CMD) but also on the MLR.

Though in the above data we clearly observe the dropoff of the LF that is related to the H-burning limit, the derivation of the implied shape of the MLR is statistically weak, in that the extension that is made relies on only 22 to 29 stars (the actual number of crosses in the two panels of Fig. 4), and on only 3 stars in the region where the MLR slope is steepest. We have therefore proposed that the next round of HST observations (Cycle 8) include second-epoch imaging of another field in NGC 6397 and of 3 fields in M4. The observations proposed would greatly increase the available sample of globular-cluster stars close to the hydrogen-burning limit.

6. Microlensing

Meanwhile, however, another HST program may make a major contribution to the subject of brown dwarfs in globular clusters. It follows up a suggestion by Paczynski (1994) that in a globular cluster that lies in front of a rich background it might be possible to carry out photometric monitoring of enough stars simultaneously that a significant number of microlensing events could be observed. These events give information, of course, about the nature of the objects that do the lensing.

The ideal cluster for such observations is M22, a globular that lies in front of the Galactic bulge, at about half the distance to the bulge. Kailash Sahu, of the Space Telescope Science Institute, has followed up on Paczynski's suggestion, and has HST observing time (GO-7615) to look for microlensing events in M22, in a 3-year project, of three 120-day observing campaigns. Calculations in the proposal that won Sahu the observing time predict that he will see 12 to 25 microlensing events, many of them caused by objects whose masses are below the H-burning limit. As we said earlier, we can expect a lot of brown dwarfs in globular clusters; here is a chance actually to detect some of them.

7. Conclusions

Brown dwarfs are almost certainly present in appreciable numbers in globular clusters, but their role in the dynamics of the cluster is probably quite minor. A few of them might actually be detected by microlensing of Galactic bulge stars by brown dwarfs in M22.

Observationally, the visible stars that are just above the hydrogen-burning limit are of considerable interest, because of the information that they offer about the mass–luminosity relation in this range of mass.

I. R. K. thanks Dr. Jay Anderson for help with the method and computations that went into Sec. 5 and Fig. 4, and acknowledges support from Grant GO-6797 from the Space Telescope Science Institute. This paper is based on observations with the NASA/ESA *Hubble Space Telescope*, obtained at the Space Telescope Science Institute, which is operated by AURA, Inc., under NASA contract NAS5-26555.

REFERENCES

Alexander, D. R., Brocato, E., Cassisi, S., Castellani, V., Ciacio, F., & Degl'Innocenti, S. 1996, A&A, 317, 90

Baraffe, I., Chabrier, G., Allard, F., & Hauschildt, P. H. 1997, A&A, 327, 1054

Cool, A. M., Piotto, G., & King, I. R. 1996, ApJ, 468, 655

Gnedin, O., & Ostriker, J. P. 1997, ApJ, 474, 223

Gould, A., Bahcall, J. N., & Flynn, C. 1997, ApJ, 482, 913

Paczynski, B. 1994, Acta Astr., 44, 235

Piotto, G., Cool, A. M., & King, I. R. 1997, AJ, 113, 1345

Vesperini, E., & Heggie, D. C. 1997, MN, 289, 898

The DENIS Very Low Mass Stars and Brown Dwarfs Results (Sample, Spectroscopy and Luminosity Function)

By X. DELFOSSE[1] AND T. FORVEILLE[2]

[1] Instituto de Astrofísica de Canarias, 38200 La Laguna, Tenerife, Spain

[2]Observatoire de Grenoble, BP 41, F-38041 Grenoble Cedex 9, France

In this paper we review the results of the DENIS survey on very low mass stars and brown dwarfs. The analysis of DENIS catalogs for 1500 square degrees has produced a sample of ~100 very-late M dwarfs and 15 L dwarfs. Spectroscopy of these objects has established a spectroscopic classification sequence for L dwarfs, and determined the underlying effective temperature scale.

We use this sample to obtain the local luminosity function of the very low mass stars and brown dwarfs, with particular attention to correcting possible error sources and Malmquist-like biases. This first DENIS luminosity function has good statistical accuracy down to the limit between M and L dwarfs.

1. Introduction

Very low mass stars and brown dwarfs can be looked for around known brighter stars, in clusters, or in the general field, with advantages and disadvantages which have been repeatedly discussed in detail (for instance, Hambly 1998). Companion searches have historically identified the coolest object known at any given time, though usually not the least massive (which are found in clusters, where they haven't yet cooled much). Companion searches in the immediate solar neighbourhood also provide the information needed to correct cluster and field samples for the contribution of unresolved companions to more distant objects, and as such they are an essential complement to both field and cluster surveys. Cluster searches benefit both from an increased source density and from the much larger luminosity of younger brown dwarfs, and as a consequence they are sensitive to much lower mass objects (e.g. Zapatero-Osorio et al. 1999). Mass segregation and selective cluster evaporation however complicate their interpretation in terms of initial mass and luminosity functions. Field searches finally, which we discuss here, have the lowest efficiency factor, since they don't take advantage of any local density enhancement. For this same reason however they provide the most direct determinations of the local density of very low mass objects, and they have also produced samples of very cool objects which can be followed-up without the hindrance of a much brighter close companion.

Given the low efficiency mentioned above and the very low luminosity of the brown dwarfs, the first requirement for a field brwon dwarf search is a wide deep sky survey, preferably in the near infrared domain, where the flux of very low mass stars and brown dwarfs peaks. The main difficulty then becomes the "needle in haystack" aspect of the search: at high galactic latitudes DENIS only detects one brown dwarf for every 5.10^5 stars. Efficient selection criteria with a a very low false alert rate are therefore essential. To date DENIS and 2MASS have identified most of the field brown dwarfs, with the SLOAN survey recently making significant contributions.

TABLE 1. Limiting distance for detection of typical very low mass stars and brown dwarfs with DENIS. For a GL229B-like object this is the limiting distance for detection in J and K only; for the other object types the distance is for a 3-colours detection.

Object	**Limiting distance of detection**
M5.5V	~140 pc
M9V	~40 pc
GD165B-like	~27 pc
GL229B-like	~5 pc

2. DENIS

2.1. *The* DENIS *survey*

The DEep Near-Infrared Survey (DENIS) will be a complete near infrared survey of the southern sky (Epchtein 1997) It aims to provide full coverage in two near-infrared bands (J and K′) and one optical band (I), using a ground-based telescope and digital array detectors. The approximate 3-σ sensitivity of the survey is I=19, J=17, K=14. The products of this survey will be databases of calibrated images, extended sources, and small objects. The survey started in January 1996 and is expected to be completed within five years. As of December 1999 over 70% of the southern sky have been observed.

The study of the lowest mass stars and brown dwarfs is one area of astronomical research in which such a database will have a particularly profound impact. With a completeness limit of I≈18.5 (Table 1), and coverage of the whole southern sky, DENIS offers the opportunity to detect numerous examples of this currently poorly understood class of objects. The optical-infrared I–J and I–K colours provided by DENIS are more sensitive probes of effective temperature for very cool stars than the infrared J-H or H-K colours, and these objects are thus relatively easily identified in the DENIS catalogues.

DENIS observations are carried out on the ESO 1m telescope at la Silla (Chile), with a purpose-built three channels infrared camera. Dichroic beam splitters separate the three channels, and focal reducing optics provides image image scales of 3" on the 256×256 NICMOS3 arrays used for the two infrared channels and 1" on the 1024×1024 Tektronix CCD detector of the I channel, and a 12' instantaneous field of view for all three channels. A focal plane microscanning mirror is used to obtain 1" sampling for the two infrared channels. The sky is scanned in a step and stare mode, along 30 degrees strips at constant right ascension which contitute the basic DENIS observing unit.

Two data analysis centres (DACs) have been created to process these data. The Paris DAC processes the raw data from the telescope into flattened and cleaned images (Borsenberger 1997). The Leiden DAC will then process these images to extract point source information (Deul et al 1995), which will be assembled into a database at the Paris DAC.

To date the PDAC works in a routine mode and the data are usually processed within a month of the observations. The LDAC has been producing data since September 1997, but not yet in a full production mode.

2.2. *The present sample*

We first carried out a "Mini-survey" with spectroscopic follow-up (Delfosse et al. 1997, 1999a) of the very low-mass star and brown dwarf candidates identified in ≈1% of the DENIS survey data (230 square degrees). The image data from the high latitude part ($|b_{II}| > 20$-30°) of 47 survey strips (for a total surface area of 230 square degree) were

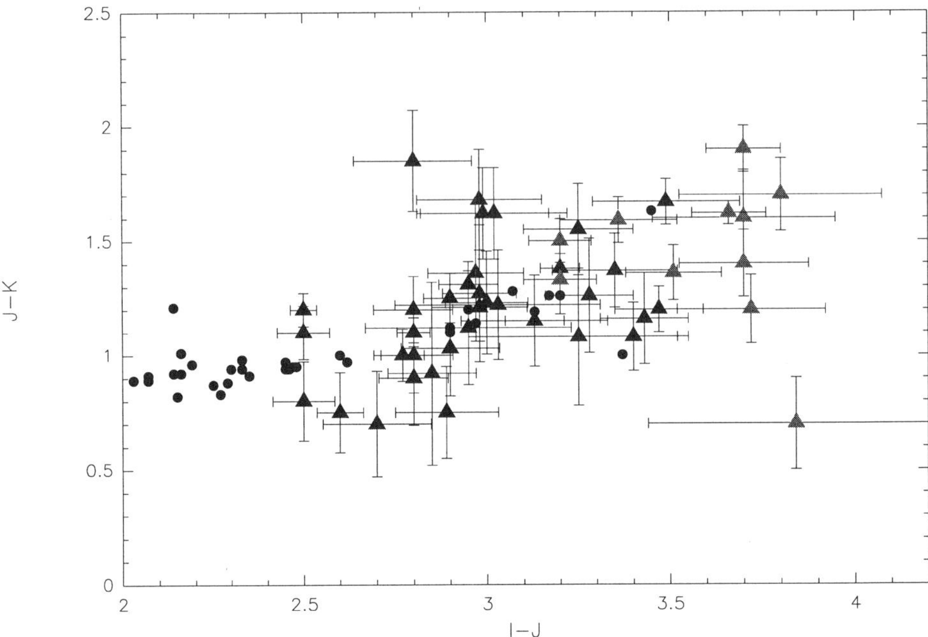

FIGURE 1. I–J:J–K diagram for the DENIS very low mass stars and brown dwarfs. The filled circle represent previously known very low mass stars, the triangle show DENIS objects.

obtained from the Paris DAC. As the Leiden source extraction pipeline was not yet operational these images were processed in Grenoble to create catalogs of I, J and K photometry, using the SExtractor package (Bertin & Arnouts 1996). The resulting source list contains 37 dwarfs redder than M7V, 5 of which have the colors of an M9-M9.5 dwarf ($3.1< I-J <3.35$) and 3 of which are as red as GD165B or redder (I-J>3.6) (Delfosse et al 1997). As the quality of the "Mini-survey" data is representative of the overall survey, those results can be scaled to evaluate the brown dwarf content of DENIS.

Since then, we search the PDAC and LDAC databases for very low mass stars and brown dwarfs as data become available. To date we have analysed DENIS data for 1500 square degrees, producing an updated sample of $\sim$ 100 objects redder than $I-J=2.8$ and $\sim$15 L dwarfs (defined below). Figure 1 shows those objects which have been detected in all three bands in an I–J:J–K diagram, with known very low mass stars added for orientation.

3. Follow-up spectroscopy

We have obtained follow-up spectroscopic observations of the DENIS very low mass stars and brown dwarfs using the AAT telescope (for infrared spectroscopy (Delfosse et al. 1999a) and low resolution visible spectroscopy (Tinney, Delfosse, Forveille & Allard 1998) and mostly the Keck telescope (for high resolution visible spectroscopy with HIRES (Martín, Basri, Delfosse & Forveille 1997; Basri et al., 2000) and visible low resolution spectroscopy with LRIS (Martín et al., 1999)).

The infrared observations (Delfosse et al., 1997, 1999a) demonstrate that infrared spectroscopy can be used to classify these very cold objet and determine their effective

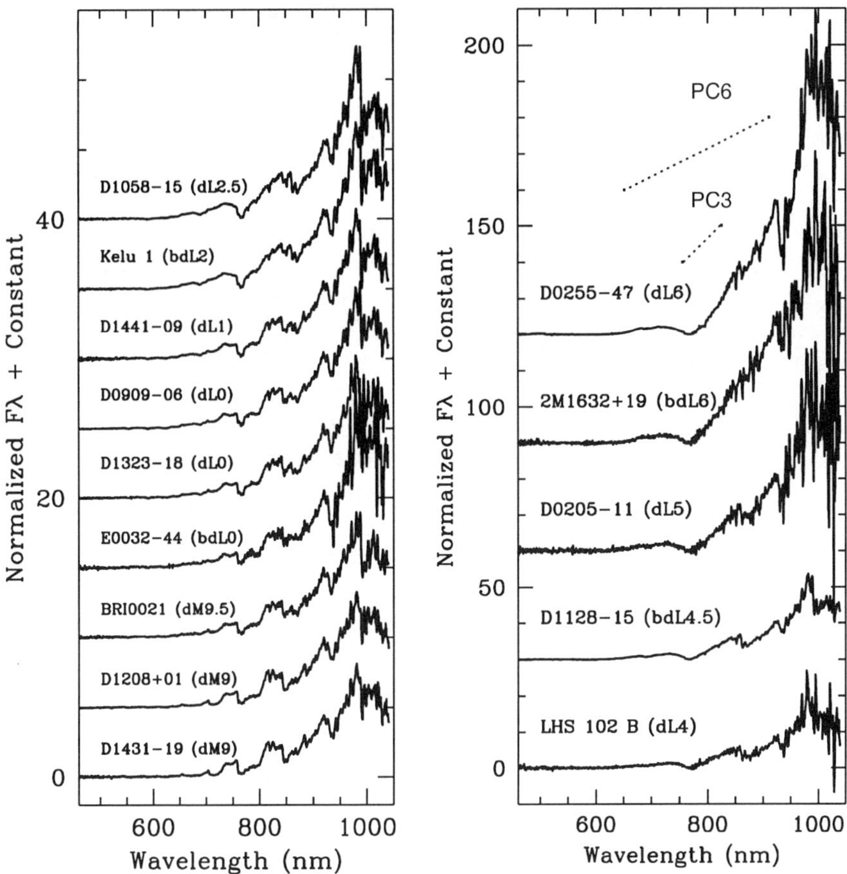

FIGURE 2. Optical spectra from a sequence of L and very late M dwarfs, from Martín et al. 1999. Our proposed L subclass assignment is based on the the PC3 and PC6 pseudo-continuum indices. The DXXXX-XX are DENIS detections and the 2MXXXX+XX are 2MASS objects.

temperatures. The slope of the 1.6μm H_2O band in particular is a very good temperature diagnostic.

The visible spectroscopy resulted in the identification of the first field brown dwarf (simultaneously with the independent detection of Kelu 1 by Ruiz, Legget and Allard 1997), from the detection of lithium in DENIS-P J1228-1547 (Martín et al., 1997; Tinney, Delfosse and Forveille 1997). This detection indicates that the mass of this fully convective object must be lower than the lithium burning threshold of ~0.06 solar masses.

The low resolution spectroscopy has revealed a whole population of objects with spectra similar to that of GD 165B, previously considered as an isolated oddity and which had occasionally been explained as the result of binary star evolution. The TiO and VO bands which constitute the defining characteristic of the M spectral class rapidly weaken with decreasing effective temperature, and finally disapear (Figure 2, from Martín et al. 1999) in the optical spectrum of very cool dwarfs. This is due to depletion of heavy elements in their cool photospheres through formation of dust grains – in particular solid VO and perovskite ($CaTiO_3$) (Allard 1998). A whole new spectral class is clearly needed, now represented by a few dozen objects mostly discovered by the DENIS, 2MASS (Kirkpatrick et al. 1999b) and now SLOAN (Fan et al. 2000) surveys. Martín, Basri, Delfosse & Forveille (1997) proposed the letter "L" to designate it, which has been adopted by consensus. Two conflicting definitions of the L subclasses exist on the other hand, the Kirkpatrick et al. (1999b) classification, based on spectra of mostly 2MASS objects, and our classification (Martín et al 1999, Figure 2), based mostly on spectra of DENIS detections. The two definitions are mutually consistent for subclasses L0 to L2, but then diverge such that L6(Martín) = L8(Kirkpatrick). We chose to define broader L subclasses to have a more significant difference between adjacent subclasses (the difference between the later Kirkpatrick et al. subclasses is rather subtle), and also to preserve space for objects to be discovered with effective temperatures intermediate between the latest known L dwarfs (T_{eff}~1600 K) and the methane dwarfs (T_{eff}~1000 K). Our classification has also been guided by the effective temperature scale of Basri et al. (2000, discussed below), and the subclasses correspond to constant T_{eff} steps of 100 K on this scale. The main justification for this is that a spectral type scale is a proxy for a temperature scale, so it makes sense to follow systematically the guidance of a T_{eff} scale, even when future revisions will likely shift it slightly. We derive this T_{eff} scale from the analysis of high resolution spectra (R=40000) of the resonance absorption lines of Cs I and Rb I, using detailed atmospheric model (Basri et al. 2000). We find the L0 dwarfs have an effective temperature of 2200 K and that for L6(Martín) dwarfs correspond to T_{eff} = 1600 K .

4. The first DENIS luminosity function

The mass function is still rather poorly determined for $M \lesssim 0.3 M_{\odot}$. This has important implications for star formation theories (through the shape of the initial mass function (IMF)) and potentially for galactic dynamics (brown dwarfs remain a viable candidate for the local dark matter). In practice mass is not measured for single stars, and the mass function is obtained in three steps: 1/ A luminosity function for stellar systems (i.e. the number of systems per unit volume and per luminosity bin) is derived from deep multicolour differential star counts, using an accurate color-luminosity relation; 2/ The system luminosity function is transformed into a stellar luminosity function, using the binarity statistics to correct for unresolved multiple systems 3/ The mass function is finally obtained by combining this stellar luminosity function with a mass-luminosity relation.

DENIS is very well suited to the first task, determining the luminosity function, as it detects large numbers of very low mass stars and measures one color index, I–J, which is an excellent luminosity estimator in this part of the HR diagram. In this section we describe the classical derivation of the luminosity function from two-colour star counts, and discuss corrections for various errors and biases that can affect such derivations.

4.1. *Error sources and biases in photometric luminosity functions*

4.1.1. *Error sources*

To derive a luminosity function from multicolour surveys like DENIS one needs two basic physical inputs:

- a colour-luminosity relation, and its intrisic dispersion due to age and metallicity;
- a good knowledge of the local structure of the Galaxy, out to the maximum distance sampled by the survey, usually parametrized by an exponential scale height for the disk.

Each of those inputs can potentially introduce large errors in the luminosity function.

The luminosity of each survey object is estimated from its colour and the colour-luminosity relation. It is used both to assign it to one given luminosity bin and to derive the maximum distance for a survey detection of that particular object. Stars are then counted as the inverse of their maximum detection volume, correcting back the survey to the case of a constant detection volume. This estimator, known as the Schmidt estimator (Schmidt 1968), takes as described here no account of density variations within the accessible volume. It will thus underestimate the local density of the brighter objects that the survey can detect beyond one galactic disc scale height. It can be readily generalised however (Felten 1976, Stobie et al. 1989, Tinney et al. 1993) to correct for this effect by simply replacing the maximum detection volume by a generalized volume:

$$V_{Gen} = \Omega \int z^2 \frac{\rho(\Omega, z)}{\rho_0} dz$$

Where Ω is the solid angle of the survey, and ρ_0 the local density. For an exponential disk (which is a good representation of the locally dominant galactic disk), $\rho(\Omega, z)/\rho_0 = \exp(-z/h)$, and at the galactic pole, one obtains:

$$V_{Gen} = \Omega h^3 [2 - (y^2 + 2y + 2)\exp(-y)].$$

with $y = Z/h$.

For a general galactic latitude b this becomes:

$$V_{Gen} = \Omega \frac{h^3}{\sin^3 b} [2 - (\xi^2 + 2\xi + 2)\exp(-\xi)] \tag{4.1}$$

where $\xi = d \sin b \;/\; h$, with d is the star distance.

The stellar distance and galactic scale height both enter at high powers in this equation, and errors in any of the two inputs (colour-luminosity relation and galactic scale height) thus propagate into large luminosity function errors, as will be discussed in more details in a forthcoming paper (Delfosse & Forveille 2000).

Simple error calculation shows that:

$$\frac{\Delta V_{Gen}}{V_{Gen}} = f_1(\xi)\frac{\Delta d}{d} + f_2(\xi)\frac{\Delta h}{h} \tag{4.2}$$

with

$$f_1(\xi) = -\frac{\xi^3}{-2\exp(\xi) + \xi^2 + 2\xi + 2} \tag{4.3}$$

$$\text{and} f_2(\xi) = -\frac{-6\exp(\xi) + \xi^3 + 3\xi^2 + 6\xi + 6}{-2\exp(\xi) + \xi^2 + 2\xi + 2} \tag{4.4}$$

For the fainter very low mass stars and the brown dwarfs ξ remains small. Distance estimation errors then dominate in the luminosity function errors, since equation 4.2 approaches the homogeneous limit:

$$\lim_{\xi \longrightarrow 0} \frac{\Delta V_{Gen}}{V_{Gen}} = 3\frac{\Delta d}{d} \tag{4.5}$$

This can be written as:

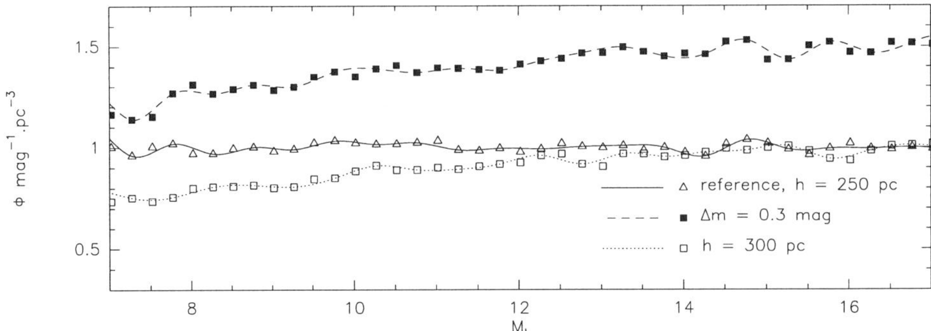

FIGURE 3. Effects on the luminosity function of systematic errors on galactic scale height and color-magnitude relation. See the text for details.

$$\lim_{\xi \longrightarrow 0} \frac{\Delta V_{Gen}}{V_{Gen}} = 0.6 \ln 10 \Delta m \simeq 1.38 \Delta m \tag{4.6}$$

showing that in this regime a 0.3 mag error on the colour-luminosity relation induces a 50% (!!) error on the corresponding bin of the luminosity function.

Figure 3 shows a more complete error simulation, where a population of K-M dwarfs is synthesised with a constant luminosity function $\Phi(M) = 1$, for a galactic scale height of 250 pc and for a nominal colour luminosity relation. Its luminosity function is then reconstructed, first with the parameters used to generate it (triangles), then using a colour-luminosity offset (in luminosity) by 0.3 mag (filled squares), and finally using a galactic scale height of 300 pc instead of 250 pc (empty squares). The largest errors are found for the faintest stars and due to the colour luminosity errors, but the errors induced by an invalid galactic scale height are not negligible either, specialy for the "brighter" stars.

The intrinsic scatter of the colour-luminosity relation is larger than 1 magnitude (e.g. Goldman et al., for the late-M and L Dwarfs). Local 0.3 mag errors (i.e. 50% luminosity function errors) can therefore quite easily creep into analytical fits to this relation, for instance when using a polynomial degree too low to reproduce some features, or on the contrary when a too small number of stars leaves a high degree fit free to oscillate. This is particularly critical at I–J∼ 2.0, where the slope of the colour-luminosity flattens as the stellar interior becomes degenerate (this corresponds to the luminosity function peak). Figure 4 provides a graphic illustration of those difficulties: the 4th degree polynomial fit does not fully reproduce the M_I=10 plateau, which the theory tells us must be present, and yet it may oscillate at the faintest magnitudes where it is little constrained. Some luminosity functions having in the past been constructed with linear colour-luminosity relations (e.g. Stobie, Ishida and Peacock 1989), there is clearly potential here for large systematic errors. Theoretical relations by contrast already includes most of the physical structure in the colour-luminosity relation. We think they represent a much better alternative, after they have been validated by observations, and perhaps with some adjustments for remaining minor deficiencies.

Commonly used values of the Galactic scale height currently range between 300 and 350 pc, but this parameter is not very well determined and the more recent measurements obtain values outside this range. Furthermore, a single exponential law doesn't describe the vertical structure of the disk well (Haywood et al. 1997a, 1997b): the stellar density in fact decreases faster than a 300 pc scale height exponential. Our use of a 50 pc error on the scale height to assess the errors steming from galactic strucure uncertainties (Figure 3) is thus *plausible*.

As can be seen from this discussion, significant work remains to be done on the two physical inputs used to derive photometric luminosity functions. The recent improvements in very low mass stars and brown dwarfs models (Allard 2000 and Baraffe 2000 in these proceedings; Chabrier et al. 2000) set the foundations of a better colour-luminosity relation, together with the first determinations of L dwarfs parallaxes (Goldman et al. 1999, Kirkpatrick et al 1999b), but a sizeable effort is still needed to piece the theory and observations together. Substantial work is also needed to directly determine the scale height of the very low mass stars and brown dwarfs. In the mean time photometric luminosity functions should be considered as potentially suspect, including the one we present below.

4.1.2. *Malmquist-like biases*

Colour-luminosity relations have a significant intrinsic scatter, which reflects the age and metallicity dispersion of the field population. Observed colours are furthermore affected by some level of observational errors (both noise and systematic calibration errors). Stars with a given observed colour therefore actually span a range of luminosities, while their distance and luminosity are determined assuming a one-to-one correspondance between colour and luminosity. As is now well known (for instance in an extragalactic distance scale context), such an use of a dispersed luminosity indicator to infer distances introduces biases, collectively known as Malmquist-like biases. In the present luminosity function context there are two separate, though interelated, effects, as discussed for instance by Stobie, Ishida and Peacock (1989) and Kroupa (1998):

- A magnitude-limited sample has a brighter average absolute magnitude at a given colour than a volume-limited sample, since the brighter stars are detectable to larger distances. This the classical Malmquist bias
- A magnitude-limited sample has a larger number of stars at a given colour than the equivalent volume-limited sample for the average colour-luminosity relation: since the volume grows as d^3, dispersion removes fewer fainter stars from the sample than it adds brighter stars.

Stobie et al. (1989) derived an analytic correction for those two aspects of the Malmquist bias, which has been used for most recent luminosity function determinations (including e.g. Leggett and Hawkins (1988), Kirkpatrick et al. (1994), Kroupa (1995)). This correction is exact for a constant gaussian dispersion in the absolute magnitude at a given colour. This is a useful first order description of the intrinsic dispersion (see below though), but it cannot describe the bias introduced by noise, which ranges from large for faint stars to negligible for bright stars. To investigate this bias we have performed some Monte Carlo simulations.

Monte Carlo simulation of the observational errors bias:

We use a simple model (Delfosse and Forveille 2000) to simulate observations of a galactic population of very low mass stars and brown dwarfs, and produce apparent magnitudes in some selected filter set. The input population has a *classical* luminosity function (which peaks around M_I=10), and an exponential vertical distribution $D(r) = \exp(-d\sin(b)/h)$ (with b the galactic latitude and h the scale height). It obeys polynomial colour-luminosity relation, which where adjusted to stars with accurate parallaxes and

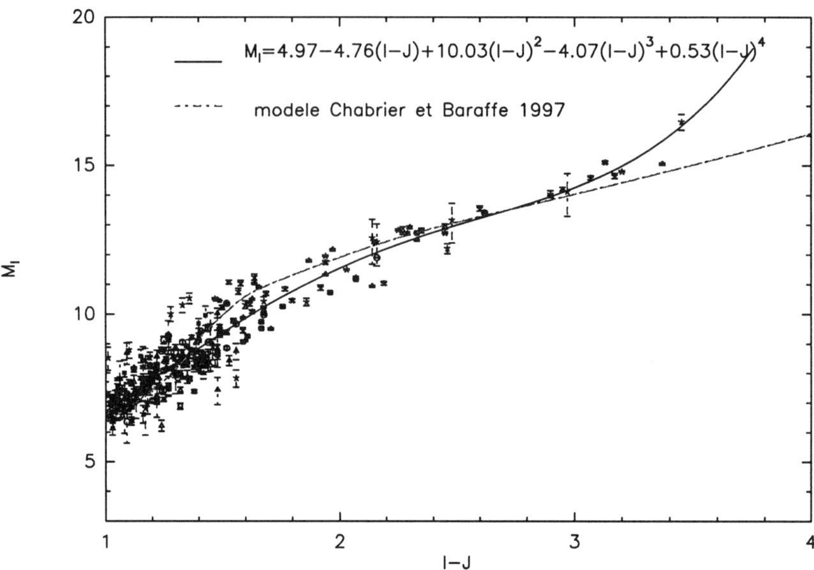

FIGURE 4. $M_I/(V-I))$ polynomial colour-luminosity used for in the Monte Carlo simulations and to construct the DENIS luminosity function. Data are from Leggett (1992) and Tinney et al. (1993b).

photometry from a sample combining Leggett's (1992) extensive compilation with Tinney et al. (1993b) for some fainter stars (Fig. 4 for $M_I/(V-I))$.

After sampling from this population, an approximate PSF is used to spread the source over a small array of detector pixels for each relevant photometric band. Gaussian noise is then added to the data counts in each of those pixels to account for background and readout noise (DENIS observations are background-limited at J and K, and usually readout noise-limited at I). Fluxes are extracted from those synthetic detector arrays with the same algorithm used to analyse the actual survey images, and converted to magnitudes. Small gaussian errors are finally added to those magnitudes, to account for (multiplicative) calibration errors. Large scale surveys such as DENIS typically compromise somewhat on photometric homogeneity to complete observations within the limits of their funding. The corresponding errors are clearly systematic within one observing night, but they become random when averaged over many nights. They can thus be modeled as a dispersion.

Calibration errors (gaussian in magnitude) tend to smooth the intrinsic stellar colour distribution, filling-in dips and scrapeing peaks (Fig 5). This respectively over-estimates and under-estimates the number of stars around dips and peaks of the intrinsic colour distribution. Even for a flat luminosity function (figure 5(a)), the intrinsic distribution has a well defined peak at $I-J \sim 2$. This peaks reflects the flattening of the colour-luminosity relation (Fig 4), and the resulting crowding in colour space, when electron degeneracy sets in for the stellar interior. The uncorrected luminosity function therefore (figure 6(a)) displays a marked bias at $M_I \sim 11.5$. The above-mentioned peak of the inverse colour-luminosity slope actually corresponds (Fig. 6(b)) to a peak of real luminosity functions, further increasing colour-space crowding at $I-J \sim 2$ and the significance of the bias (Figs. 5(b) and 6(b)). The large calibration errors used here for illustrative purposes are (fortunately) clearly unrealistic. Actual calibration are of the order of $\sigma = 0.05$mag, and produce maximum errors of $\sim$20%.

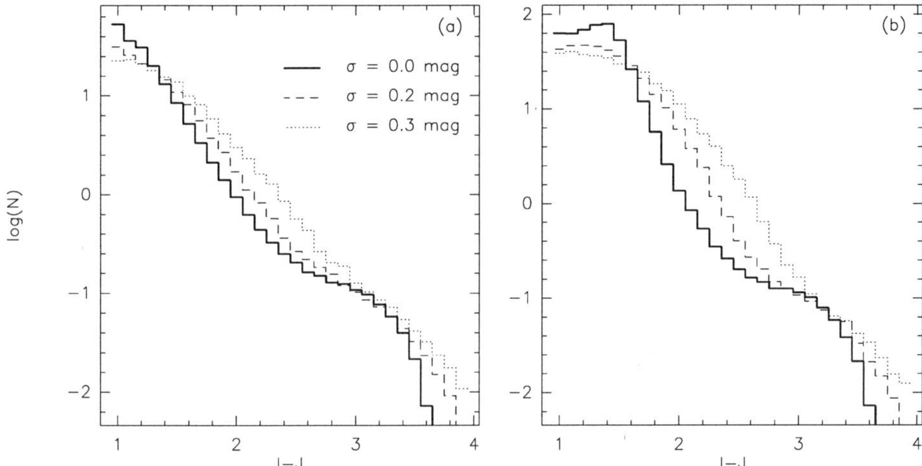

FIGURE 5. I-J colour distribution of the detected stars for a simulation of a galactic pole field with a h = 300 pc galactic scale height, and limiting magnitudes of I_{lim} = 19.0, J_{lim} = 17.0 and K_{lim} = 14.7 (close to the DENIS limits). (a) is for a flat luminosity function and (b) for a "realistic" luminosity function (as shown in Fig. 6). Only stars detected in all three bands are counted. These particular simulations have no noise, to separately illustrate the effect of (multiplicative) calibration error. They were performed for exagerated gaussian dispersions of σ =0.0, 0.2, 0.3 to emphasize this bias.

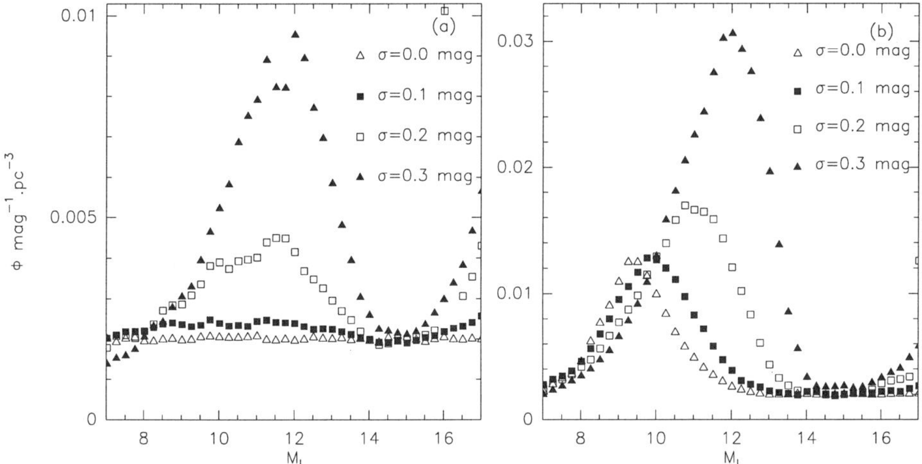

FIGURE 6. Uncorrected luminosity function for the simulation of Figure 5, showing the biases introduced by calibration errors for (a) a flat and (b) a "realistic" luminosity function.

The biases from detector and background noise similarly couple with the shapes of the colour-luminosity relations and of the luminosity function itself, though noise has a skewed, non-gaussian, distribution when looking at magnitudes. As noise affects the fainter stars more, the resulting luminosity function bias depends on the magnitude cutoff adopted for the analysis. Figure 7 shows that this bias remains significant even for relatively conservative cutoffs (20% errors when cutting 1 magnitude above the detection limit).

Clearly observational errors introduce significant biases, which are quite complex since they result from an interplay with the shape of the luminosity function istself and with

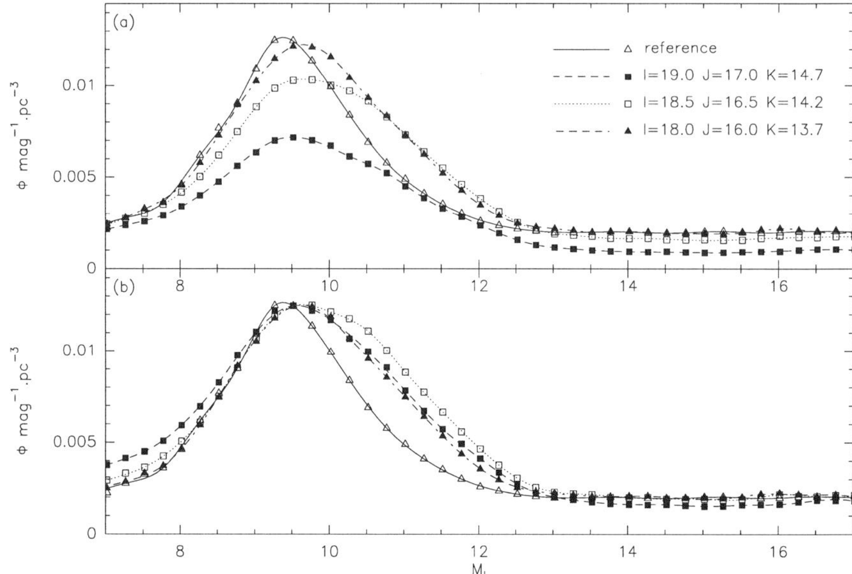

FIGURE 7. Luminosity function bias due to noise. The limiting magnitudes are I_d=19.0, J_d=17.0 and K_d=14.7. Sources are considered detected if they have 5 contiguous pixels 1.5σ above background, and the analysis used the sources detected at I and J. The simulations were analysed with cutoffs at the detection limit, 0.5 above this limit and 1 magnitude above this limit. The luminosity function is affected by detection incompleteness for cutoffs close to the limiting magnitude. The luminosity function in (a) has not been corrected for this incompleteness, while that in (b) has been to emphasize the other biases that noise can cause.

the slope of the colour-luminosity relation. One should note that changes in the slope of the colour-luminosity relation similarly affect the classical Malmquist bias due to the intrinsic scatter of this relation. The interplay between intrinsic dispersion and features of the luminosity function has been studied analytically by Stobie et al. (1989), but only for a linear colour-luminosity relation. Such a study is not easily generalised to a more general colour-luminosity relation with variable dispersion, which is best handled through montecarlo simulations.

To minimise the overall bias, it be desirable to select a colour with as large a dynamic range as possible over the relevant luminosity interval, such as I-J, and preferably one that also provides a constant colour-luminosity slope. This last goal unfortunately can be reached at best approximately for very low mass stars and brown dwarfs, because several important physical transitions occur over this mass range and affect the slope of the colour-luminosity relations:

- H_2 recombines in the atmosphere for early-M dwarfs,
- dust forms in the photosphere for late-M and L dwarfs,
- electrons become degenerate in the stellar interior at $\sim 0.1\ M_\odot$.

4.1.3. *Multiplicity bias:*

Deep wide field survey only resolve some (small) fraction of the multiple systems that they detect. The others are too close or/and have too large a luminosity contrast to be separated by the survey, and are thus counted as single stars. If not accounted for, the unresolved systems bias the luminosity function in three ways:

- Unresolved multiple stars are brighter than single stars, and therefore detected at larger distances (by a factor of $\sqrt{2}$ for an equal mass binary) and thus in larger numbers. This will overestimate the luminosity function.

• The colour of an unresolved multiple system is not that of a single star with the same luminosity (with a difference which depends on the mass ratio). It will therefore be attributed to an incorrect luminosity bin.

• The unresolved companions are not counted. This underestimates the luminosity function, particularly for the faintest luminosity bins since those objects are lost to a larger fraction of potential companions.

One should note that the last effect also affects cluster luminosity functions.

The multiplicity bias in photometric luminosity functions has been discussed in detail by both Kroupa (1995) and Reid & Gizis (1997), with opposite conclusions and a loud controversy. The two groups essentially agree on the basic mechanism, but they choose completely different distribution functions for the binary parameters (proportion of multiple systems, period distribution, mass ratio distribution). This leads them to conclude either that the three mechanisms mentioned above cancel out (Reid), or on the contrary that they have a large net effect (Kroupa). The net multiplicity bias depends strongly on the binary parameters distribution, and the bottom line in our opinion is that none of their two conflicting choices can yet be excluded on an observational basis, as the multiplicity statistics for M dwarfs are rather poorly constrained (Delfosse et al., 1999b). We should soon be in a good position to determine the multiplicity fraction from a volume limited sample of ~100 nearby M dwarf systems, which we have been studying with adaptive optics imaging and radial radial velocity monitoring over the last ~5 years (Delfosse et al. 1999b; Delfosse et al 2000, poster in these proceeding). A larger sample will however be needed to assess the period and mass ratio distributions.

4.2. DENIS *Luminosity function*

We have used the well characterized DENIS mini-survey data (230 squares degrees) to produce a very low mass stars and brown dwarfs luminosity function

Above $M_I = 13$ our analysis only retained objects detected in the three DENIS bands and brighter than $I = 17$, $J = 15$ and $K = 13$. These conservative cutoffs ensure 100% completness and reliability. To minimize the consequences of galactic structure uncertainties we also rejected all stars with photometric distances larger than 300 pc. We then use the generalised Schmidt estimator, with the colour-luminosity relation of Fig. 4, to build the luminosity function.

Fainter than $M_I = 13$ we cannot afford the luxury of conservative cutoffs, and retain the whole mini-survey sample (Delfosse et al. 1999a). Its reliability is guaranteed by complete spectroscopic observations, and its completeness is estimated on a strip by strip basis (Delfosse et al. 1999a, for details). We retain all objects detected at I and J above the magnitude for 50% completeness, and correct for the remaining incompleteness (i.e. a star at the 50% completeness limit carries an extra weight factor of 2). We use polynomial colour-luminosity relation of Figure 4 above the luminosity of GD 165B. This polynomial fit is essentially unconstrained for fainter L dwarfs and therefore diverges. We therefore provisionally (and certainly incorrectly) assume that redder objects all have the same luminosity as GD 165B. This only affects the last two bins of this DENIS luminosity function (Fig. 8), which also have large statistical errors.

We correct for the Malmquist-like biases resulting from the intrinsic dispersion of the colour-luminosity relation using the analytical formulation of Stobie et al. (1989), and use the Monte-Carlo simulation described above to correct those which result from the observational errors. The resulting luminosity function (Fig. 8) has very good statistics

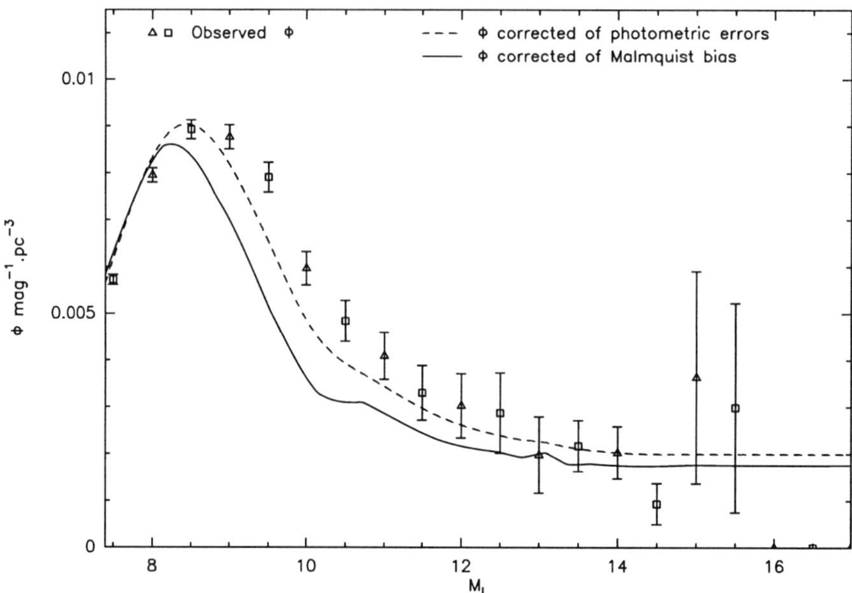

FIGURE 8. The first DENIS luminosity function for very low mass stars and brown dwarfs. The empty squares and triangles represent the uncorrected luminosity function, the dashed line the luminosity function corrected for the observation errors bias, and the full line is the luminosity function corrected for the biases introduced by both photometric errors *and* the intrinsic dispersion of the colour-luminosity relation.

for all M dwarfs. It shows the well known "Wielen" peak at $M_I \sim 8.25$ (due to the onset of H_2 recombination in the photosphere) and then decreases until $M_I \sim 12$. It then seems to flatten for fainter objects, out to the last bin which contain the L dwarfs and is quite uncertain. To confirm this flattening we will very soon analyse a much larger fraction of the DENIS data, using the updated theoretitical colour-luminosity relation of Chabrier et al. (2000) and a new determination of the intrinsic dispersion around this relation (Delfosse et al., in preparation).

5. Conclusions

To date the analysis of $\sim$ 1500 square degrees of DENIS data has identified a sample of $\sim$100 very late-M dwarfs and 15 L dwarfs. After suggesting the letter L to designate this new class (Martín et al 1997) of very faint dwarfs, we recently defined a spectral classification within the class (Martín et al 1999). We also determined the effective temperature scale of the L dwarfs from high resolution spectroscopy of the alkali optical resonance lines (Basri et al. 2000).

We also present here for the first time a DENIS stellar luminosity function for the solar neighbourhood, derived from the DENIS mini-survey data (230 squares degrees). We show that the Luminosity Function is very sensitive to relatively small errors on the colour-luminosity relation and Galactic scale height used as input. We furthermore demonstrate through Monte Carlo simulations that observational errors (measurement noise and calibration errors) can introduce large biases which couple with the real features of the colour-luminosity relation and of the luminosity function itself.

To improve on this Luminosity Function we will soon analyse a much larger fraction of the DENIS data to obtain better statistics in the fainter bins, and will use updated colour-luminosity relations. It will also be important to re-examine the Malmquist bias corrections, since the existing analytic corrections assume a linear average colour-luminosity relation, and a constant gaussian dispersion around this relation. The actual colour-luminosity relation clearly is *not* linear, and the underlying age and metallicity dispersions are unlikely to convert into a constant luminosity dispersion at all colours around this relation.

The ultimate goal of this work is an accurate Stellar Mass Function. In addition to a Luminosity Function, this requires a Mass-Luminosity relation. We have measured some M dwarf masses with 1-3% accuracy (Forveille et al. 1999, Delfosse et al. 1999c) from radial velocity and adaptive optics observations. With forthcoming additional accurate mass determinations (Delfosse et al., these proceedings) and some literature measurement from HST we plan to rapidly produce a more accurate Mass-Luminosity relation (Delfosse et al., in preparation) and therefore to newly derive the local Stellar Mass Function.

REFERENCES

ALLARD, F. 1998 *in Brown Dwarfs & Extra-solar Planets, Rebolo, Martín & Zapaterio-Osorio (eds), ASP Conf.* **Vol 134**, 370.

ALLARD, F. 2000 *in this proceding*

BARAFFE, I. 2000 *in this proceding*

BASRI, G., SUBHANJOY, M., ALLARD, F., HAUSCHILDT, P., DELFOSSE, X., MARTÍN, E. L., FORVEILLE, T., GOLDMAN, B. 200 *Submitted to ApJ*

BERTIN, E. & ARNOUTS, S. 1996 *A&AS* **117**, 393.

BORSENBERGER, J. 1997 *in Proceedings of the 2nd* DENIS *Euroconference The impact of large scale near-infrared surveys, F. Garzon, N. Epchtein, A. Omont, W.B. Burton, P. Persi (eds) Kluwer Ac. Publishers, Dordrecht*, p. 181.

CHABRIER, G., BARAFFE, I., ALLARD, F., HAUTSCHILDT, P. H., 2000 *A&A* **in preparation**

DELFOSSE, X., TINNEY, C. G., FORVEILLE, T., et al. 1997 *A&A* **327**, L25.

DELFOSSE, X., TINNEY, C. G., FORVEILLE, T., EPCHTEIN, N. BORSENBERGER, J., FOUQUÉ, P., KIMESWENGER, S., TIPHÈNE, D. 1999a *A&ASS* **135**, 41.

DELFOSSE, X., FORVEILLE, T., BEUZIT, J.-L., UDRY, S., MAYOR, M., PERRIER, C. 1999b *A&A* **344**, 897.

DELFOSSE, X., FORVEILLE, T., UDRY, S., BEUZIT, J.-L., MAYOR, M., PERRIER, C. 1999c *A&A* **350**, L39.

DELFOSSE, X. & FORVEILLE, T. 2000 *A&A* **(in preparation)**

DELFOSSE, X., FORVEILLE, T., BEUZIT, J.-L., UDRY, S., SEGRANSAN, D., MAYOR, M., PERRIER, C. 2000 *in this proceding*

DEUL, E. et al. 1995 *in Euroconference on Near Infrared Sky Surveys, P. Persi, B. Burton, N. Epchtein, & A. Omont (eds), Memorie della Società Astronomica Italiana* **66**, 549

EPCHTEIN, N. 1997 *in Proceedings of the 2nd* DENIS *Euroconference The impact of large scale near-infrared surveys, F. Garzon, N. Epchtein, A. Omont, W.B. Burton, P. Persi (eds) Kluwer Ac. Publishers, Dordrecht*, 15.

FAN, X., ET AL. 2000 *ApJ* **in press**

FELTEN, J. E. 1976 *ApJ* **207**, 700.

FORVEILLE, T., BEUZIT, J.-L., DELFOSSE, X., SEGRANSAN, D., BECK, F., MAYOR, M., PERRIER, C., TOKOVININ, A., UDRY, S. 1999 *A&A* **351**, 619.

GOLDMAN, B., DELFOSSE, X., FORVEILLE, T., ET AL. 1999, *A&A* **351**, L5.

HAMBLY, N. C. 1998, *in "Brown Dwarf and Extrasolar Planets Workshop", eds Rebolo, Martín,*

Zapatero Osorio, ASP Conference Series **134**, 11.

Haywood, M., Robin, A. C., Crézé, M. 1997a *A&A* **320**, 428.

Haywood, M., Robin, A. C., Crézé, M. 1997b *A&A* **320**, 440.

Kirkpatrick, J. D., McGraw, J. T., Hess, T. R., Liebert, J., McCartht Jr, D. W. 1994 *ApJSS* **94**, 749.

Kirkpatrick, J. D., Reid, I. N., Liebert, J., Cutri, R. M., Nelson, B., Beichman, C. A., Dahn, C. C., Monet, D. G., Gizis, J. E., Skrutskie, M. F. 1999b *ApJ* **519**, 802.

Kroupa, P. 1995 *ApJ* **453**, 358

Kroupa, P. 1998 The Stellar Mass Function *in "Brown Dwarf and Extrasolar Planets Workshop", eds Rebolo, Martín, Zapatero Osorio, ASP Conference Series* **134**, p483.

Leggett, S. K., Hawkins M. R. S. 1988 *MNRAS* **234**, 1065.

Leggett, S. K. 1992 *ApJS* **82**, 351.

Martín, E. L., Basri, G., Delfosse, X., Forveille, T. 1997 *A&A* **327**, L29

Martín, E. L., Delfosse, X., Basri, G., Goldman, B., Forveille, T., Zapatero Osorio, M. R., 1999 *AJ* **11**, 2466

Reid, I. N., Gizis, J. E. 1997 *AJ***113**, 2246.

Ruiz, M.T., Leggett, S.K., Allard, F. 1997 *ApJ* 491, L107.

Schmidt, M. 1968 *ApJ* **151**, 393.

Stobie, R.S., Ishida, K., Peacock, J.A. 1989 *MNRAS* **238**, 709.

Tinney C.G., Reid I.N., Mould J.R. 1993 *ApJ* **414**, 254.

Tinney C.G., Mould J.R., Reid I.N. 1993b *AJ* **105**, 1045.

Tinney, C.G., Delfosse, X., Forveille, T. 1997 *ApJL* **490**, L95.

Tinney, C.G., Delfosse, X., Forveille, T., Allard, F. 1998 *A&A* **338**, 1066.

Zapatero-Osorio, M.R., Bejar, V.J., Rebolo, R., Martín, E.L., Basri, G. 1999 *ApJ* **524**, L115.

Preliminary Results from the 2MASS Core Project

By J. LIEBERT[1], I. N. REID[2], J. D. KIRKPATRICK[3], R. CUTRI[3], B. NELSON[3], D. G. MONET[4], C. C. DAHN[4], M. SKRUTSKIE[5], J. E. GIZIS[5], AND M. D. FISHER[1]

[1]Steward Observatory, University of Arizona, Tucson, AZ 85721, USA

[2]California Institute of Technology, 105-24, Pasadena, CA 91125, USA

[3]IPAC, Caltech 100-24, Pasadena, CA 91125, USA

[4]U.S. Naval observatory, P.O. Box 1149, Flagstaff AZ 86002, USA

[5]University of Massachusetts, Amherst MA 01003, USA

The initiation of the DENIS and 2MASS surveys are resulting in the discoveries of dozens of field brown dwarf candidates, and the need to develop a new spectral class of L dwarfs. The L dwarfs are perhaps only a few hundred degrees cooler than the latest M dwarfs (M9–9.5 V), and are many hundreds of degrees warmer than the brown dwarf Gliese 229B. However, the formation of dust removes TiO and VO from the atmospheres, resulting in qualitatively different red spectra, with for example a strong, pressure-broadened K I resonance doublet. In contrast, the infrared spectra show the same (H_2O and CO) molecular features as in late M dwarfs, with no evidence of methane. The detection of the Li I 6707 Å resonance doublet shows that many L dwarfs, at least, have substellar masses. Based on analysis of point sources from the first 1% of sky, well over 1,000 L dwarfs should be detected in the 2MASS survey. Our results suggest that several might exist within 5 pc of the Sun.

1. Introduction

After waiting three decades since Kumar (1963) proposed their existence, we are gratified to see literally dozens of candidates probably or definitely below the stellar mass limit being found in young clusters and associations. Here one has the big advantages that the age, the distance and luminosity of a cluster member are generally known. In this presentation, complementary to the topic of this meeting, we report the finding of a large number of candidates in one of the first infrared surveys of the field population. For these objects, we do *not* know the ages, nor initially their distances and luminosities, and must put up with the difficulty of finding cooler and fainter objects, because they are generally older. However, surveys of large areas of sky should enable us to find the nearest such neighbors to the Sun, assuming they form as single objects in the field. At a given temperature or luminosity, these will be the brightest such objects, and the most amenable to detailed follow-up studies. The older brown dwarfs are indeed cooler, yet it turns out that the 1–1.5μ wavelength region – roughly the J and H bands – includes relatively transparent intervals of the spectrum. Models of Burrows et al. (1997) show that brown dwarfs down to below 1,000 K emit a relatively large amount of flux there.

Fortunately, two groups have recently initiated the first substantial sky surveys at near-infrared wavelengths. The DEep Near Infrared Sky survey (DENIS; see Delfosse & Forveille, these proceedings) was started in January 1996 by a consortium of European investigators for the southern hemisphere. The survey covers the I, J, and K bands. The Two Micron All Sky Survey (2MASS, Skrutskie et al. 1997) began data acquisition in May 1997 at Mt. Hopkins AZ for the northern sky, and at Cerro Tololo Interamerican

Observatory in March 1998 for the southern sky. The 2MASS infrared camera is a unique design: Dichroic optics split the radiation into three beams for simultaneous measurements of $J(1.2)\mu$, $H(1.6)\mu$ and the "short" $K_s(2.2)\mu$ bands (Milligan et al. 1996). The telescope is scanned in declination, covering 8.5 arc minute by 6 degree "tiles." The nominal 10 sigma detection limiting magnitudes are J <15.8, H <15.1 and K < 14.3, substantially fainter than those of DENIS.

In the paper by Delfosse et al. (1997), DENIS announced the discovery of the first brown dwarf candidates from an infrared survey of the general sky. These objects showed spectra indicating they are cooler than stars assigned the M9–M9.5 V subtype. In a companion paper, Martín et al. (1997) reported the detection of lithium in one of the DENIS candidates. We do not need to go through the general arguments for this audience that – for an object this cool – passing the "lithium test" (Rebolo et al. 1992; Magazzù, Martín & Rebolo 1993) proves unambiguously that DENIS-P J1228.2-1547 is substellar. It is also in the Martín et al. paper that there first appears in a refereed journal the brief proposal that the objects later than M9.5 V be called L dwarfs. These authors in turn acknowledged the suggestion by Kirkpatrick at last year's Tenerife meeting (1998, in "Brown Dwarfs and Extrasolar Planets", eds. Rebolo, Martín, Zapatero Osorio, *ASP*, Vol. 134).

Of course the much cooler Gliese 229B, while not observed at short enough wavelengths for the Li I 6707Å resonance doublet, is the first unambiguous brown dwarf – it passes the "methane test." The other distant-companion brown dwarf candidate, GD 165B, lacks enough flux at these red wavelengths for a measurement of lithium. As we shall see, GD 165B sits near the middle of the L spectral sequence that has been developed.

The first DENIS discoveries underscore the critical role the lithium test plays for a field population in determining whether a given candidate is substellar, provided the effective temperature (T_{eff}) is below about 2,000 K. Note, however, that since Li undergoes nuclear reactions at a lower central temperature than hydrogen, the highest mass brown dwarfs (~0.065-0.075$M_\odot$) are able to deplete lithium.

Nearly concurrently with the early DENIS work, three other field stars were found which pass the lithium test. From UK and ESO Schmidt plates, Thackrah, Jones & Hawkins (1997) found an M6 dwarf which shows a distinct Li I 6707Å detection. Depending on the age, T_{eff}, implied Li abundance, and the uncertainties due to models, this object still appears to be warm enough to straddle the stellar boundary. A second, much cooler object is Kelu 1, found in the Chilean proper motion survey by Ruiz, Leggett & Allard (1997). The model atmospheres generated by Allard suggest that T_{eff} is about 1,900 K. Its spectrum places this object among the earlier L dwarfs discussed here. The presence of Li makes it highly probable that this object is substellar. A third object of spectral type M9 V, has been found from an old Luyten proper motion catalog, LP 944-20 (Tinney 1998). Tinney (1998) argues that the mass is near 0.06 $M_\odot$, and the age 475-700 Myr. As with the first example, T_{eff} may be a little too high to be confident about the mass. The sequence of Li observations in the Pleiades – where ages and luminosities are known – lends credence to the model-dependent interpretations of the field objects (Martín et al. 1998).

2. Selection criteria

Our goal has been to define systematic survey criteria so that a complete sample of late M dwarfs and now L dwarfs can be assembled. At the time of the meeting, 493 survey tiles had been analyzed, plus enough data taken earlier with prototype cameras

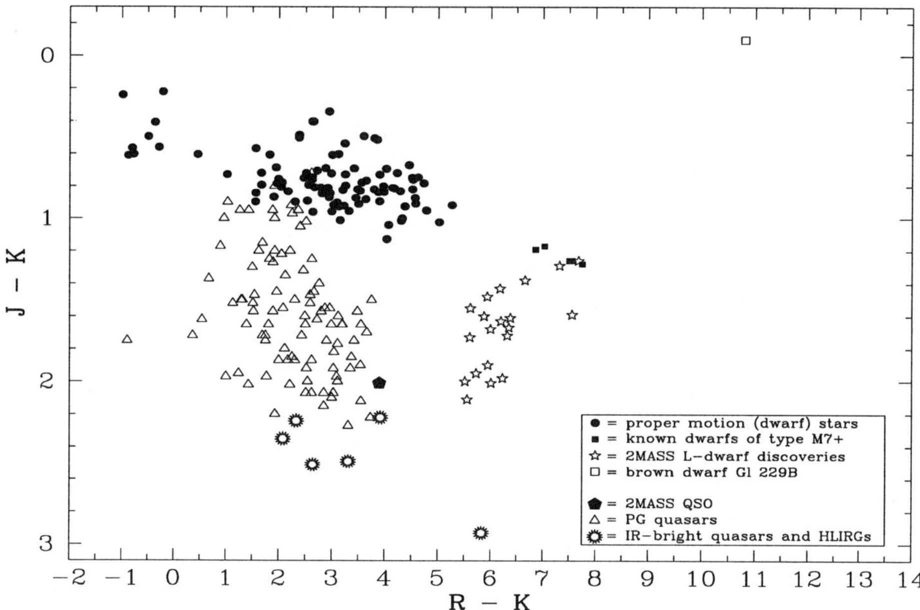

FIGURE 1. The two-color diagram used for identifying L dwarf and late M dwarf candidates from all 2MASS point sources (symbols as indicated). Red and reddened extragalactic objects are separated nicely with these colors. The L dwarfs are not detected on the POSS red plates, so the plotted positions are lower limits to the R-K colors. Note in particular the unique position of GL 229B.

to bring the total area searched to about 400 square degrees. This is close to one per cent of the sky.

The selection criterion which has been shown to have a high degree of success in identifying very late M and L dwarfs is that J-K > 1.3, while R-K >5.5, where "R" is the photographic red magnitude from the Palomar Observatory Sky Survey (POSS 1 or 2). Since it is now appreciated that the onset of methane in cooler objects causes J-K in GL 229B to be blue, we also searched for objects with J-K < 0. We did not find any similar counterparts in our fields. However, since brown dwarfs with the luminosity of GL 229B would not be detectable with 2MASS beyond about 8 parsecs, the volume surveyed would have been only about 30 cubic parsecs.

In Fig. 1 the initial set of 2MASS point sources satisfying the above criteria are shown as open stars. *None* of these are detected on the POSS red plate, but to minimize clutter on the graph, the plotted positions are the lower limits to the R-K color. Also shown for comparison is a sequence of cool stars (filled circles), and the much cooler brown dwarf GL 229B. In this initial analysis, 20 L dwarfs were found and these are discussed in Kirkpatrick et al. (1998), hereafter K98. In all 25 L dwarfs have been studied, including new Keck observations of the three DENIS objects of Delfosse et al. (1997), of Kelu 1 and GD 165B.

3. The L dwarfs

A complete classification system has been developed in K98; some principal features are summarized here. In Fig. 2 spectrophotometry is shown of one late M dwarf, one middle

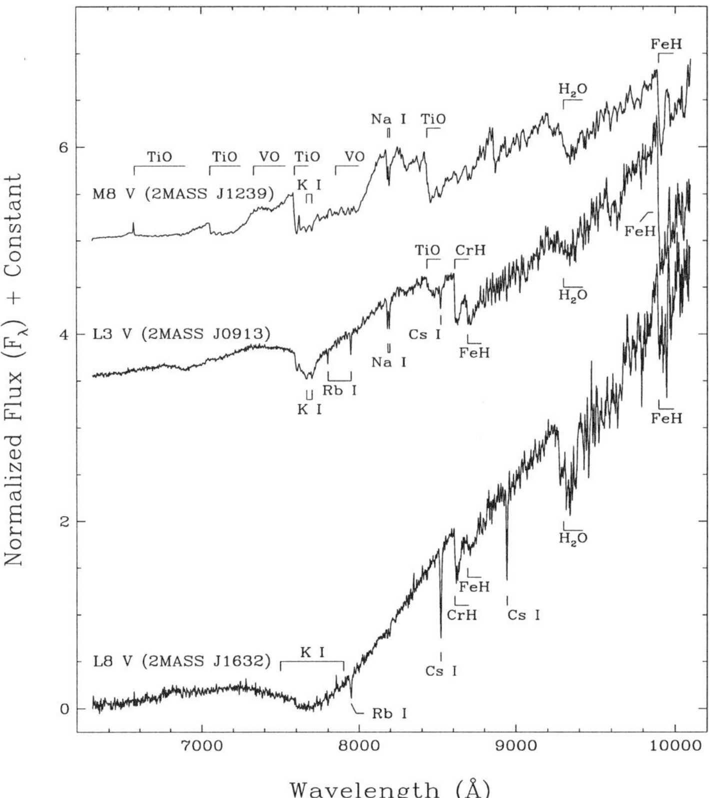

FIGURE 2. Red spectra of an M8 V dwarf (top) an L3 V dwarf (middle) and very late L8 V dwarf (bottom). Particularly in the last spectrum, note the dominating strength and width of the K I resonance doublet, and the other resonance lines of rarer alkalis, along with the disappearance of the TiO and VO bands.

and one very late 2MASS L dwarf. These are at "far red" wavelengths (6200-10000Å), obtained with the Keck II low resolution spectrograph (LRIS).

The strongest features at these wavelengths in late M dwarfs are the prolific band systems of titanium and vanadium oxides, just as water and carbon monoxide dominate at longer wavelengths. So strong are these that nearby "pseudo-continuum" peaks tower high above adjoining absorption troughs, though at no wavelength is the true continuum level reached. By the latest M types, however, the TiO strengths have peaked out and begun to weaken. At the slightly cooler T_{eff} where VO begins to weaken as well, this is defined as the beginning of the L sequence (L0 V). With progressively increasing L type (and decreasing T_{eff}), the TiO and VO bands weaken (in the L3 V dwarf of Fig. 2) and then disappear entirely (the L8 V object).

Our understanding of the physics of the above phenomena is as follows: The M dwarf temperature scale is not known to better than 10% or so, but it is believed that below about 2500 K, first the TiO and then VO molecules in the atmospheres precipitate out as dust grains (Tsuji et al. 1996ab, Allard 1998, Burrows and Sharp 1998). By removing these principal sources of opacity from the atmospheres, dwarfs a few hundred degrees cooler than late M stars have much more transparent atmospheres at these wavelengths. Among the remaining absorbers are hydride bands with limited wavelength coverage. Otherwise, there is little continuum opacity, due to the extremely low electron density

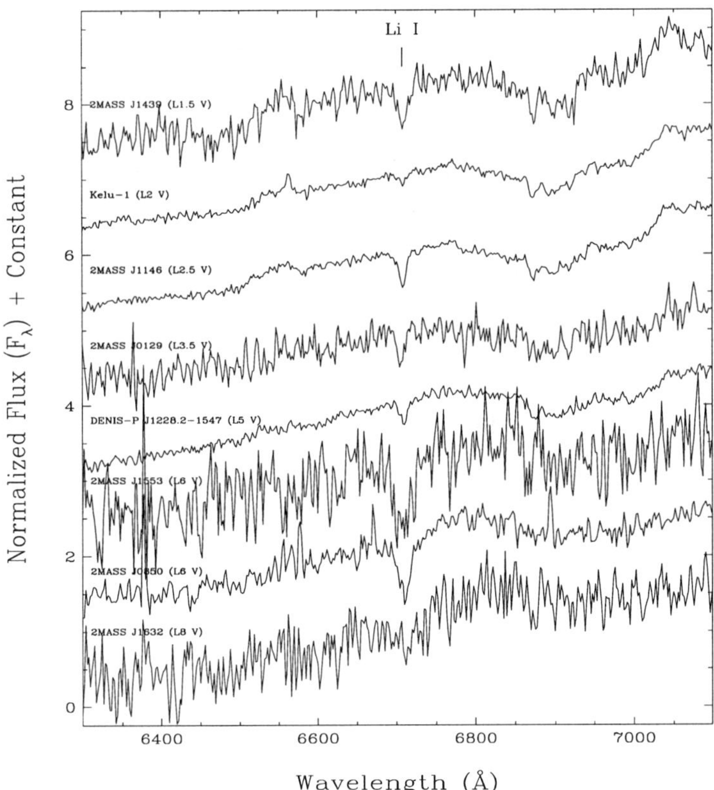

FIGURE 3. Spectra of a sequence of L dwarfs, increasing in type from to to bottom, centered on the Li I 6707Å resonance doublet. The line is generally detected, but highly variable in strength. The Hα line is also included, but appears convincingly in emission only for Kelu 1.

(for H^- and H_2^-). Thus, the atomic resonance lines of the neutral alkalis become quite strong for an abundant species like K I, and easily detected for rare species like Rb I, Cs I, and if undepleted, Li I. That is, the crucial spectral identifier (Li I) of brown dwarfs below about $0.065M_\odot$ appears strongly enough at an undepleted abundance to be detectable on low resolution spectra, rather than requiring the high resolution needed for M dwarfs – see Fig. 3, discussed below. At late L types, these trends carry to extreme. As is evident for the L8 V dwarf in Fig. 2, the K I doublet with great width and strength becomes the dominant feature of the red spectrum. (The Na I resonance doublet would be even stronger, if our spectra extended to short enough wavelengths.)

With these red spectra featuring pressure-broadened K I, the late L dwarfs may resemble cool white dwarfs more than they do M dwarfs. On the other hand, the L dwarfs show strong CO and H_2O like M dwarfs in the near infrared. The effective temperatures of the coolest L dwarfs are not known, but in any case they are not cool enough for the CO molecule to give way to CH_4, as it is predicted to occur near 1500 K (Burrows and Sharp 1998). Indeed, our infrared spectra of the latest objects (K98) show no evidence of CH_4 bands. The only known "methane dwarf" remains GL 229B.

The onset of methane in the near-infrared spectrum will reverse the trend in J-K color with decreasing temperature (Burrows et al. 1997). GL 229B has a very blue J-K= -0.1. Thus the selection criteria must be modified to search for objects with blue or neutral J-K color, increasing enormously the numbers of apparent point sources which must be screened. At the time of this writing, the search has uncovered no methane dwarfs with blue J-K colors like GL 229B. However, we have not yet searched effectively for methane dwarfs between the T_{eff} of an L8 V dwarf and GL 229B (~1,000 K), since these would likely have neutral J-K colors characteristic of the vast majority of 2MASS point sources.

The yield of the L dwarfs is substantial enough to conclude that they are present in significant numbers in the field around the Sun. It is elementary to do the math: 20 were found in the first 400 square degrees of sky, which means that the entire sky may yield at least 1,000 objects to the 2MASS survey depth. (This neglects the likelihood that the current color selection techniques do not find 100% of the L dwarfs, not to mention methane dwarfs.) It is too early for a real estimate of the space density of even just the L dwarfs, since the distances and luminosities are not yet known (not to mention the ages and masses). For a sense of the likely answer, one might assume that the $M_K = 11.65$ of GD 165B is appropriate to all L dwarfs, all brighter than K=14.5. In this case the volume surveyed would be about 2100 cubic parsecs, or a space density of 0.009 pc^{-3}, suggesting that 4-5 might exist within 5 pc of the Sun.

We can show that many are substellar. In Fig. 3, a handful of our objects are illustrated in spectra centered on the Li I 6707Å resonance doublet. The candidates observed with Keck II fall into three categories: (1) those with Li I detections at equivalent widths of several Angstroms, due to the high atmospheric transparency discussed earlier; (2) those with good enough spectra to show that Li I is depleted and undetected; and (3) those too faint for a decent spectrum to be achieved at these wavelengths, even with a 10 meter telescope. The statistics suggest that most of the L dwarfs are indeed brown dwarfs, but the reader is referred to K98 for justification of this conclusion.

4. Flare activity in a 2MASS object

The displayed interval for spectra of Fig. 3 also include the Hα line (6563Å). Only the high signal-to-noise ratio spectrum of Kelu 1 (classified L2 V) shows a definite emission feature. It has not been clear what happens to magnetic dynamo activity – as manifested in coronal X-rays or chromospheric emission lines – at very low masses (see, for example, Stauffer et al. 1991).

2MASS J0149090+295613 (hereafter 2M0149) is unique among the very late M/L dwarf counterparts to *2MASS* point sources in having shown, in the first spectrum taken (1997 Dec 7), a diverse emission line spectrum at red wavelengths, featuring an Hα equivalent width of 300Å. On four repeat observations, however, the spectrum was that of a more ordinary dMe object, albeit with a late spectral type of M9.5 V.

Our interpretation is that the object underwent an extreme magnetic flare event. The continuum flux at the short wavelength end of the spectrum was also several times stronger during the apparent flare than in quiescence. In Fig. 4, we display its flare spectrum with that of another known M9.5 emission line object PC 0025+0047 (PC0025, Schneider et al. 1991; Graham et al. 1992). These objects differ decisively in that (1) the emission line spectrum of 2M0149 is more diverse, at both higher and lower ionization states, and (2) the PC object always seems to exhibit the same spectrum – that is, it appears to be maintaining the same state, rather than undergoing flare events.

The *2MASS* object adds to the evidence that magnetic dynamo eruptions may release more energy relative to L_{bol} as the stellar mass decreases. If the total flare luminosity scales with Hα in the same way as, for example a multi-wavelength observed event of the middle-M dwarf AD Leo (Hawley & Petterson 1991), 2M0149 may have had a flare luminosity which approached or exceeded its quiescent L_{bol} during the brief (1000 second) impulse phase characteristic of flare stars. In contrast, its quiescent state shows

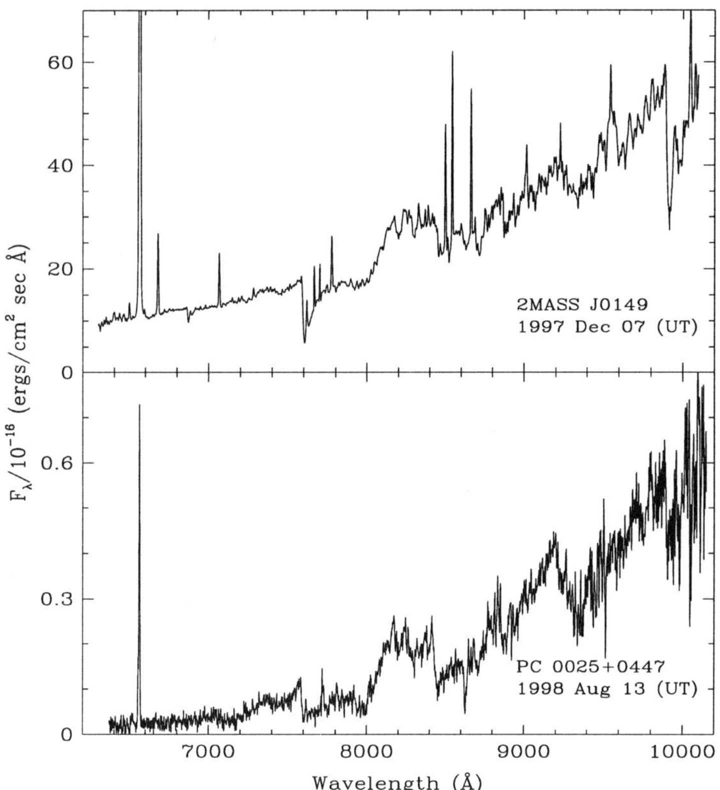

FIGURE 4. A Keck LRIS spectrum recently obtained of PC0025 is shown alongside the "flare" spectrum of 2M0149 to contrast their emission line properties. In addition to the dominant Hα, 2MASS0149 shows He I 6678Å, the resonance doublet of K I. and a continuum excess that fills in the absorption features particularly at the shorter wavelengths. The absorption spectrum of 2M0149 in quiescence is, however, very similar to that of PC0025.

an unremarkable level of chromospheric activity, which leads us to question whether this object is unique at all; rather, it turn out that many of the ultracool M and L dwarfs found by *DENIS* and *2MASS* undergo similar flares. A more complete discussion of 2M0149 is given in Liebert et al. (1999).

This work was supported by NASA's Jet Propulsion Lab through contract number 961040NSF, a core science grant to the Two Micron All Sky Survey science team. The 2MASS Project has been supported though NASA Cooperative Agreement NCC , JPL Contract 960847, and NSF grant AST 93-17456. JL also acknowledges support from the NSF through grant AST 92–17961.

REFERENCES

Allard, F. 1998, in *Very Low Mass Stars and Brown Dwarfs in Clusters and Associations*, La Palma, Spain, 11-15 May 1998.

Burrows, A. & Sharp, C.M. 1998, submitted to *Ap.J.*.

Burrows, A., Marley, M., Hubbard, W.B., Lunine, J.I., Guillot, T., Saumon, D., Freedman, R., Sudarsky, D., & Sharp, C. 1997, ApJ, 491, 856.

Delfosse, X., Tinney, C.G., Forveille, T., Epchtein, N., Bertin, E., Borsenberger, J., Copet,

E., de Batz, R., Fouque, P., Kimeswenger, S., Le Bertre, T., Lacombe, F., Rouan, D., & Tiphene, D. 1997, Astr. Ap., 327, L25.

Graham, J.R., Matthews, K., Greenstein, J.L., Neugebauer, G., Tinney, C.F., & Persson, S.E. 1992, A.J., 104, 2016.

Hawley, S.L. and Petterson, B.R. 1991, Ap.J., 378, 725.

Kirkpatrick, J.D., Reid, I.N., Liebert, J., Cutri, R.M., Nelson, B., Beichman, C.A., Dahn, C.C., Monet, D.G., Skrutskie, M.F., & Gizis, J. 1999, preprint. (K98)

Kumar, S. 1963, Ap.J., 137, 1121.

Liebert, J., Kirkpatrick, J.D., Reid, I.N., & Fisher, M.D. 1999, Ap.J. in press.

Magazzu, A., Martín, E.L., & Rebolo, R. 1993, Ap.J.Let., 404, L17.

Martín, E.L., Basri, G., Delfosse, X., & Forveille, T. 1997, Astr. Ap., 327, L29

Martín, E.L., Basri, G., Zapatero Osorio, M.R., Rebolo, R., & Garcia Lopez, R.J. 1998, in *Very Low Mass Stars and Brown Dwarfs in Clusters and Associations*, La Palma, Spain, 11-15 May 1998.

Milligan, S., Cranton, Brian W., Skrutskie, Michael F. 1996, in *Development of a three-channel SWIR camera, for ground-based astronomical imaging*, Proceedings of the SPIE, 2863, pp 2-13.

Rebolo, R., Martín, E.L., & Magazzu, A. 1992, Ap.J.Let., 389, L83.

Ruiz, M.T., Leggett, S.K., & Allard, F. 1997, Ap.J.Let., 491, L107.

Schneider, D.P., Greenstein, J.L., Schmidt, M., & Gunn, J.E. 1991, A.J., 102, 1180.

Skrutskie, M.F., et al. 1997, in *The Impact of Large Scale Near-IR Sky Surveys*, eds. F. Garzon et al., Kluwer (Netherlands), pp 25-32.

Stauffer, J.R., Giampapa, M.S., Herbst, W., Vincent, J.M., Hartmann, L.W., & Stern, R.A. 1991, Ap.J., 374, 142.

Thackrah, A., Jones, H., & Hawkins, M. 1997, Mon.Not.R.A.S., 284, 507.

Tinney, C.G. 1998, Mon.Not.R.A.S., 296, L42.

Tsuji, T., Ohnaka, K., & Aoki, W. 1996, Astr. Ap., 305, L1.

Tsuji, T., Ohnaka, K., Aoki, W., & Nakajima, T. 1996, Astr. Ap., 308, L29.

II.
SPECTROSCOPIC PROPERTIES, FUNDAMENTAL PARAMETERS AND MODELLING

Properties of M dwarfs in Clusters and the Field

By SUZANNE L. HAWLEY[1], I. NEILL REID[2], AND JONATHAN G. TOURTELLOT[1]

[1]Department of Physics and Astronomy, Michigan State University, East Lansing, MI 48824, USA

[2]Palomar Observatory, Caltech, Pasadena, CA 91125, USA

We report on magnetic activity and luminosity function results from our field and cluster surveys of low mass stars. Magnetic activity in M dwarfs has several notable effects on the colors, magnitudes and molecular bandstrengths. The presence of activity only up to a limiting mass (color, magnitude) in a coeval population can be used as an age indicator. We have calibrated several age-activity relations using new observations of dMe stars in M67 to anchor the relations at large age. The changes in activity strength along the M dwarf sequence are discussed. The luminosity functions for several clusters show evidence for mass segregation and two clusters appear to have lost their low mass population. Unusually rapid dynamical evolution or a skewed initial mass function could account for these results. Either explanation would have implications for the number of brown dwarfs and very low mass stars expected in the field at the present epoch.

1. Introduction

We have been carrying out large surveys of M dwarfs in the field (Reid *et al.* 1995, hereafter PMSU1, Hawley, Gizis & Reid 1996, hereafter PMSU2) and in nearby open clusters (Hawley, Tourtellot & Reid 1998, hereafter HTR98). Although these stars, on the whole, might not qualify as "very low mass" (VLM) stars for this conference, they are interesting to study in order to understand the properties that might affect stars even further down the main sequence. I will divide the talk into two main parts, the first focussing on magnetic activity in these stars, how it affects what we observe, and how we can use it as an age indicator. In one cluster, the Hyades, we did observe several VLM and brown dwarf candidates and I will comment briefly on those. In the second part, I will concentrate on our new membership surveys for low mass stars in several less-observed clusters, and discuss the present day luminosity functions and implications for mass segregation and the initial mass function.

2. Magnetic activity

We measure magnetic activity on an M dwarf star by looking for the Hα line in emission, which indicates the presence of a strong chromosphere. X-ray emission is another way to find activity but most M dwarfs are too faint and far away (especially in clusters) to have good measurements as yet. Figure 1a shows the red part of the optical spectrum which we observe in our surveys, for the active dMe star AD Leo. Several molecular bands are marked including CaOH, CaH and TiO. We measure the band strength by taking the ratio of the band depth to a nearby continuum point, as shown on the figure.

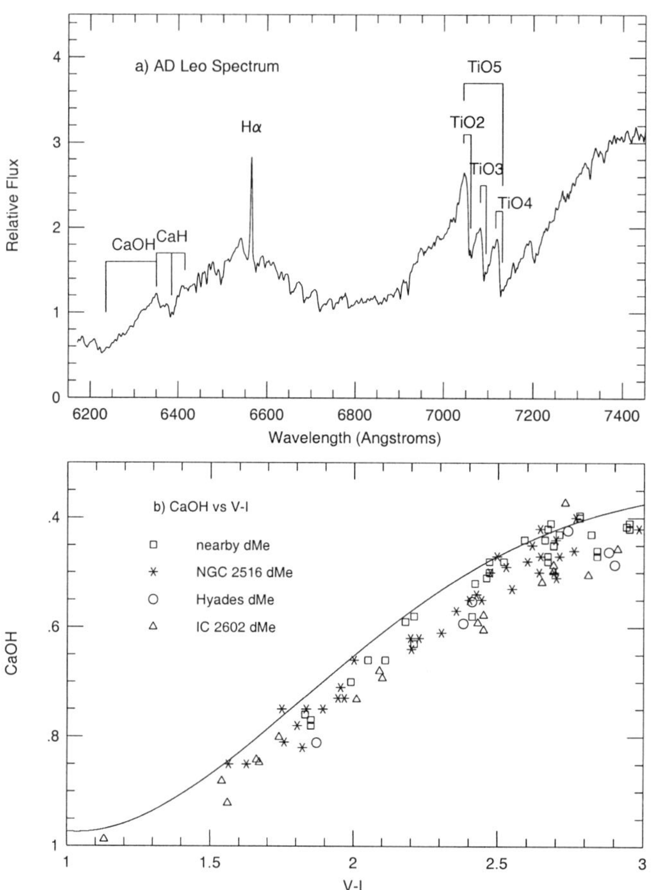

FIGURE 1. a) AD Leo spectrum showing molecular bands and Hα emission; b) CaOH vs. V-I for active stars from the field and several clusters, solid line is the mean relation for inactive stars.

2.1. *Activity effects on bandstrengths, colors and magnitudes*

The first point to emphasize is that some of these bandstrengths vary systematically with the presence of magnetic activity (Hα emission) on the star. We showed in PMSU2 that TiO2 was deeper and TiO4 was shallower in active stars, compared to inactive stars at the same spectral type or V-I color. In HTR98 we showed that CaOH was similarly affected, appearing shallower in active stars at the same V-I color as inactive ones. Figure 1b illustrates this effect for nearby dMe field stars, and for dMe stars from several nearby clusters. The inactive dM star relation is shown as the solid line; typical scatter about this line is about ±0.05 in the CaOH bandstrength. Clearly the temperature of the star is not being accurately reflected by both the CaOH bandstrength and the V-I color. The questions is, which is causing the active stars to deviate from the inactive star relation?

A second relevant observation is found from looking at an M_V vs. V-I HR diagram,

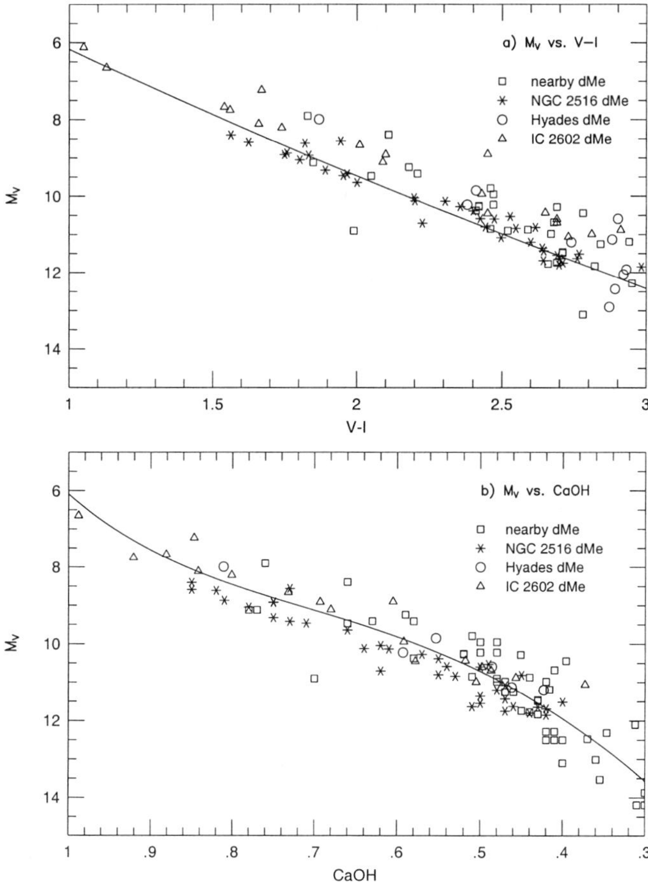

FIGURE 2. a) M_V vs. V-I for active stars from the field and several clusters, solid line is the mean relation for inactive stars; b) M_V vs. CaOH, same symbols as in (a). Note that using CaOH on the temperature axis brings the active and inactive stars into agreement.

as shown in Figure 2a. Symbols are the same as in Figure 1b, and the scatter in M_V is about ± 0.5 mags. The dMe stars show a definite tendency to lie brighter/redder than the dM main sequence.

Combining these results in Figure 2b (M_V vs. CaOH), we see that when CaOH is used as the temperature indicator, the dMe stars fall into agreement with the dM star main sequence, and the discrepancy is resolved. The tendency for the NGC 2516 stars to lie below the main sequence in this figure is probably a result of the low cluster metallicity. Based on this figure, it appears that the V-I color is being most affected by the magnetic activity. However, these are only empirical results and the next step will be to model an M dwarf photosphere and chromosphere in detail, to understand the physical processes that lead to these color, magnitude and bandstrength effects.

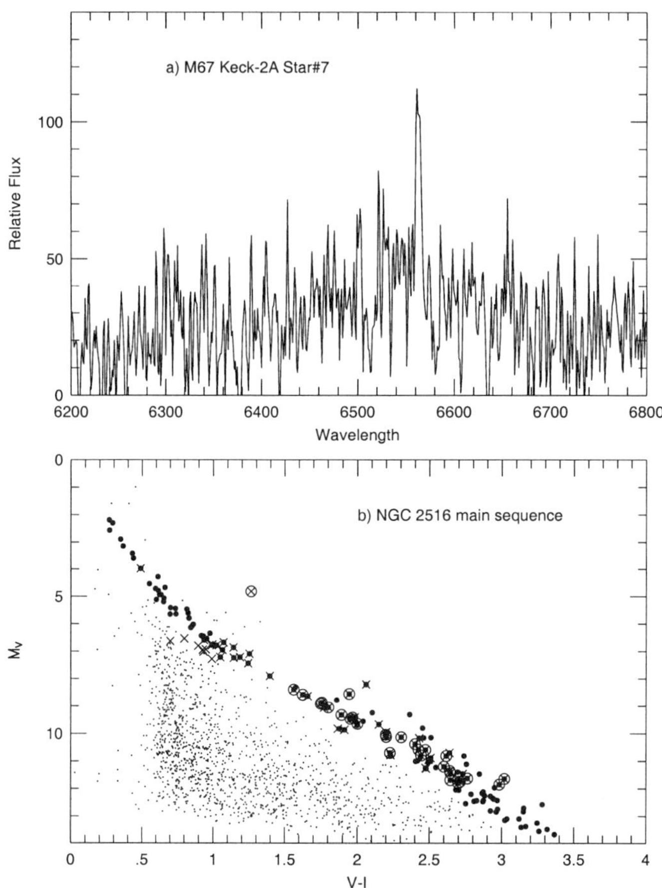

FIGURE 3. a) Spectrum of a photometric candidate member of M67 showing Hα in emission; b) Color-magnitude diagram for NGC 2516 showing photometric candidates (enlarged points), spectroscopically observed stars (crosses), and Hα emission stars (circled).

2.2. *Age and activity*

A primary goal of our open cluster survey was to quantify and calibrate an age-activity relation for M dwarfs. Such a relation has been suspected in the past on various grounds: 1) F,G and K stars show chromospheric activity which decays with age (Skumanich 1972); 2) kinematic studies indicate that dMe stars are a younger population than dM stars in the field (Wielen 1975); and 3) observations of dMe stars in clusters (e.g. the Pleiades and Hyades) have shown that the color at which Hα emission becomes common is redder in older clusters (Stauffer *et al.* 1991). In PMSU2 we carried out a kinematic study of the dM stars in the field and found evidence that the lower mass dMs were an older population than the higher mass dMs. We suggested therefore that the age-activity relation appropriate for M dwarfs differed from the usual F,G and K star relations. The emission does not decay with age but turns off at a given mass depending on age. This

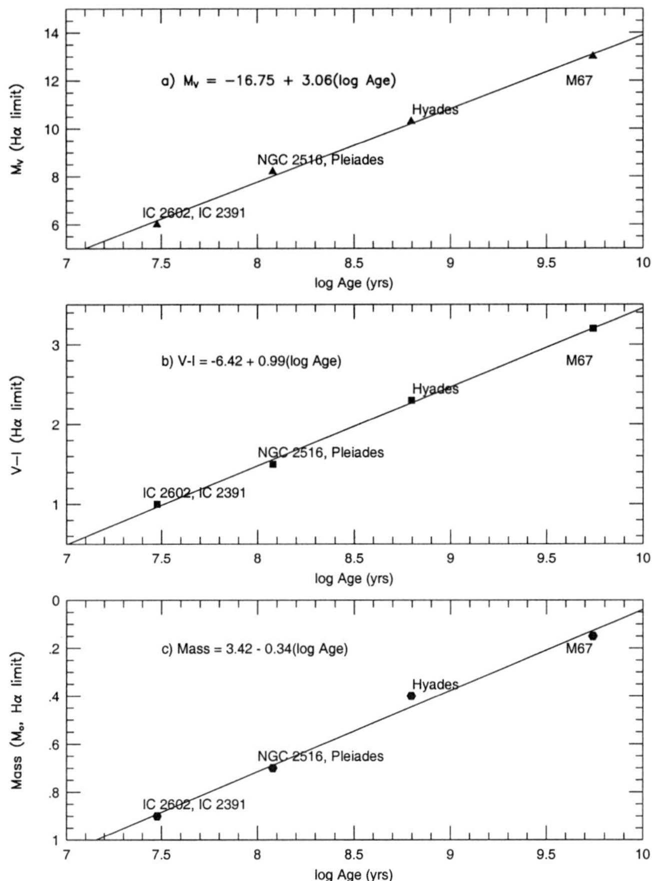

FIGURE 4. Age-Activity relations for M dwarfs using Hα limit measurements in clusters. Panels show: a) M_V vs. log age; b) V-I vs. log age; and c) Mass ($M_\odot$) vs. log age.

explains the prevalence of dMe stars at late spectral types (they mostly haven't turned off yet) and also fits in with the cluster mass-Hα results. We call the mass at which the cluster M dwarfs become mostly dMe stars the "Hα limit" mass.

To calibrate the Hα limit – mass (color, absolute magnitude) vs. age relations, we required observations of dMe stars in a fairly old cluster. We chose M67 as the nearest old (and fairly uncontaminated) cluster, and obtained deep color-magnitude diagrams, and subsequently spectra of photometrically-selected candidates. Our observations were made at the Keck Observatory using the LRIS spectrograph. A spectrum of the first dMe star we found is shown in Figure 3a. It has R=21, spectral type dM4e, and Hα EW = 8Å. A second run turned up two more dMe stars at about the same color. Earlier type stars showed no emission.

Another cluster that we have spectroscopic data for is NGC 2516, which has an age similar to the Pleiades. Figure 3b shows an HR diagram for this cluster, with

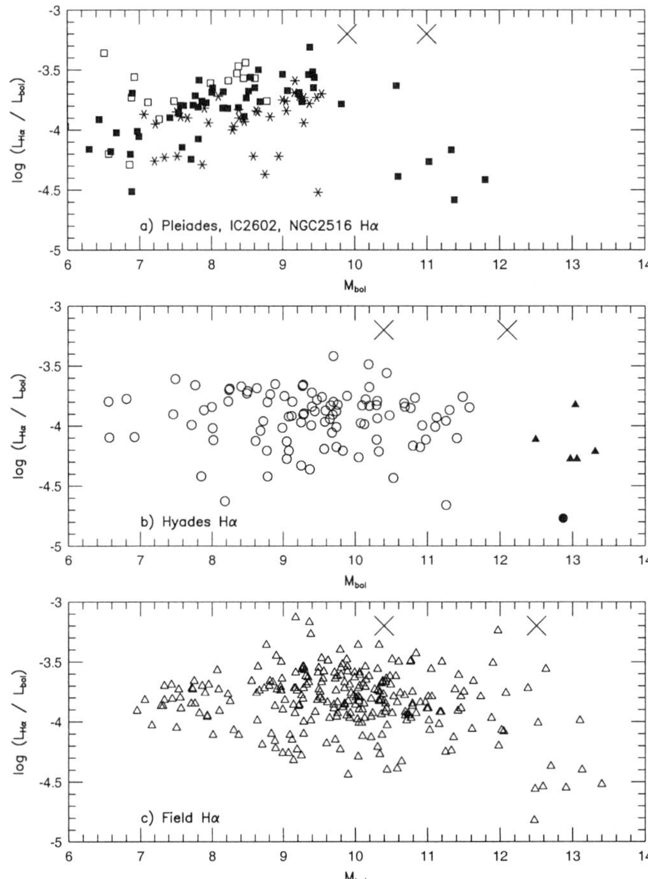

FIGURE 5. Activity strength for dMe stars in a) young clusters – Pleiades are closed squares, IC2602 are open squares, NGC 2516 are asterisks; b) the Hyades (see text for explanation of closed symbols); and c) the field. Large X's mark the positions of (left) 0.2 $M_\odot$ and (right) 0.1 $M_\odot$ stars for the age of each population.

photometrically-selected candidates that have Hα emission indicated. This figure illustrates the concept of the Hα limit color, which we take to be at V-I = 1.5 for this cluster. Essentially all of the redder cluster stars that we observed have emission. (Note that not all stars were observed in this cluster, see figure caption.) Using similar data for several other clusters and our data for M67, we constructed the age-activity relation shown in Figure 4. Cautionary note: the Hα limit is subjectively determined and there are also uncertainties in the cluster ages. Therefore these relations should be regarded as suggestive rather than definitive.

Nevertheless, it certainly appears that linear color-magnitude-mass vs. log age relations fit the available data and that a clear physical explanation should be sought to explain these results. It is not obvious that age-related spindown of a rotation driven dynamo will be adequate, since our relation does not show *decay* with age as in the earlier type stars. There are other pieces of evidence against rotation being the most important parameter governing activity in these stars, which can be found in e.g. PMSU2 or Hawley *et al.* (1998).

2.3. *Activity strength and Hyades VLM candidates*

Figure 5 shows the activity strength, log $L_{H\alpha}$ / L_{bol}, (see PMSU2 for discussion of why we use this ratio) vs. M_{bol} for three young clusters, the Hyades and the field. As promised, some of the VLM candidates in the Hyades are shown in Figure 5b. We observed the 12 reddest stars noted as possible candidate members in Leggett & Hawkins (1989), and found that five of the stars (shown as closed triangles) are probably background, pre-main sequence brown dwarfs. Six others (not shown) are foreground stars, and only one (closed circle) remains as a Hyades VLM/BD candidate. These results are described in Reid & Hawley (1998). Unfortunately, the lack of confirmed VLM members of the Hyades means that we cannot test whether the activity strength declines at ~ 0.1 $M_\odot$ as it appears to in the Pleiades (Stauffer *et al.* 1994) and in the field (PMSU2), as shown in Figures 5a,c. Basri & Marcy (1995) pointed out that there are VLM stars in this mass range with fast rotation and no evidence of activity; again suggesting that rotation is no longer instrumental in the production of magnetic fields on these stars. Note that in Figure 5a, there are four NGC 2516 stars (asterisks) with quite low activity strength at higher mass. These may provide a clue to the lower mass, low activity stars.

A second interesting feature of Figure 5 is the *rise* in activity strength that occurs between $6 < M_{bol} < 10$ in the young clusters. This rise is also seen in the X-ray activity strength (log L_X/L_{bol}) in that magnitude range. Among the earlier type, active M dwarfs, the less massive stars generally emit more luminosity in Hα (and X-rays). This effect has not yet been explained adequately, although suggestions such as saturation, rotation (again), field strength and heating efficiency are all floating around the literature (see HTR98).

3. Luminosity functions and mass segregation

We showed in a previous paper (Reid & Hawley 1996) that the old clusters M67 and NGC 2420 show evidence of mass segregation (lower mass stars have a larger core radius than high mass stars), but still have well populated main sequences down to at least $M_V \sim 10 - 12$, with the magnitude limit being imposed by the depth of our surveys. Similarly NGC 2516, shown in Figure 3b, has an obvious main sequence down to $M_V > 12$ and many of the stars are confirmed members due to their Hα emission. We found evidence of significant mass segregation in NGC 2516, with several stars in the offset field (one degree away) showing emission. A few other clusters with low mass members have also been studied (e.g. the Pleiades - Raboud & Mermilliod (1998) and the Hyades - Reid (1992)), and all show some evidence of mass segregation.

We have carried out a photometric survey on the Swope telescope at Las Campanas Observatory, comprising ten southern clusters covering a large range in age. The observational techniques and data analysis are described in HTR98. One of the clusters which has not been previously searched for low mass stars, is the $\sim$ 4 Gyr old cluster Melotte 66. The main field and an offset field to assess field star contamination are shown in Figures 6a and 6b. We identify candidate photometric members, shown as enlarged points, around a fiducial main sequence using the known cluster distance, reddening and metallicity (Kassis *et al.* 1997). Figure 6c shows the luminosity functions for the main and offset fields, with a clear overdensity of stars visible at all magnitudes. This cluster

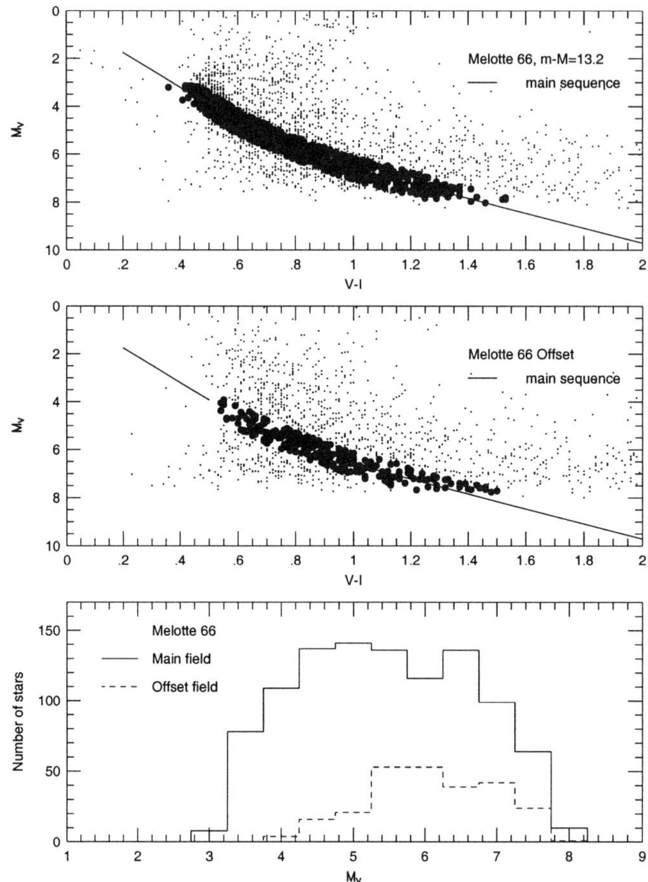

FIGURE 6. Photometric data for Melotte 66 showing the position of the main sequence in the main and offset fields, and the candidate members as enlarged points. The bottom panel is the present day luminosity function, with a well populated main sequence down to the limit of our observations.

obviously has a well-populated main sequence to the limit of our observations. Although it is rather far away, it would be interesting to search for mass segregation even among the G and K stars in the cluster.

Two clusters in our survey are notable for their distinct lack of low mass stars, possibly indicating mass segregation to the point of extinction. NGC 2287 is a fairly young cluster ($\sim$ 200 Myrs) with a large number of bright, early type stars but few or no stars with $M_V > 5$. Figure 7 shows the color magnitude diagrams of a) the main field (near the center, but chosen to avoid the brightest stars) and b) the offset field one degree away. Again we identify candidate cluster members using the known cluster distance, reddening and metallicity (Harris *et al.* 1993). Figure 7c shows the resulting luminosity function of the candidates in the cluster and offset fields. There is little or no overdensity of cluster

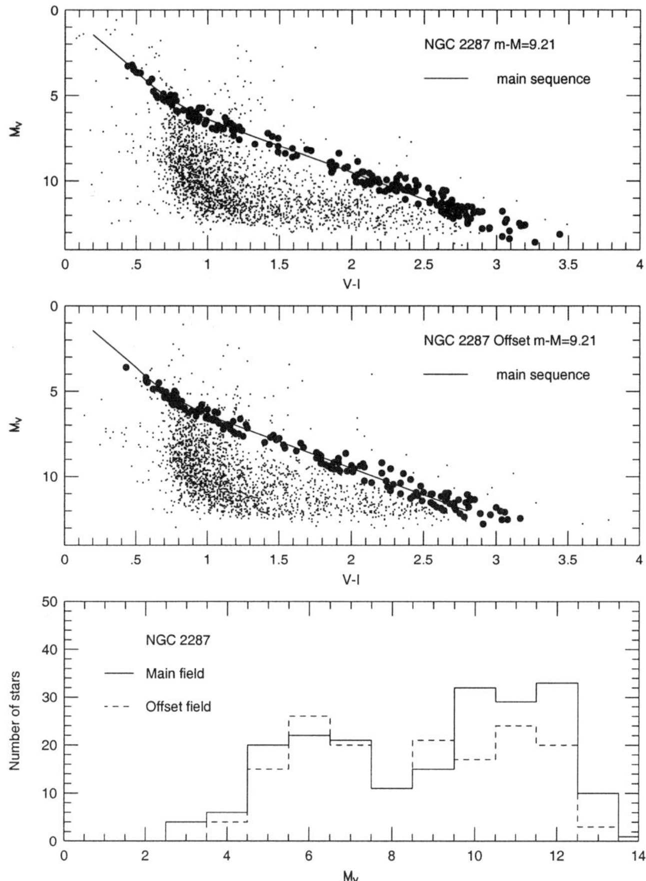

FIGURE 7. Photometric data for NGC 2287 showing the position of the main sequence in the main and offset fields, and the candidate members as enlarged points. The bottom panel is the present day luminosity function, with very few low mass stars visible over the background level.

stars at faint magnitudes. The cluster is quite young so extensive dynamical evolution and evaporation of low mass stars must have taken place. Alternatively, the initial mass function could have been skewed toward high mass stars, rather than following a Salpeter or Scalo form.

We found a similar result in the $\sim$ 2 Gyr old cluster NGC 3680. Those data are described in detail in HTR98. These two clusters are interesting for future detailed study to uncover the reasons for their current dearth of low mass stars. Of course these results also have implications for determining the number of low mass stars and brown dwarfs that have either evaporated from the cluster and are now floating freely in space – or were never born in the first place.

We thank the staffs of Las Campanas Observatory and the Keck Observatory for their help in obtaining the observations described here. We acknowledge useful discussions with John Gizis and Brendan Byrne. SLH and JGT were partially supported by NSF (NYI) grant AST94-57455.

REFERENCES

BASRI, G. & MARCY, G.W. 1995, *Astronomical J.* **109**, 762.

HARRIS, G.L.H., FITZGERALD, M.P.V, MEHTA, S. & REED, B.C. 1993, *Astronomical J.* **106**, 1533.

HAWLEY, S. L., GIZIS, J. E. & REID I. N. 1996, *Astronomical J.* **112**, 2799. (PMSU2)

HAWLEY, S. L., REID, I.N., GIZIS, J.E. & BYRNE, P.B. 1998, Proceedings of the Armagh conference on Solar and Stellar Activity, ed. G. Doyle and J. Butler, ASP conference series, in press.

HAWLEY, S.L., TOURTELLOT, J.G. & REID, I.N. 1998, *Astronomical J.*, submitted. (HTR98)

KASSIS, M., JANES, K.A., FRIEL, E.D. & PHELPS, R.L. 1997, *Astronomical J.* **113**, 1723.

LEGGETT, S.K., & HAWKINS, M.R.S. 1989, *Mon. Not. R.A.S.*, **238**, 145.

RABOUD, D. & MERMILLIOD, J.C. 1998, *Astr. & Astrophys.* **329**, 101.

REID, I.N. 1992, *Mon. Not. R.A.S.* **257**, 257.

REID, I.N. & HAWLEY, S.L. 1998, *Astronomical J.*, in press.

REID, I.N. & HAWLEY, S.L. 1996, in ASP Conference Series No. 109, eds. R. Pallavicini and A. Dupree, Page 383.

REID, I. N., HAWLEY, S. L. & GIZIS, J. E. 1995, *Astronomical J.* **110**, 1838. (PMSU1)

REID, I.N., HAWLEY, S.L. & MATEO, M. 1995, *Mon. Not. R.A.S.* **272**, 828.

SKUMANICH, A. 1972, *Astrophys. J.* **171**, 565.

STAUFFER, J.R., LIEBERT, J., GIAMPAPA, M., MACINTOSH, B., REID, N. & HAMILTON, D. 1994, *Astronomical J.* **108**, 160.

STAUFFER, J.R., GIAMPAPA, M.S., HERBST, W., VINCENT, J.M., HARTMANN, L.W. & STERN, R.A. 1991 *Astrophys. J.* **374**, 142.

WIELEN, R. 1975, Highlights in Astronomy ed. G. Contopolous,(D. Reidel - Dordrecht-Holland), **3**, 395.

Spectroscopy of Very Low Mass Stars and Brown Dwarfs in Young Clusters

By E. L. MARTÍN

[1]Department of Astronomy, University of California, Berkeley, CA 94720, USA

Spectroscopy provides key information about the membership of brown dwarf candidates in clusters, and allows the study of important evolutionary and structural properties like atmospheric composition, chromospheric activity, lithium depletion and rotation. Indeed, spectroscopy is the technique that has allowed the unambiguous confirmation (mainly via the lithum test) of all the known brown dwarfs. In this review, the spectroscopic observations that have been taken to date on very low-mass stars and brown dwarf candidates in clusters are summarized. Particular attention has been paid to the information that we are obtaining on the early evolution of very low-mass objects.

---

1. Introduction

1.1. *Substellar terminology*

What do we understand by "Brown Dwarf"? What is a transition object? How do we distinguish between brown dwarfs and planets? The answers to all these questions rely on conventions. With the discoveries of the first unambiguous substellar objects, language ambiguities that may lead to widespread confusion and misunderstanding should be avoided. Here, some definitions are favored for sake of simplicity, even though there is not yet a general consensus among researches in the field. Throughout this paper, I will use the following terminology which relies on clear-cut mass ranges:

Brown Dwarf (BD): A gaseous object with enough mass to kindle nuclear reactions in its core (H, Li and/or D burning), but these reactions are never sufficiently energetic to halt gravitational contraction. A BD never reaches a main sequence equilibrium state, and cools forever. This definition has been advocated by Basri & Marcy (1997) and Oppenheimer et al. (1999). For solar metallicity, BDs are confined to the mass range **0.075–0.013 $M_\odot$**. For lower metallicities this range shifts slightly to higher masses.

Extrasolar Giant Planet (EGP): A gaseous body that is less massive than a BD, and thus never ignites any nuclear reaction. The maximum mass of a solar-metallicity EGP is hence 0.013 $M_\odot$ or 13.6 $M_{Jupiter}$ Note that this definition is independent of the formation mechanism. In particular, there could be free-floating planets formed in relative isolation out of small molecular cores, or ejected from unstable multiple systems. The minimum mass of an EGP is not known, but it could be about 0.02 $M_{Jupiter}$ if EGPs require a minimum critical mass (e.g., Pollack et al. 1996).

Substellar Mass Object (SMO): A gaseous body that never stabilizes its luminosity by H burning. Both BDs and EGPs are SMOs. Hence, SMO is the most generic word to use if we are not sure about the mass of an object.

Transition Stellar/Substellar Object (TSO): A gaseous body with a mass close to the borderline between stars and BDs. TSOs derive a substantial fraction of their luminosity by nuclear reactions for a long time (D'Antona & Mazzitelli 1985). The mass range of TSOs is **0.09 $M_\odot$–0.065 $M_\odot$**, where lithium is completely depleted at some point of the evolution (Magazzù et al. 1993; Nelson et al. 1993). TSOs straddle the low-mass end of stars and the high mass end of the BDs. The term TSO is useful for objects for which we are not sure of their substellar or stellar nature due to observational and/or theoretical uncertainties.

Very low-mass (VLM) star: A star with a mass approaching the substellar limit. I arbitrarily define this mass range as that in which stars are always fully convective, i.e. **0.3–0.075 $M_\odot$**.

1.2. *Background*

In the 1994 *"The Bottom of the Main Sequence and Beyond"* ESO workshop held in Garching (Tinney 1995), there was general consensus that no SMOs had been unambiguously discovered. In the Summary Remarks, Virginia Trimble wrote: "Clearly, existing data do not require that BDs form anywhere." She also noted that if the temporal spacing of conferences on faint stars would remain constant, the next one would take place sometime between 2003 and 2011.

Trimble's prediction was (happily) not fulfilled. In 1997 a meeting devoted specifically to *"Brown Dwarfs and Extrasolar Planets"* was held in Tenerife (Rebolo et al. 1998). Several discoveries had revolutionized the field in only 2 years. A cool substellar companion had been imaged around the nearby early M-type star Gl 229 (Nakajima et al. 1995; Oppenheimer et al. 1995). This is still the only SMO known to show CH_4 bands in the spectrum. Periodic radial velocity variations had been detected in the solar-type star 51 Peg, indicating the presence of an EGP (Mayor & Queloz 1995; Marcy et al. 1997). A number of other stars with EGP-induced radial velocity curves quickly followed (Butler et al. 1997). The third breakthrough came from the discovery of free-floating BDs in the Pleiades open cluster (see references in Section 3). I recommend the review of Nigel Hambly as a good summary of the status of BD searches in open clusters about 1.5 years ago. I also direct the reader to the papers of Kulkarni (1997) and Rebolo (1998) for more general reviews.

In a short time the substellar arena has changed from a barren desert to a fertile land where new candidate SMOs are being revealed almost every month. Soon after I finish writing this review, it will be outdated. Nevertheless, my aim will be to summarize the present status of the topic of spectroscopy of VLM stars and BDs in young clusters, and to discuss possible future developments. I will discuss individual regions following an order of cluster age. From the extremely young, possible open cluster progenitors, ρOph and the Orion nebula cluster (ONC), through α Per, the Pleiades, the Hyades and finally Praesepe. Then, I will put together the results of different clusters in order to have a perspective of the temporal evolution of VLM stars and BDs.

2. Star-forming regions: ρ Oph and Orion

Rieke & Rieke (1990) performed an imaging survey of the ρ Oph molecular cloud and reported the discovery of 3 BD candidates. They used the Steward Observatory 2.3 m telescope to cover 200 arcmin^2 and to reach limiting magnitudes of m_H=17.5 and m_K=15.0. For two of their candidates they obtained low-resolution (R$\sim$50-100) K-band spectroscopy, which they used for constraining the effective temperatures (T_{eff}). Comerón et al. (1993) presented a more detailed analysis of the same data and derived luminosities and masses using the evolutionary tracks of D'Antona & Mazzitelli (1985). Seven objects were found with masses below the substellar limit (0.075 $M_\odot$). However, Williams et al. (1995) revised the analysis of Comerón et al. using the models of Burrows et al. (1993) and found generally higher masses. Williams et al. also obtained moderate resolution K-band spectra (R$\sim$800) of 7 objects. Despite the low signal-to-noise (S/N) ratios, they were able to recognize bands characteristic of cool star photospheres,

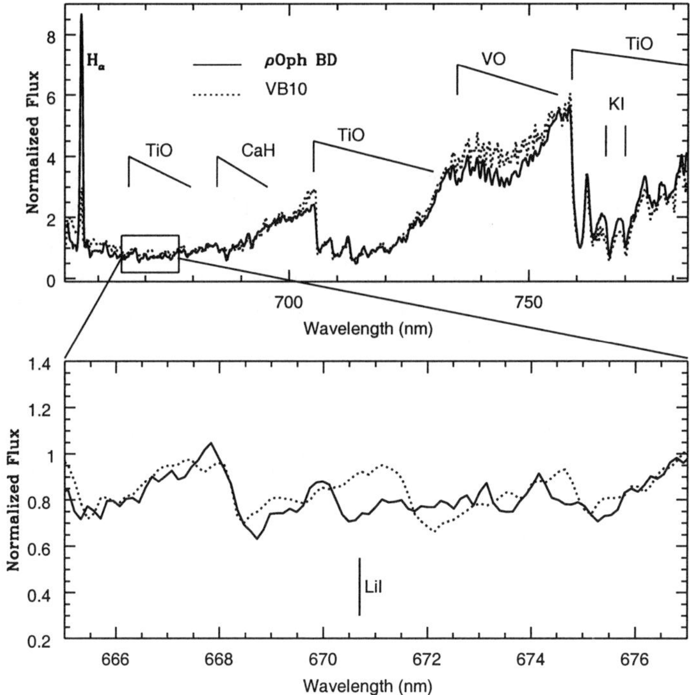

FIGURE 1. Keck LRIS spectrum of the BD in ρ Oph compared with the field M8 dwarf VB10. Identification of the main spectral features is given in the top. A zoom of the lithium feature is shown in the lower panel. A boxcar smoothing of 3 pixels has been applied to the spectra.

in particular the first ^{12}CO overtone bandhead at λ=2.29 μm and longer wavelengths. Williams et al. considered that only one of the BD candidates was safely in the substellar range given the uncertainties of the method. In order to confirm any of these objects as a BD, higher resolution and S/N spectroscopy needs to be obtained so that a detailed comparison with other VLM stars and BDs can be made.

Ironically, the first BD confirmed in ρ Oph is an object discovered by Rieke & Rieke (1990), but considered by them a likely nonmember of the cluster. The full name of this object is ρ Oph 162349.8-242601, hereafter abbreviated to ρ OphBD for simplicity. Luhman, Liebert & Rieke (1997) obtained low-resolution (R$\sim$400) optical spectra of this object with the Multiple Mirror Telescope on Mount Hopkins. They derived a spectral type of M8.5$\pm$0.5 and strong H$_\alpha$ emission (W$_{H_\alpha}$=-60 Å). They also argued that ρ OphBD has spectral characteristics intermediate between dwarfs and giants, supporting a surface gravity of a very young object. Recently, I have obtained, in collaboration with G. Basri, Keck Low-Resolution Imaging Spectrograph (LRIS) medium resolution (R$\sim$3,500) spectra of ρ OphBD. We confirm the very strong H$_\alpha$ emission and the presence of low-gravity signatures such as weak K I resonance doublet (Figures 1 and 4). We also probably detect a Li I feature (lower panel of Figure 1) that supports the SMO nature of ρ OphBD (see Section 5 for more details).

Luhman et al. (1997) placed ρ OphBD in an HR diagram and compared with recent theoretical models. They derived an age of 3-10 Myr and a mass of 0.035$\pm$0.025 M$_\odot$ (could it be an EGP?). The distance to the ρ Oph clouds has recently been revised from 160 pc to 120 pc using the Hipparcos database (Knude & Høg 1999). The closer distance implies that ρ OphBD is less luminous by 0.25 dex than was estimated by Luhman et al. (1997). The luminosity should then be log $L/L_{\odot}$=-2.83 for zero reddening, which places the object on the 10 Myr isochrone in Fig. 2 of Luhman et al. (1997). This is somewhat too old for the age of ρ Oph core (1-3 Myr). A plausible solution to this paradox is that ρ OphBD does have some extinction.

Wilking et al. (1999) have obtained K-band spectroscopy (R$\sim$300) of 20 BD candidates in ρ Oph selected mostly from the work of Comerón et al. (1993). They derived spectral types from steam bands, and in one case they obtained an optical spectrum which yielded an spectral type consistent with the infrared spectrum. For five of their targets, they assigned spectral types between M7 and M8.5, implying that they are likely BDs.

Hillenbrand (1997) compiled the CCD surveys of the ONC made by several people and obtained new CCD images. Using the KPNO 4 m and 3.5 m WIYN telescopes and the RCSP and HYDRA spectrographs, she obtained low-resolution spectroscopy (R$\sim$1,000) of several hundred objects. Five of them had spectral types M7.5 or later, implying that they are good BD candidates. Hillenbrand did not discuss these objects in detail, delaying it for a future paper which has not yet been published. Another interesting region in Orion is the σOri cluster. A CCD survey and follow-up spectroscopy of VLM candidate members was presented in this meeting by Béjar et al. They derive a spectral type of M8.5 for their faintest objects, making them likely BDs. Very recent results of low-resolution optical and infrared spectroscopy of BD candidates in L1495E and IC 348 are reported in Luhman's contribution to these proceedings. Three of the IC 348 sources have optical spectral types in the range M7.5–M8 (Luhman et al. 1998).

3. Young open clusters: α Per and the Pleiades

The "lithium test" for brown dwarfs was first proposed in a paper that dealt with a BD candidate in α Per (Rebolo et al. 1992). Using HIRES on the Keck II telescope, Basri & Martín (1999) have obtained high resolution spectroscopic observations (R$\sim$31,000) of all the BD candidates discovered by Rebolo et al. (1992) and Prosser (1994). Only one of them, AP 270, shows lithium. Basri & Martín use the method of "lithium dating" for determining a cluster age in the range 65–90 Myr, which is older than the classical turnoff age of 50 Myr. AP 270 is an example of a TSO. We will not know if it is stellar or substellar until the cluster age can be determined more precisely. Several groups are presently conducting deeper CCD surveys in α Per that will allow to nail down the cluster age and reveal the BD population of this interesting cluster. A poster showing the first results of a new CCD survey was presented in this meeting by Stauffer et al.

Rebolo et al. (1995) reported the discovery of a proper motion Pleiades member with such a low luminosity and late spectral type that it had to be a BD for any reasonable cluster age. This object, namely Teide 1, was subsequently confirmed to be a bona fide BD via the lithium test (Rebolo et al. 1996). The lithium confirmation of PPl 15, a BD candidate found by Stauffer et al. (1994), was made before that of Teide 1 (Basri et al. 1996). Nevertheless, I consider Teide 1 to be the first discovered free-floating BD in the Pleiades because the evidence presented in Rebolo et al. (1995) was overwhelming, and also because this object was actually present in the first deep CCD imaging survey of this cluster (Jameson & Skillen 1989). These authors did not recognize Teide 1 because it was close to the CCD edge. Nevertheless, their observations were kept in the La Palma data archive and they proved to be crucial for obtaining the proper motion measurement reported by Rebolo et al. (1995).

The Pleiades is an important cluster because we know the photometric and spectral sequence across the substellar mass limit, and we can compare other clusters with it. Table

TABLE 1. Pleiades BD surveys and spectroscopic follow-ups

Survey	Year −1900	Area arcmin^2	Ref. deg^{-1}	Ncand	Nconf	Nrej	Nleft
INT	86	170	JS89	106	0	4	1
Palomar	89	870	S89,S94	12	2	1	0
UKIRT-UH	89	200	SB92	396	0	22	0
Teide	94-96	750	R95,M98a	10	2	0	0
Calar	94	350	ZO97a	72	1	6	0
NOT	95-96	850	F98a	25	3	1	0
CFHT	96	9000	B98	7	3	1	9
ITP	96	3600	ZO97b,ZO99	35	9	0	26
KPNO	96	18000	P99	5	0	38	45
MHO	96	3600	S98a	6	1	0	5
Total	**86-96**	**37390**			**21**	**73**	**86**

1 summarizes the results of spectroscopic follow-up of Pleiades BD candidates found in different surveys. The last four columns are Ncand=Number of BD candidates identified in the survey, normalized to an area of 1 deg^2; Nconf=Number (not normalized to the area) of BD candidates discovered in a given survey and confirmed as substellar cluster members via spectroscopy (BDs detected in different surveys have been counted only once); Nrej=Same as before, but for objects not confirmed as BD cluster members via photometric (all the KPNO rejections are based on infrared photometry) or spectroscopic follow-ups; and Nleft=Similar to the previous two quantities, but for BD candidates for which no follow-up observations have been made yet. The references to the surveys are the following: JS89=Jameson & Skillen (1989); S89=Stauffer et al. (1989); S94=Stauffer et al. (1994); SB92=Simons & Becklin (1992); R95=Rebolo et al. (1995); M98a=Martín et al. (1998a); ZO97a=Zapatero-Osorio et al. (1997a); F98a=Festin (1998a); B98=Bouvier et al. (1998); S98a=Stauffer et al. (1998a); ZO97b=Zapatero-Osorio et al. (1997b); ZO99=Zapatero-Osorio et al. (1999); P99=Pinfield et al., this Euroconference.

Table 1 shows how important spectroscopic follow-up really is for determining the nature of the photometric BD candidates, and that there is still a lot of work to be done. The first surveys were hampered by the presence of many contaminating objects (mainly field red dwarfs and faint galaxies). All the surveys that have claimed more than 50 BD candidates per square degree were suffering from heavy confusion with contaminating objects. With time the survey results have greatly improved because they were deeper, wider and they used the first discovered BDs (Calar 3, PPl 15 and Teide 1) as benchmarks. I consider as confirmed BDs those that have spectroscopic observations that support their membership to the cluster. The level of confidence of these confirmations varies from object to object. The highest confidence is for objects that have radial velocity, spectral type, gravity indicators and photospheric lithium absorption consistent with their Pleiades BD status. The list of these BDs presently includes 10 objects: Calar 3, CFHT-Pl-11, CFHT-Pl-12, CFHT-Pl-15, MHObd3, PPl 1, PPl 15, Roque 13, Teide 1 and Teide 2. The spectroscopic observations were presented in the following papers: Basri et al. (1996); Martín et al. (1996, 1998a); Rebolo et al. (1995, 1996); and Stauffer et al. (1998b). A lower level of confidence has to be placed to objects with some but not all the membership data (Cossburn et al. 1997; Zapatero Osorio et al. 1997b; Festin 1998b; Martín et al. 1998b). Some of the BD candidates are so faint (I>20) that it will be difficult to obtain high resolution spectroscopic data in the near future. It is important to measure their proper motions as a further membership check.

A new spectral class, namely "L-type", for dwarfs cooler than M-type has been proposed in order to account for new objects discovered in the field (Delfosse et al. 1997; Martín et al. 1997; Ruiz et al. 1997; Tinney et al. 1998; Liebert et al. in this volume). The optical characteristics of the class are: very weak or absent TiO bands, strong hydride bands (CaH, CrH, FeH), broad Na I and K I resonance doublets, strong alkali lines (Cs I, Rb I and Li I if undepleted). The first L-type dwarf known in an open cluster is Roque 25 (Martín et al. 1998b). This object indicates that for the Pleiades age, the transition between M-type and L-type takes place at a temperature of about 2200 K and a mass of about 0.04 $M_\odot$. For older/younger ages the transition between these two spectral classes presumably takes place at higher/lower masses. Recent models indicate that VLM stars at the bottom of the main sequence may cool down to about 1800 K (Burrows et al. 1997). Thus, all L-type dwarfs are not BDs. The lithium test is a useful tool for constraining the age and mass of field L-type dwarfs. In Figure 2, an updated magnitude versus spectral type diagram of Pleiades BDs is shown (data has been taken from the references cited above). The L-type subclasses are tentative (Roque 25 has been assigned a subclass of L1). The masses have been estimated using models for VLM stars and BDs for an age of 120 Myr kindly provided by I. Baraffe (see Baraffe & Chabrier contribution to this proceedings book). These evolutionary tracks and isochrones use the NextGen atmosphere models of Allard et al. (1997).

4. Intermediate-age open clusters: Hyades and Praesepe

The Hyades is the closest young open cluster to the Sun. However, its age of ~5 times older than the Pleiades makes Hyades BDs fainter that their Pleiades cousins. Moreover, the very large area covered by the Hyades hampers the task of searching for BDs. With the advent of large mosaic CCDs, it should be possible to discover Hyades BDs in a reasonable amount of telescope time. The all sky near-infrared survey 2MASS will barely reach the substellar limit in this cluster. New results of spectroscopic follow up of the Hyades BD candidates found by Leggett & Hawkins (1989) have been presented by Reid & Hawley (1999). Five of these objects are surprisingly identified as very young BDs based on lithium detections (2 objects), gravity indicators and trigonometric parallaxes. They do not belong to the Hyades or to any known molecular clouds. Only one object remains as a possible Hyades BD candidate.

Two different groups have carried out recent deep CCD surveys of the Praesepe open cluster reaching I~21. This cluster is interesting because it has a similar age than the Hyades but it is more distant and compact. Pinfield et al. (1997) identified 19 BD candidates and estimated a mass function rising into the BD range. However, no follow-up spectroscopy has been reported yet. Magazzù et al. (1998) reported on the discovery of one BD candidate for which they presented a low-resolution optical spectrum. They estimated a spectral type of M8.5, which is consistent with cluster membership. Praesepe is indeed a promising site for locating intermediate-age VLM stars and BDs.

5. Evolution of VLM stars and BDs

Putting together the results coming from clusters of different ages allows to get a grip on the time evolution of stars and BDs. We are starting to have some information in the range of ages 1 Myr to about 1 Gyr. It is interesting to look at the trends that begin to appear in the data, and compare them with our naive expectations.

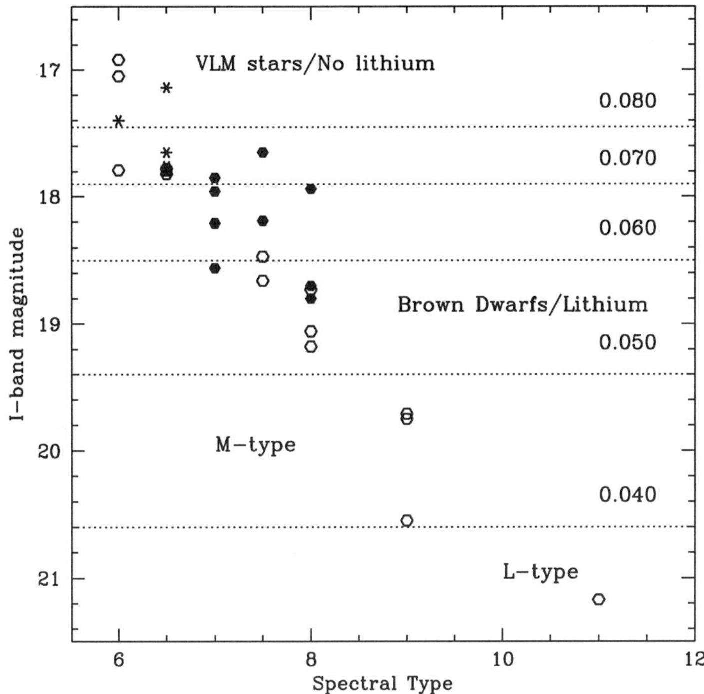

FIGURE 2. The Pleiades spectral sequence from the stellar to the substellar regime, and from M-type to L-type. Masses in solar units are indicated. Open symbols denote objects with low-resolution spectroscopy (not high enough to detect lithium), filled symbols denote objects with lithium detections, star-like symbols denote objects with lithium non-detections.

5.1. *Activity*

Since VLM stars and BDs are fully convective, they might be expected to generate strong magnetic fields through a mechanism of turbulent dynamo similar to the quiet Sun (e.g. Schmitt et al. 1996). They may also heat chromospheres and coronae, and have winds and photospheric spots. Optical spectra include several lines that have been observed to appear in emission in the spectra of field stars as cool as M8. The strongest lines are H_α and emission reversals in the core of the broad Na I and K I resonance doublets (Martín 1999). Generally, the most available chromospheric indicator is H_α because it can be detected even in low-resolution spectra.

The extremely young BDs in ρ Oph and σOri have very strong H_α emission equivalent widths ($W_{H_\alpha} \geq 30$Å). They are so young that it is quite possible that such emission partly arises in a boundary layer, resulting from the interaction of the BD and a hot accretion disk. VLM stars and BDs in α Per have all moderate W_{H_α} down to spectral type of about M6.5 (Basri & Martín 1999). Zapatero Osorio et al. (1996) have shown that in this cluster the W_{H_α} has a maximum at around spectral types M3–M4, and decreases for cooler stars. In the older Pleiades cluster, Hodgkin et al. (1995) found a similar turnover of W_{H_α} for stars cooler than M4. However, it is not clear whether W_{H_α} continues decreasing into the BD regime. For Pleiads with spectral types M6–L1, the W_{H_α} values range from 20Å to less than 5Å(Martín et al. 1996, 1998a,b; Zapatero Osorio et al. 1997b; Festin 1998b). The L_{H_α}/L_{bol} ratio does seem to decrease in the

Pleiades BDs with respect to the VLM stars. It would be useful to monitor the temporal variability of H_α emission in Pleiades BDs in order to investigate if strong H_α emission is mainly associated with flare events.

5.2. *Rotation*

Rotational broadening can be measured in spectra of sufficiently high spectral resolution. A resolution of R$\sim$30,000 is enough to measure $v\sin i$ larger than about 5 km s^{-1}. The HIRES spectrograph on the Keck I telescope is the first instrument that allows to obtain high resolution spectra of the faint objects around the substellar limit in young open clusters. Rotational velocities have been derived for two VLM cluster members in α Per and five in the Pleiades (Basri & Marcy 1995; Oppenheimer et al. 1997; Martín et al. 1998a; Basri & Martín 1999). The $v\sin i$ measurements are the following: AP 270 (24 km s^{-1}), AP J0323+4853 (other name AP 275, 33 km s^{-1}), HHJ 3 (37 km s^{-1}), HHJ 6 (50 km s^{-1}), HHJ 10 (52 km s^{-1}), PPl 1 (18.5 km s^{-1}) and Teide 2 (13.0 km s^{-1}). They all have similar spectral types (M6–M6.5) and masses (0.08–0.07 $M_\odot$). One of them (AP J0323+4853) has a photometric rotation period of 7.6 hours (Martín & Zapatero Osorio 1997), which is consistent with the $v\sin i$ value. Fast rotation appears to be a general property of the TSOs with ages 70–120 Myr. There is probably some dispersion in the rotational velocities of TSOs of similar age and mass, but the spread does not seem to be as large as in the K-type stars. Basri & Marcy (1995) and Martín & Zapatero Osorio (1997) discussed the evolution of rotation velocities from the T Tauri stars to the Hyades. They found that there is little angular momentum loss. This is consistent with current ideas that the timescale for rotational braking is very long in VLM stars.

The relationship between chromospheric activity and rotation in the VLM members of young open clusters is shown in Figure 3. There is not any evidence for a higher level of activity in faster rotators. The rotation-activity connection present in higher mass stars has broken down (or saturated) for these VLM objects. This suggests that the chromospheric heating is largely independent of the rotation rate in these objects.

5.3. *Gravity sensitive spectral indicators*

As BDs contract gravitationally with time, their surface gravity increases. For example, recent evolutionary models calculated by the Lyon group (e.g. Baraffe et al. 1998) give the following evolution of log g for 0.04 $M_\odot$: 4.52 (30 Myr), 4.70 (60 Myr), 4.88 (150 Myr). The same models give a log g and $T_{\rm eff}$ of 5.01 and 2900 K for a VLM star with 0.09 $M_\odot$ and age of 150 Myr. A BD model of 0.06 $M_\odot$ and age 30 Myr has similar temperature (2850 K) but lower gravity (log g=4.55). Thus, gravity can in principle be used as an age indicator for BDs, and to discriminate between young BDs and old VLM stars. Davidge & Boeshaar (1991) recognized that many features in the spectra of cool dwarfs are sensitive to gravity and that they could be used for identifying BDs. However, the real problem is to accurately calibrate gravity sensitive spectral indicators so that they can actually be used quantitatively. This is a complicated task because it requires that we can produce good fits of observed spectra with synthetical spectra.

Qualitatively, we can already see gravity-related effects in the spectra of VLM stars and BDs. Steele & Jameson (1995) and Martín et al. (1996) claimed that the Na I doublet at 818.3,819.5 nm is systematically weaker in the Pleiades VLM stars and BDs than in field stars of the same spectral type. The latter authors also noticed an effect in the VO bands. Luhman et al. (1997) and Reid & Hawley (1999) have found that the spectra of very young M7–M9 BDs have characteristics intermediate between dwarfs and a giants. The gravity effects are most conspicuous in the K I and Na I lines, and in CaH molecular bands. I have compared in Figure 4 the spectra of 3 objects with similar

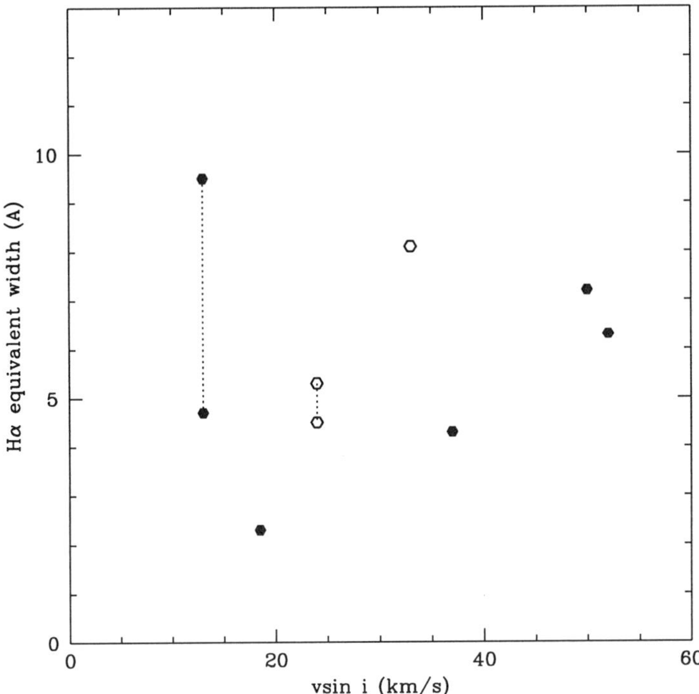

FIGURE 3. A plot of H_α equivalent widths versus $v\sin i$ for M6–M6.5 members of α Per (empty symbols) and the Pleiades (filled symbols). The dashed lines join different values of H_α for the same object, indicating variability. I have included only measurements coming from high-resolution spectra.

spectral types (M8–M8.5) but different gravities. All of them were observed with Keck II LRIS using the 1200 l/mm grating. The resolution is about R=3,500. The strongest K I resonance doublet corresponds to the oldest object (VB10), and the weakest corresponds to the youngest object (ρ OphBD). Teide 1 is intermediate between the two, but closer to VB10. The Rb I line at 780.0 nm follows the behaviour of the K I lines. The sequence of line strengths shown in Figure 4 corresponds to a sequence of ages and gravities. For ρ OphBD, assuming an age of 5 Myr and mass 0.040 $M_\odot$, one gets a gravity of log g=3.8; for Teide 1, the age and mass can be 120 Myr and 0.055 $M_\odot$, leading to log g=4.9; and for VB10, the age and mass are unknown but it is probably not very old (Martín 1999 finds that it is probably younger than VB8). If VB 10 has a mass of 0.08 $M_\odot$ and an age of 1 Gyr, the gravity is log g=5.1. The strength of the K I and Rb I lines in Fig. 4 appears to be nicely correlated with the estimated gravities. Gravity sensitive indicators can be very useful in the future, particularly for very faint objects where the lithium test will be quite difficult to perform.

5.4. *Lithium dating of young open clusters*

The lithium test for BDs was first proposed by Rebolo et al. (1992), and developed by Magazzù et al. (1993) in all its significant aspects (calculation of lithium depletion as a function of age and mass; formation of the resonance Li I doublet in cool atmospheres and first lithium search in BD candidates). After some failures (Marcy et al. 1994; Martín et al. 1994), the lithium test finally met with success for the first time in the

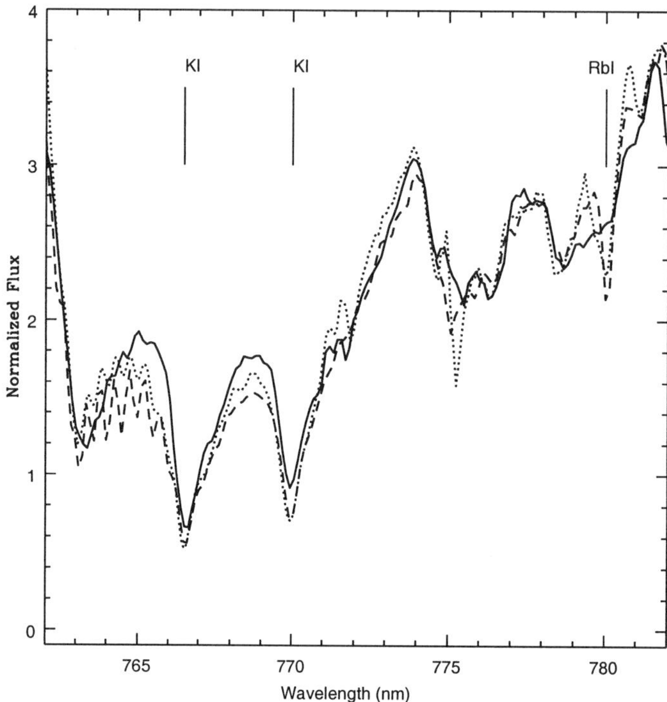

FIGURE 4. Comparison of LRIS spectra of VB10 (dashed line), Teide 1 (dotted line) and ρ OphBD (solid line). Note the different strengths of the K I and Rb I lines.

Pleiades BDs Calar 3, PPl 15 and Teide 1 (Basri et al. 1996; Rebolo et al. 1996). Since then, it has proved to be the best means for confirming BDs in clusters and the field. Of all the known BDs, all but one (Gl 229B) have been identified using the lithium test. With the lithium observations of TSOs and BDs in the Pleiades, it has become clear that any dwarf with a spectral type M7 or later and with detectable lithium must be a BD (Martín et al. 1996, 1998a). However, the main caveat of the lithium test is that TSOs do deplete this light element if they are sufficiently old. Hence, the absence of lithium in an M7 or later dwarf does not rule out a substellar mass, although it constrains it to be larger than 0.06 $M_{\odot}$.

In young open clusters, a lithium chasm develops with age. The term "lithium chasm" refers to the gap between stars with shallow convective regions that preserve lithium only in the outermost layers and the fully convective VLM objects that preserve it because they have not reached sufficiently high central temperatures. The formation and evolution of this chasm can be clearly seen in Figs. 15 and 16 of D'Antona & Mazzitelli (1994). At an age of 5 Myr, the chasm is very narrow because only the early M-type stars have had time to deplete lithium. The chasm widens up quickly and reaches the early K-type stars and late-M stars for an age of 30 Myr. The cool side of the chasm is very steep. The luminosity and temperature at which lithium reappears again in an open cluster is a very sensitive function of the age. Thus, the age of a cluster can in principle be obtained from observations of lithium and luminosity in the VLM members. This method was first applied by Basri et al. (1996) in the Pleiades, and they obtained an age of 115 Myr, which is considerably older than the canonical turnoff age of 70 Myr. The object used by Basri et al. (PPl 15) turned out to be the first spectroscopic double-lined BD-BD binary (Basri

TABLE 2. Some young clusters where lithium dating is currently feasible

Cluster	Distance (pc)	Turnoff age (Myr)	m_I (SB)	Lithium age (Myr)	m_I (SB)
ρ Oph	120	1	13.8		
IC 348	300	3	16.2		
σOri	350	10	17.5		
IC 2391	150	30	16.6		
IC 2602	150	30	16.6		
α Per	175	50	17.5	65	17.7
Pleiades	125	70	17.1	120	17.6

& Martín 1998). Nevertheless, new observations of lithium in VLM Pleiades members (Martín et al. 1998a; Stauffer et al. 1998) have basically confirmed the results of Basri et al. (1996). Lithium observations of VLM members of α Per give an age for this cluster $\geq$65 Myr (Basri & Martín 1999), that is also older than the canonical turnoff age of 50 Myr. A plausible explanation for this age discrepancy between the massive stars and the VLM stars is that additional mixing has to be included in the high-mass evolutionary models. Several mechanisms are possible (turbulent diffusion, overshooting, rotationally induced mixing) that would increase the lifetime of massive stars by supplying extra-fuel to their cores (e.g. Meynet & Maeder 1997).

The physics of the fully convective VLM stars are simpler than their high-mass counterparts where the interplay between the convective core and the radiative envelope is difficult to theorize. Therefore, the cluster ages obtained from lithium dating should have a higher degree of confidence than the turnoff ages. The obvious drawback is that the VLM stars are too faint to be observed at large distances. Thus, the lithium dating method can only be applied to nearby open clusters. Our hope is that using the lithium dating constraints, the high-mass turnoff method can be recalibrated in the nearest clusters and subsequently applied to distant clusters. I have estimated the location of the substellar boundary in some young clusters using the Lyon group models and zero reddening. The results are given in Table 2. The apparent I-band magnitudes of the substellar boundary for the turnoff age and lithium age (if available) are given in the 4th and 6th columns respectively. Taking into account that the cool side of the lithium chasm is located at masses larger than the substellar limit for ages younger than $\sim$120 Myr, it is clear that lithium dating is feasible with current instrumentation in several young open clusters. In very young clusters (age$<$5 Myr) all members are expected to preserve lithium. Nevertheless, lithium observations are still useful in order to confirm that the objects are indeed very young. An age spread of several Myr might be present even in the youngest clusters. Lithium observations will constrain the possible age dispersion.

I thank the Scientific Organizing Committee for their invitation to deliver a talk in this meeting, and the Local Organizing Committee for their hard work. I am grateful to Gibor Basri, Simon Hodgkin, Kevin Luhman and Rafael Rebolo for valuable discussions. I have been supported by the Spanish MEC under their FPI/Fullbright program. This research has made use of the Simbad database, operated at CDS, Strasbourg, France.

REFERENCES

ALLARD, F., HAUSCHILDT, P. H., ALEXANDER, D. R. & STARRFIELD, S. 1997 Model atmospheres of very low-mass stars and brown dwarfs *ARAA* **35**, 137–177.

BARAFFE, I., CHABRIER, G., ALLARD, F. & HAUSCHILDT, P. H. 1998 Evolutionary models for solar metallicity low-mass stars: mass-magnitude relationships and color-magnitude diagrams *A&A* **337**, 403–412.

BASRI, G. & MARCY, G. W. 1995 A surprise at the bottom of the main sequence: Rapid rotation and no H_α emission *AJ* **109**, 762–773.

BASRI, G., MARCY, G. W. & GRAHAM, J. R. 1996 Lithium in brown dwarf candidates: the mass and age of the faintest Pleiades stars *ApJ* **458**, 600–609.

BASRI, G. & MARCY, G. W. 1997 Early hints on the substellar mass function. In *Star Formation Near and Far* (ed. S. S. Holt and L. G. Mundy), AIP Press, New York, pp. 228–240.

BASRI, G. & MARTÍN, E. L. 1998 PPl15: the first binary brown dwarf system?. In *Brown dwarfs and extrasolar planets* (ed. R. Rebolo, E. L. Martín and M. R. Zapatero Osorio), San Francisco: Astronomical Society of the Pacific, Conference Series, Vol. 134, pp. 284–287. .

BASRI, G. & MARTÍN, E. L. 1999 The mass and age of very low-mass members of the open cluster Alpha Persei *ApJ* in press.

BOUVIER, J., STAUFFER, J. R., MARTÍN, E. L., BARRADO Y NAVASCUES, D., WALLACE, B. & BÉJAR, V.J.S. 1998 Brown dwarfs and very low-mass stars in the Pleiades cluster: a deep wide-field imaging survey *A&A* **336**, 490–502.

BURROWS, A., HUBBARD, W. B., SAUMON, D. & LUNINE, J. I. 1993 An expanded set of brown dwarf and very low mass star models *ApJ* **406**, 158–171.

BURROWS, A., MARLEY, M., HUBBARD, W. B., LUNINE, J. I., GUILLOT, T., SAUMON, D., FREEDMAN, R., SUDARSKY, D. & SHARP, C. 1997 A nongray theory of extrasolar giant planets and brown dwarfs *ApJ* **491**, 856–875.

BUTLER, P. R., MARCY, G. W., WILLIAMS, E., HAUSER, H. & SHIRTS, P. 1997 Three new "51 Pegasi-type" planets *ApJ* **474**, L115–L118.

COMERÓN, F., RIEKE, G. H., BURROWS, A. & RIEKE, M. J. 1993 The stellar population in the rho Ophiuchi cluster *ApJ* **416**, 185–203.

COSSBURN, M. R., HODGKIN, S. T., JAMESON, R. F. & PINFIELD, D. J. 1997 Discovery of the lowest mass brown dwarf in the Pleiades *MNRAS* **288**, L23–L27.

D'ANTONA, F. & MAZZITELLI, I. 1985 Evolution of very low-mass stars and brown dwarfs. I. The minimum main-sequence mass and luminosity *ApJ* **296**, 502–513.

D'ANTONA, F. & MAZZITELLI, I. 1994 New pre-main sequence tracks for $M \leq 2.5 M_\odot$ as tests of opacities and convection model *ApJS* **90**, 467–500.

DAVIDGE, T. J. & BOESHAAR, P. C. 1991 The near-infrared spectrum of the suspected brown dwarf binary system Gliese 473 *AJ* **102**, 267–271.

DELFOSSE, X. ET AL. 1997 Field brown dwarfs found by DENIS *A&A* **327**, L25–L28.

FESTIN, L. 1998a Brown dwarfs in the Pleiades. II. A deep optical and near infrared survey *A&A* **333**, 497–504.

FESTIN, L. 1998b Spectroscopy of New Pleiades Brown dwarfs *MNRAS* **298**, L34–L36.

HAMBLY, N. 1998 Deep searches in open clusters. In *Brown dwarfs and extrasolar planets* (ed. R. Rebolo, E. L. Martín and M. R. Zapatero Osorio), San Francisco: Astronomical Society of the Pacific, Conference Series, Vol. 134, pp. 11–19.

HILLENBRAND, L. A. 1997 On the stellar population and star-forming history of the Orion Nebula Cluster *AJ* **113**, 1733–1768.

HODGKIN, S. T., JAMESON, R. F. & STEELE, I. A. 1995 Chromospheric and coronal activity of low-mass stars in the Pleiades *MNRAS* **274**, 869–883.

JAMESON, R. & SKILLEN, I. 1989 A search for low-mass stars and brown dwarfs in the Pleiades *MNRAS* **239**, 247–253.

KNUDE, J. & HØG, E. 1999 Interstellar reddening from the Hipparcos and Tycho Catalogues I. Distances to nearby molecular clouds and star forming regions *A&A* in press.

KULKARNI, S. R. 1997 Brown dwarfs: a possible missing link between stars and planets *Science* **276**, 1350–1354.

LEGGETT, S. K. & HAWKINS, M. R. S. 1989 Low mass stars in the region of the Hyades cluster *MNRAS* **238**, 145–153.

LUHMAN, K. L., LIEBERT, J. & RIEKE, G. H. 1997 Spectroscopy of a young brown dwarf in the ρ Ophiuchi cluster *ApJ* **489**, L165–L168.

LUHMAN, K. L., RIEKE, G. H., LADA, C. J. & LADA, E. A. 1998 Low mass star formation and the initial mass function in IC 348 *ApJ* **508**, in press.

MAGAZZÙ, A., MARTÍN, E. L. & REBOLO, R. 1993 A spectroscopic test for substellar objects *ApJ* **404**, L17–L20.

MAGAZZÙ, A., REBOLO, R., ZAPATERO OSORIO, M. R., MARTÍN, E. L. & HODGKIN, S. T. 1998 A brown dwarf candidate in the Praesepe open cluster *ApJ* **497**, L47–L50.

MARCY, G. W., BASRI, G. & GRAHAM, J. R. 1994 A search for lithium in brown dwarf candidates using the Keck hires echelle *ApJ* **428**, L57–L60.

MARCY, G. W., BUTLER, P. R., WILLIAMS, E., BILDSTEN, L., GRAHAM, J. R., GHEZ, A. M. & GARRET JERNIGAN, J. 1997 The planet around 51 Pegasi *ApJ* **481**, 926–935.

MARTÍN, E. L., REBOLO, R. & MAGAZZÙ, A. 1994 Constraints to the masses of brown dwarf candidates from the lithium test *ApJ* **436**, 262–269.

MARTÍN, E. L., REBOLO, R. & ZAPATERO OSORIO, M. R. 1996 Spectroscopy of new substellar candidates in the Pleiades: toward a spectral sequence of young brown dwarfs *ApJ* **469**, 706–714.

MARTÍN, E. L., BASRI, G., DELFOSSE, X. & FORVEILLE, T. 1997 Keck HIRES spectra of the brown dwarf DENIS-P J1228.2-1547 *A&A* **327**, L29–L32.

MARTÍN, E. L. & ZAPATERO OSORIO, M. R. 1997 The rotation period of a very low-mass star in alpha Persei *MNRAS* **286**, L17–L20.

MARTÍN, E. L., BASRI, G., GALLEGOS, J., REBOLO, R., ZAPATERO OSORIO, M. R. & BEJAR, V.J.S. 1998a A new Pleiades member at the lithium substellar boundary *ApJ* **499**, L61–L64.

MARTÍN, E. L., BASRI, G., ZAPATERO OSORIO, M. R., REBOLO, R. & GARCÍA LÓPEZ, R. 1998b The first L-type brown dwarf in the Pleiades *ApJ* **507**, L41–L44.

MARTÍN, E. L. 1999 Utrecht echelle spectroscopy of a flare in VB8 *MNRAS* in press.

MAYOR, M. & QUELOZ, D. 1995 A Jupiter-mass companion to a solar-type star *Nature* **378**, 355–357.

MEYNET, G. & MAEDER, A. 1997 Stellar evolution with rotation. I. The computational method and the inhibiting effect of the mu-gradient *A&A* **321**, 465–476.

NAKAJIMA, T., OPPENHEIMER, B. R., KULKARNI, S. R., GOLIMOWSKI, D. A., MATTHEWS, K. & DURRANCE, S. T. 1995 Discovery of a cool brown dwarf *Nature* **378**, 463–465.

NELSON, L.A., RAPPAPORT, S. & CHIANG, E. 1993 On the Li and Be tests for brown dwarfs *ApJ* **413**, 364–367.

OPPENHEIMER, B., KULKARNI, S. R., MATTHEWS, K. & NAKAJIMA, T. 1995 Infrared spectrum of the cool brown dwarf Gl 229B *Science* **270**, 1478–1479.

OPPENHEIMER, B. R., BASRI, G., NAKAJIMA, T. & KULKARNI, S. R. 1997 Lithium in very low-mass stars in the Pleiades *AJ* **113**, 296–305.

OPPENHEIMER, B. R., KULKARNI, S. R. & STAUFFER, J. R. 1999 Brown Dwarfs. In *Protostars and Planets IV*, in press.

PINFIELD, D. J., HODGKIN, S. T., JAMESON, R. F., COSSBURN, M. R. & VON HIPPEL, T. 1997 Brown dwarf candidates in Praesepe *MNRAS* **287**, 180–188.

POLLACK, J. B., HUBICKYJ, O., BODENHEIMER, P., LISSAUER, J. J., PODOLAK, M. & GREENZWEIG, Y. 1996 Formation of the giant planets by concurrent accretion of solids and gas *Icarus* **124**, 62–85.

PROSSER, C. F. 1994 Photometry and spectroscopy in the open cluster Alpha Persei *AJ* **107**, 1422–1431.

REID, I. N. & HAWLEY, S. L. 1999 Brown dwarfs in the Hyades and beyond? *AJ* in press.

REBOLO, R., MARTÍN, E. L. & MAGAZZÙ, A. 1992 Spectroscopy of a brown dwarf candidate in the Alpha Persei open cluster *ApJ* **389**, L83–L86.

REBOLO, R., ZAPATERO OSORIO, M. R. & MARTÍN, E. L. 1995 Discovery of a brown dwarf in the Pleiades star cluster *Nature* **377**, 129–131.

REBOLO, R., MARTÍN, E. L., BASRI, G., MARCY, G. W. & ZAPATERO OSORIO, M. R. 1996 Brown dwafs in the Pleiades cluster confirmed by the lithium test *ApJ* **469**, L53–L56.

REBOLO, R., MARTÍN, E. L. & ZAPATERO OSORIO, M. R. 1998 eds. Brown Dwarfs and Extrasolar Planets *San Francisco: Astronomical Society of the Pacific, Conference Series* Vol. 134.

REBOLO, R. 1998 Spectroscopy of brown dwarfs. In *The Tenth Cambridge Workshop on Cool Stars, Stellar Systems and the Sun* (ed. R. A. Donahue and J. A. Bookbinder), San Francisco: Astronomical Society of the Pacific, Conference Series, Vol. 154, pp. 47–57.

RIEKE, G. H. & RIEKE, M. J. 1990 Possible substellar objects in the rho Ophiuchi cloud *ApJ* **362**, L21–L24.

RUIZ, M. T., LEGGETT, S. K. & ALLARD, F. 1997 Kelu 1: A free-floating brown dwarf in the solar neighborhood *ApJ* **491**, L107–L110.

SCHMITT, D., SCHUSSLER, M. & FERRIZ MAS, A. 1996 Intermittent solar activity by an on-off dynamo *A&A* **311**, L1–L4.

SIMONS, D. A. & BECKLIN, E. E. 1992 A near-infrared search for brown dwarfs in the Pleiades *ApJ* **390**, 431–439.

STAUFFER, J. R., HAMILTON, D., PROBST, R. G., RIEKE, G. & MATEO, M. 1989 Possible Pleiades members of about 0.07 solar mass - Identification of brown dwarf candidates of known age, distance and metallicity *ApJ* **344**, L21–L24.

STAUFFER, J. R., HAMILTON, D. & PROBST, R. G. 1994 A CCD-based search for very low mass members of the Pleiades cluster *AJ* **108**, 155–159.

STAUFFER, J. R., SCHILD, R., BARRADO Y NAVASCUES, D., BACKMAN, D. E., ANGELOVA, A. M., KIRKPATRICK, J. D., HAMBLY, N. & VANZI, L. 1998a Results of a deep imaging survey of one square degree of the Pleadies for low-luminosity cluster members *AJ* **504**, 805–820.

STAUFFER, J. R., SCHULTZ, G. & KIRKPATRICK, J. D. 1998b Keck spectra of Pleiades brown dwarf candidates and a precise determination of the lithium depletion edge in the Pleiades *ApJ* **499**, L199–L203.

STEELE, I. A. & JAMESON, R. F. 1995 Optical spectroscopy of low-mass stars and brown dwarfs in the Pleiades *MNRAS* **272**, 630–646.

TINNEY, C. G. 1995 ed. The bottom of the main sequence - and beyond *ESO Astrophysics Symposia* Springer.

TINNEY, C. G., DELFOSSE, X., FORVEILLE, T. & ALLARD, F. 1998 Optical spectroscopy of DENIS mini-survey brown dwarf candidates *A&A* **338**, 1066–1072.

WILLIAMS, D. M., COMERÓN, F., RIEKE, G. H. & RIEKE, M. J. 1995 The low-mass IMF in the ρ Ophiuchi cluster *ApJ* **454**, 144–150.

WILKING, B. A., GREENE, T. P. & MEYER, M. 1999 Spectroscopy of brown dwarf candidates in the ρ Ophiuchi molecular core, preprint.

ZAPATERO OSORIO, M. R., REBOLO, R., MARTÍN, E. L. & GARCÍA LÓPEZ, R. 1996 Stars approaching the substellar limit in the α Persei open cluster *A&A* **305**, 519–526.

ZAPATERO OSORIO, M. R., REBOLO, R. & MARTÍN, E. L. 1997a Brown dwarfs in the Pleiades cluster: a CCD-based R,I survey *A&A* **317**, 164–170.

ZAPATERO OSORIO, M. R., REBOLO, R., MARTÍN, E. L., BASRI, G., MAGAZZÙ, A., HODGKIN, S., JAMESON, R. & COSSBURN, M. R. 1997b New brown dwarfs in the Pleiades cluster *ApJ* **491**, L81–L84.

ZAPATERO OSORIO, M. R., REBOLO, R., MARTÍN, E. L., HODGKIN, S., COSSBURN, M. R., MAGAZZÙ, A., STEELE, I. & JAMESON, R. 1999 Brown dwarfs in the Pleiades cluster. III. A deep I,Z survey *A&A Supp* in press.

High Resolution Spectra of L Type Stars and Brown Dwarfs

By GIBOR BASRI[1], FRANCE ALLARD[2], PETER HAUSCHILDT,[3] AND SUBHANJOY MOHANTY[1]

[1]Astronomy Dept., Univ. of California, Berkeley, CA 94720, USA

[2]Centre de Recherche Astronomique de Lyon (UMR 142 CNRS), Ecole Normale Superieure, 69364 Lyon Cedex 07, France

[3]Dept. of Physics and Astronomy & Center for Simulational Physics, University of Georgia, Athens, GA 30602-2451, USA

The first brown dwarfs were confirmed only three years ago. Already, however, a library of echelle spectra of objects of a variety of temperatures has been accumulated. This process has been greatly aided by the discovery of relatively nearby free-floating brown dwarfs and companions to M dwarfs. Their spectra show the rapidly increasing importance of dust formation in the atmosphere, and its concomitant decrease of the TiO molecular features which define the M spectral class. This has lead to the proposal of a new spectral class, L, for cooler objects. The primary atomic features visible in red spectra of L and late M stars are resonance lines of alkali metals (Na, K, Rb, Cs, and sometimes Li). Here we present a sample of line profiles from mid-M to mid-L objects, which include both very low mass stars and confirmed brown dwarfs. We compare the line profiles in the alkali lines to very recent models which include effects of dust formation. We show that the models can already make a reasonable representation of the observations, and begin to set a temperature scale for these new very cool objects. There are certainly issues remaining to be addressed, however.

---

1. Introduction

Since the announcement of the first brown dwarfs in 1995, the field has been moving very quickly. The number of known brown dwarfs has increased rapidly, along with a growing collection of stars at the bottom of the main sequence. Though these are all very intrinsically faint, there are now a substantial number of them which are accessible to a 10-m telescope at echelle resolutions. The advantage of high spectral resolution is that one can study the kinematics of the object: rotation and radial velocity, and obtain actual line profiles to be compared with detailed stellar model atmospheric calculations. In principle, a single line profile contains information about the temperature and temperature structure of the atmosphere, turbulent and bulk velocity fields, and information on the composition and gravity of the object.

The first brown dwarfs to be confirmed as substellar were either so young that they are late M stars: eg. the Pleiades objects PPl 15 (Basri, Marcy & Graham 1996) and Teide 1 (Rebolo *et al.* 1996), or so old that they are too faint for echelle spectroscopy: Gl 229B (Nakajima *et al.* 1995). It was not until the discovery of nearby field brown dwarfs: DENIS-P J1228.2–1547 (Delfosse *et al.* 1997) and Kelu 1 (Ruiz, Leggett & Allard 1997) that objects cooler than M were found which are bright enough to study at high spectral resolution. The first line profiles from such objects were presented by Martín *et al.* (1997). These authors acted on a suggestion by Kirkpatrick (1998) and proposed the letter "L" for the new class of objects cooler than the M spectral class. Only one such object was known before 1997: the brown dwarf candidate GD 165B (Zuckerman & Becklin 1992), and only a poor low resolution spectrum existed for it. Thus, it is exciting to now be able to begin to apply all the time-tested techniques of model atmospheric analysis to the new spectral class and begin to learn about the physical state of their atmospheres.

In this paper, we present observations of the Rb I and Cs I resonance lines in a sample of L-type objects. After briefly describing the observations, we first show the evidence for the disappearance of the M-type spectral diagnostics. We then discuss the new set of models for objects in the 1500–2500 K effective temperature regime (see also the chapter by Allard) in terms of the high resolution appearance of these models. We compare them to the observations, showing that the models are a good start in describing L-type atmospheres. We discuss the physical parameters which emerge for our actual objects; in particular their temperatures and rotations. Finally we discuss some of the discrepancies that remain, and point the way to future work.

2. Observations and spectral analysis

The observations were all made at the Keck I telescope by GB using the HIRES echelle. They encompass 15 red spectral orders. The single 2048^2 chip does not cover the full echelle format. The chip is binned 2×2 in order to increase the S/N on these faint sources; the resulting resolution is about 33000. In some cases we have observed the object more than once, and the spectrum shown is the summed one. In Table 1 we list all the objects presented in this paper. Reductions are done in a standard way (Basri & Marcy (1995)).

To demonstrate the need for the L spectral class, we show in Fig. 1 the observational disappearance of the TiO molecular bands that are the defining characteristic of M stars. They are essentially gone after M9 at low spectral dispersions, but faint hints of them remain in the early L stars at our resolution. Nonetheless, it is clear that the L spectra are easily distinguished as not arising in M stars. One should not mistake the L classification for substellar status; that must be determined separately for early L stars. Beyond an (as yet undetermined) early to mid L spectral subclass, however, all objects are indeed brown dwarfs. Only the two coolest objects on our list are definitely BDs: DENIS-P J1228–1547 because it shows lithium and DENIS-P J0205–1159 because it is very likely too cool to be a star (though that is not yet indisputable).

We have not determined the radial velocities of the objects for this paper, but rather aligned each spectral feature under discussion to a common centroid (fixed by the model spectra) for all stars. We must, however, obtain an estimate of the rotation velocities (v sin$i$) in order to compare the observed spectra to models which contain no rotation. We do not estimate the v sin$i$ from the breadth of the atomic features, since that would assume that there is no important pressure broadening, which is something we would like to check. Instead we obtained estimates of v sin$i$ by looking at our redmost order, which contains molecular features in the range of 861–872 nm, even for our cool cases. These features are probably due to CrH (eg. Tinney *et al.* 1998). We only estimate a crude rotation (within $\pm$10km/s or so) which suffices for this purpose. These are listed in Table 2. We will publish a more detailed journal article on these spectra in the near future.

We do not flux calibrate the spectra, but just normalize them order by order (avoiding obvious lines and bands). A spike-removal algorithm is applied in the upward direction (mostly to remove cosmic rays), and the observed spectra are boxcar smoothed by 5 pixels. To produce comparable model spectra, we apply a rotational broadening, then an instrumental profile convolution (found from the ThAr lines), then resample to the observed pixel scale, then continuum normalize in the same manner as the observations. We note that these procedures have strong effects on the apparent strength of the atomic lines; which are measured against a pseudocontinuum produced by the amount of molec-

TABLE 1. The stars for which spectra are presented. These include very low mass main sequence objects, objects near the substellar boundary whose status is unclear, and confirmed brown dwarfs. The L-class objects are all from the DENIS survey.

Name	Spectral Type	I mag	Est. Temp.
Gl 406	M6	13.5	2700
LHS 2924	M9	17.5	2500
DENIS-P J0909–0658	L0?	17.9	2400
DENIS-P J1058–1548	L3?	18.5	1900
DENIS-P J1228–1547	L5?	18.3	1700
DENIS-P J0205–1159	L7?	18.6	1600

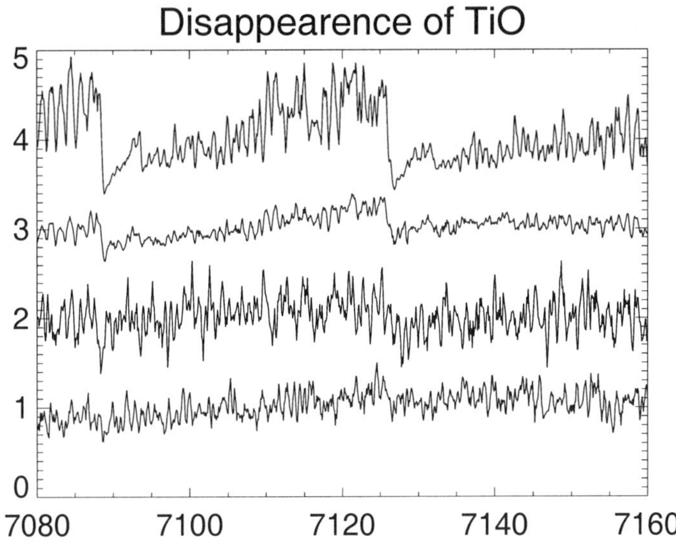

FIGURE 1. Observations of the TiO bandheads near 710 nm. From top to bottom, the stars displayed are Gl 406, LHS 2924, DENIS-P J0909–0658, DENIS-P J1058–1548. The spectra are smoothed with a boxcar of 3 pixels. The S/N for the DENIS objects is substantially lower.

ular absorption lines near the strong atomic line. The molecular strengths themselves are a strong function of temperature, and of the treatment of dust (see below). But the effective smoothing can also produce an apparent continuum with rather different effects depending on the amount of the smoothing. It should thus be kept in mind that the equivalent widths of the absorption features are not absolute, but depend on this smoothing (some instrumental and some intrinsic to the star). The molecular lines themselves are not always treated in full detail in the models anyway, as discussed below.

3. The model atmospheres

The model atmospheres are the subject of a separate contribution to this book by Allard. We do not discuss them here except to say that we have considered three variants of models, differing by their treatment of dust. For all our cases we have assumed solar metallicity and a log(g) of 5.0. In each case the opacities over the full spectrum have been treated in great detail, and the atmospheric temperature structure computed with full radiative/convective equilibrium. The "standard" case is essentially that given in the NextGen (NG) models by Hauschildt, Allard & Baron (1999), which do not consider dust formation at all. We note that the molecules VO, CaH and FeH are treated in the JOLA ("just-overlapping lines") approximation, rather than with a full list of individual molecular lines. This is because trustworthy detailed lists are not available. In particular, in the spectral regions near the Rb and Cs resonance lines we are studying, there are spectral features due to approximated bands in the models that are clearly not appropriate, given the observations. They lead to large flux jumps that are simply not seen, and to ambiguities about estimation of the pseudocontinua. We have tried to avoid these in our analysis, removing them by dividing out a linear fit to the band structure and ignoring the spectral region affected if possible.

The second (and favored) treatment is what we call CD (condensed dust) models, in which the condensation of dust is treated fully (as well as can be done right now) but the dust serves only to remove atoms from the available mix of atomic and molecular species. No opacity due to the resulting dust is included; physically this is the same as assuming all the dust forms into large particles which sink below the photosphere and do not participate in the structure of the atmosphere or the formation of the observed spectrum. We are not claiming that this is what actually happens, or even that the CD treatment is the right one to be using. This is just a preliminary exploration of what is likely to be a very complex actual situation in which dust forms with varying efficiency into particles with a distribution of sizes, whose presence in the photosphere depends on convective and turbulent velocity fields, radiative levitation, further chemistry, and even perhaps meteorology.

The third "Dusty" case is like the CD case, except that the opacity due to the dust that forms is fully included. A size distribution for the particles is assumed and the particles are taken to be fully mixed into the photospheric plasma. This and the CD case should bracket the actual situation, and serve to illustrate what the limiting cases are probably like. The three cases produce rather different surface fluxes in the red part of the spectrum. This is partly due to the heating of the upper atmosphere due to the dust in the Dusty case, the removal of many relevant molecular lines due to TiO and VO in the CD case, and the presence of these lines in the NG case. These effects lead to very different levels of pseudocontinua for the three cases, which are not apparent in the normalized spectra. Thus, the apparent strength of the molecular features adjacent to the atomic lines can change in a counterintuitive manner. In Fig. 2 we illustrate some of these effects. The Rb profiles are similar for the NG and Dusty models; the molecular lines in the NG model play the role of the dust in the Dusty model. In fact, the NG spectrum is greatly suppressed by the strong molecular lines in relative flux compared to the other two. When we normalize it to be similar, the apparent strength of the molecular lines looks at best as strong as in the Dusty case, but that is deceptive. The CD case has the weakest molecular lines, and that allows the line opacity to be more important, resulting in the broader and stronger looking Rb line.

4. Alkali resonance lines

The Na D lines (at 590 nm) are a very familiar resonance doublet due to a neutral alkali. They are already prominent in the Sun (hence the Fraunhofer designation), and become increasingly prominent in cooler stars. This is of course because they arise from

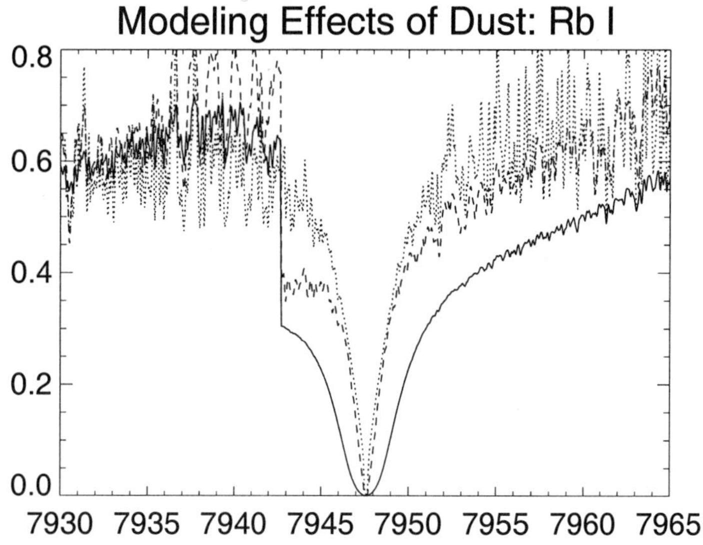

FIGURE 2. Models of the spectral region of the Rb I lines at 1900 K. The solid line shows the CD case, the dotted line shows the Dusty case, and the dashed line is the NG model. JOLA molecular bands have not been removed here, and the only normalization is to put the spectra in units of the local Planck function (except the NG model is multiplied by a factor of 6). No rotational broadening has been applied.

the ground state of a neutral species which is easily ionized. The alkalis are distinguished by having some of the redmost atomic resonance lines. Another fairly common alkali is K I, with a resonance doublet at 766,770 nm. Toward the cool end of the M spectral type, these lines are also becoming quite prominent, while the Na D lines are approaching the strength of the Ca H,K lines in solar type stars. Collisional damping wings allow the lines to cover an increasingly broad part of the spectrum. The heavier alkali metals are among the rarer of the cosmic elements, but as the other elements either form molecules or drain to low-lying levels with ultraviolet transitions, Li, Cs, and Rb eventually become the source of very strong optical atomic absorption lines. This does not occur until the late half of the M spectral class, but is true throughout the L spectral class. The lithium is eventually removed into molecules like LiCl (the brown dwarfs begin to be salty) somewhere at the cool end of the L sequence (or cooler). Even an object as cool as Gl 229B shows strong Cs I absorption (Oppenheimer *et al.* 1998).

Here we report the first detailed modeling of lines due to Rb I and Cs I in very cool dwarfs. It was suggested by Basri & Marcy (1995) that these lines grow in a simple way with decreasing temperature, and might serve as a "thermometer" for very cool stars and brown dwarfs. The lines are only visible at high spectral resolution in mid-M stars, but become visible even at low dispersion by early L. Their pseudo-equivalent widths (PEqW) are given in Table 2. We use this term because the apparent continuum used to define the units of line strength has little to do with the "true continuum", a construct which has no reality in such cool stars. Table 2 also contains the strengths of a comparable set of profiles from the model calculations, and shows that the strength of the line is a monotonic function of temperature. Indeed, a rule of thumb valid in most cases is that the PEqW of the lines doubles for every 200 K the temperature decreases (in

TABLE 2. Observations and models for the alkali lines. The rotational velocity is indicated in $\mathrm{km\,s^{-1}}$, followed by measured strengths for each star. These are followed by models of the given temperature and the measured strengths for the lines in the model. The models tabulated do not correspond directly to our estimated stellar temperatures (see Table 1 for those). Errors in the measured PEqWs are about 10%, or 15% in the coolest two objects.

Name	***v* sin*i***	**Cs PEqW**	**Rb PEqW**
Gl 406	<3	0.24	0.44
LHS 2924	9	0.44	0.53
DENIS-P J0909–0658	20	0.70	1.0
DENIS-P J1058–1548	35	2.1	2.95
DENIS-P J1228–1547	30	3.6	5.5
DENIS-P J0205–1159	20	3.9	7-9

Model (T)	***v* sin*i***	**Cs PEqW**	**Rb PEqW**
2800	–	0.13	0.31
2600	–	0.27	0.67
2400	–	0.44	1.21
2200	–	0.81	1.84
2000	–	1.65	2.6
1800	–	3.3	5.1

this temperature range). It follows that one needs to establish the absolute calibration of line strength with temperature (either by trusting these models, or obtaining checks on the temperature from low spectral dispersion or luminosity determinations) and then in principle could assign the temperature of an object by observation of only one line. In practice, even if the calibration were well determined, one would still have to worry about the gravity and metallicity of individual objects.

4.1. *Neutral cesium*

Looking first at the Cs I line at 852.1 nm, we see in Fig. 3 this line in all the objects of Table 1. The line begins with a PEqW of about 0.5Å in the M stars, and increases to almost 4Å in our coolest target. Our spectra for the two coolest objects are rather noisy, but there is no doubt that the line has become very deep and very broad. A set of model profiles of this line is shown in Fig. 4, covering the temperature range 1600 K-2800 K in increments of 200 K. Comparing the model strengths to the observed strengths (Table 2), we see that the models span a similar range of strengths. One can indeed derive fairly consistent temperatures for the objects by matching line strengths for Cs I and Rb I, and it is these that are given in Table 1. It is possible that the observed line strengths for the cooler objects include molecular lines in the wings (eg. the possible feature at 852.4 nm) which are not properly treated in the models. Of course, it would desirable to increase the S/N of these observations.

The v sin$i$ used to correct the models is a constant 30 $\mathrm{km\,s^{-1}}$. It would be better to compare model strengths computed with each individual v sin$i$. This matters most for the hotter (weaker) cases; by the time the temperature is below 2000 K, the line is so strong that collisional broadening dominates rotational broadening, and it does not matter very much what the v sin$i$ is. Of course, the rotations are all high relative to early or mid M stars. This is due in part to the fact that the angular momentum loss rate is slower in these objects, and in part to the fact that at least the BDs are preferentially young. The trend noted by Basri *et al.* 1996 (see article by Bouvier in this book) that

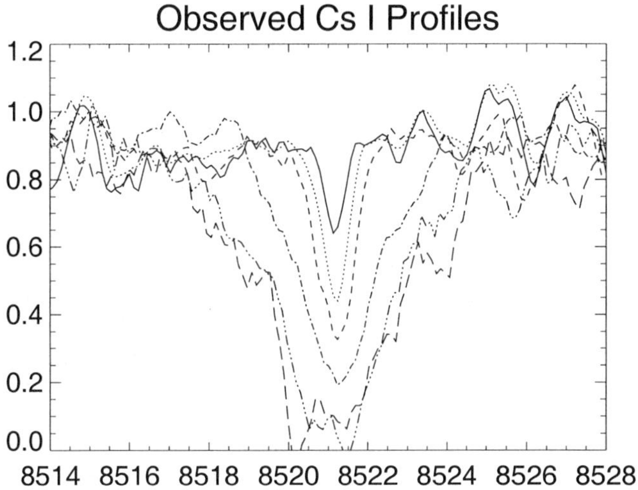

FIGURE 3. The observed Cs I lines. The identification of the stars can be inferred from the strengths of the lines, which increase as the stellar temperature decreases.

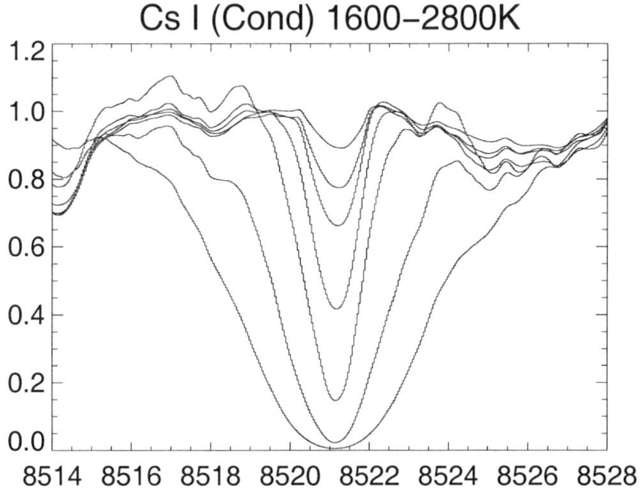

FIGURE 4. Model Cs I lines. The CD models shown are all solar metallicity with log(g)=5. A model is shown for every 200 K interval, in the range indicated. The model temperature decreases monotonically with increasing line strength.

very low mass objects tend to suffer much less angular momentum loss than higher mass convective stars is supported here.

The best profile fits are found with the CD models. Particularly at the cooler temperatures, the CD lines are broader and rounder at the bottom, more in accord with what we observe. The NG and Dusty profiles are similar to each other and have cores that are too sharp to match observations (even after rotational broadening). The NG models do not properly reproduce the molecular features: the individual lines tend to have amplitudes that are too large at high dispersion (after continuum normalization), and the strength of the molecular bands is far too strong when considered at low dispersion. The temperatures derived from fits to the CD line profiles are also in agreement (more or less) with those derived from spectral energy distribution fits to the models at much lower dispersion.

4.2. *Neutral rubidium*

For Rb I, the overall agreement between observations (Fig. 5) and models (Fig. 6) is also promising. We have plotted a restricted wavelength range which hides the problems with the JOLA molecular bands, but these should not matter where the line opacity is larger than the molecular opacity. The only visible problem here is for the two coolest model profiles, where the red wing is asymmetric and broader than the blue wing. This is due to the presence (see Fig. 2) of a VO JOLA band at 794.3 nm whose strength is important for these cases. This has been largely (but not completely) removed in Fig. 6 by dividing out a fit to it. It occurs right in the inner blue wing of the line and then weakens smoothly across the whole blueward part of the line. There is no evidence for such a feature in the observations, and the molecular band it represents should be treated with more realism in the future.

The Rb lines are generally stronger than the Cs lines in both the observations and the models, which reflects their relative cosmic abundances. Once again, it may be that the coolest line profiles are contaminated with molecular lines that have also grown in strength. The S/N for DENIS-P J0205–1159 is rather low, and we hope to increase our exposure level on this object over time. Its observed profile is about as broad as the broadest model profile, while for Cs the coolest model is broader than any of the observed profiles.

The temperature scale found from the two lines is quite compatible, agreeing to within 100 K or so. The hottest object shown is M6, and the approximate temperature of M6 objects is 2800 K (although the temperature scale for M dwarfs is actually not very well calibrated for the later half of the spectral subclasses). It seems that 1600 K is a lower limit to the temperature of our coolest object as inferred from our analysis. It is possible that the treatment of dust plays a role here; Fig. 2 demonstrates that the breadth of the model atomic lines certainly depends on the background opacity. Finally, it is not clear that we really know how to treat collisional broadening of these lines under these extreme conditions. Although there are theoretical reasons to believe that the alkali lines should be fairly well described by current theory, that has not really been checked empirically. The comparisons we are making here give some cause for hope.

5. Conclusions

We have shown strong atomic line profiles for objects ranging in temperature from about 2700 K down to 1700 K. These correspond to late M through late L spectral classes. These are the first high resolution profiles from such cool objects. The lines shown are resonance lines of neutral alkali metals, specifically Cs and Rb. We compare them to model profiles generated at various temperatures, and with different treatments of the dust formations processes which surely take place in these cool atmospheres. We find satisfactory agreement between the condensed dust (CD) models, which removes atoms that should be taken up in dust but does not allow the dust to contribute its own local opacities. This implies that dust is forming but there is at least partial settling (and perhaps condensation into clouds). On the other hand, the CD models do not do a good

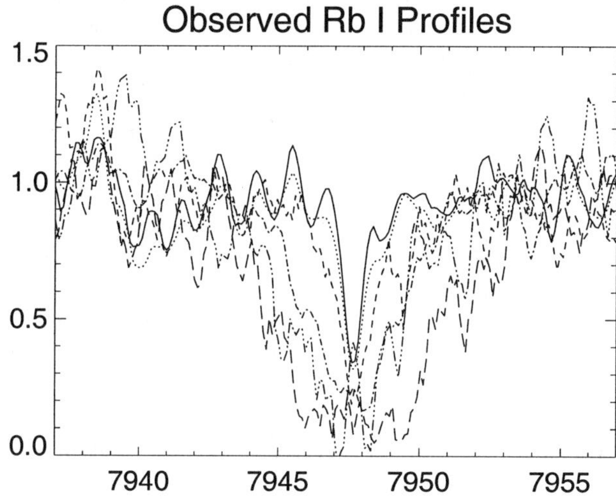

FIGURE 5. The observed Rb I lines (similar to Fig. 3).

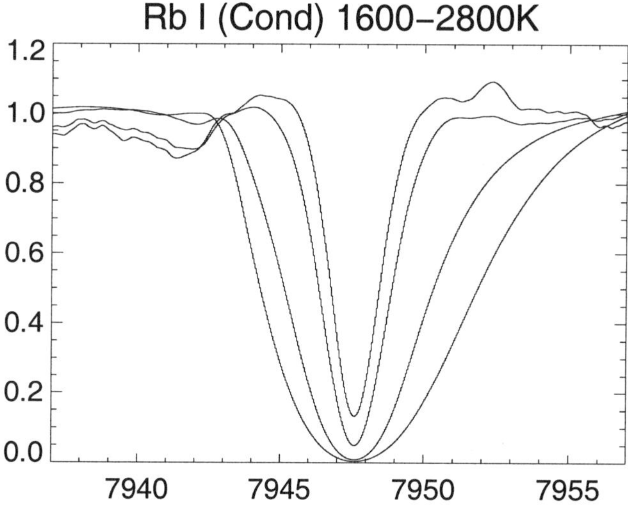

FIGURE 6. Model Rb I lines (similar to Fig. 4).

job of predicting the infrared spectral energy distribution, and some dust opacity should clearly be included. This is the major unresolved issue facing the models at the moment, and it would be premature to draw too many firm conclusions until more progress has been made.

The range of line strengths predicted is compatible with the observations, and the profile shapes are in reasonable agreement. There is a clear case to be made that one can use the growth of these lines towards cooler temperatures to assign spectral subclasses and temperatures using them in L stars. The temperature found here for Gl 406 is in good agreement with that from Jones *et al.* 1996 (Table 3). Our estimate for LHS 2924 is a little hotter than typically assigned (see above reference). If the coolest M star is really closer to 2200 K, then our coolest objects should fit in the restricted temperature range from there down to 1700 K. More likely there is an offset in the absolute temperature scales for very cool objects, and our L stars may actually be a little cooler. In that case, the coolest objects would be brown dwarfs purely on the basis of their temperature. It seems prudent to hold off in assigning the full range of L subclasses to the set of objects that have been observed so far.

We see that the treatment of both molecular and dust formation is crucial to understanding these very cool atmospheres, and that the details of the molecular and dust opacities are critical in understanding the high (and low) dispersion spectra of these objects. The molecules should be treated by using direct and detailed line lists rather than using the JOLA approximation if one wants to reproduce echelle spectra in detail. There is a need for much more theoretical and laboratory work by physicists and chemists to achieve this goal. The question of which dust species are forming, and which remain in the atmosphere (and in what state), will clearly occupy researchers for the next few years. Still, there is already a heartening agreement between the atomic line profiles modeled and observed, and we have already begun to extract the physical information inherent in them.

This research is based on data collected at the W. M. Keck Observatory, which is operated jointly by the University of California and California Institute of Technology. GB was a Miller Research Professor during the time this research was conducted, and SM acknowledges the support of NSF through grant AST96-18439. PHH acknowledges support by NSF grant AST-9720704, NASA ATP grant NAG 5-3018 and LTSA grant NAG 5-3619. We thank the DENIS team, X. Delfosse in particular, for facilitating our observation of their targets. Some of the calculations presented in this paper were performed on the IBM SP2 and the SGI Origin 2000 of the UGA UCNS, at the San Diego Supercomputer Center (SDSC) and the Cornell Theory Center (CTC), with support from the National Science Foundation, and at the NERSC with support from the DoE. We thank all these institutions for their generous allocation of computer time.

REFERENCES

Allard, F., Hauschildt, P.H., Alexander, D.R., Starrfield, S. 1998 Model atmospheres of very low mass stars and brown dwarfs *Ann. Rev. Ast. Ap.* **35**, 137–177

Baraffe, I., Chabrier, G., Allard, F., Hauschildt, P.H. 1999 New evolutionary tracks for very low mass stars *Ast.&Ap.*, submitted

Basri, G. & Marcy, G. W. 1995 A surprise at the bottom of the main sequence: Rapid rotation and no H_α emission *A.J.* **109**, 762–773

Basri, G., Marcy, G., Oppenheimer, B., Kulkarni, S.R., Nakajima, T. 1996 in Cool Stars, Stellar Systems, and the Sun, 9th Cambridge Workshop *A.S.P. Conf. Ser.* **109**, 587–588

Basri, G., Marcy, G., & Graham, J. R. 1996 Lithium in brown dwarf candidates: the mass and age of the faintest Pleiades stars *Ap.J.* **458**, 600–609

Delfosse, X., Tinney, C.G., Forveille, T., Epchtein, N. *et al.* 1997 Field brown dwarfs found by DENIS *Ast.&Ap.* **327**, L25–28

Hauschildt, P., Allard, F. & Baron, E. The NextGen Model Atmosphere grid for 3000<Teff<10000 K 1999 *Ap.J.*, in press.

Jones, H.R.A., Longmore, A.J., Allard, F. & Hauschildt, P.H. 1996 Spectral Analysis of M dwarfs *M.N.R.A.S.* **280**, 77-94

Kirkpatrick , J.D. 1998 Spectroscopic properties of ultra-cool dwarfs and brown dwarfs in Brown Dwarfs and Extrasolar Planets Workshop, Tenerife *A.S.P. Conf. Ser.* **134**, 405–415

Martín, E.L., Basri, G., Delfosse, X., Forveille, T. 1997 Keck HIRES spectra of the brown dwarf DENIS-P J1228.2-1547 *Ast.&Ap.* **327**, L29–32

Nakajima, T., Oppenheimer, B.R., Kulkarni, S.R., Golimowski, D.A., Matthews, K. & Durrance, S.T. 1995 *Nature* **378**, 463-465

Oppenheimer, B.R., Kulkarni, S.R., Matthews, K. & van Kerkwijk, M.H. 1998 The Spectrum of the Brown Dwarf Gliese 229B *Ap.J.* **502**, 932–943

Rebolo, R., Martín, E. L., Basri, G., Marcy, G. W., & Zapatero Osorio, M. R. 1996 Brown dwarfs in the Pleiades cluster confirmed by the lithium test *Ap.J.* **469**, L53–56

Ruiz, M.T., Leggett, S.K.& Allard, F. 1997 Kelu 1: a free-floating brown dwarf in the solar neighborhood *A.pJ.* **491**, L107–110

Tinney, C. G., Delfosse, X., Forveille, T. & Allard, F. 1998 Optical spectroscopy of DENIS mini-survey brown dwarf candidates *Ast.&Ap.* **338**, 1066–1072.

Zuckerman,B. & Becklin, E.E. 1992 Companions to white dwarfs: very low-mass stars and the brown dwarf candidate GD 165B *Ap.J.* **386**, 260–264

Modelling Very Low Mass Stars and Brown Dwarfs Atmospheres: The Importance of Dust Formation

By FRANCE ALLARD†

Dept. of Physics, Wichita State University, 1845 Fairmount, Wichita, KS-67260, USA

We review the current theory of very low mass stars model atmospheres including the coolest known M dwarfs, M subdwarfs, and brown dwarfs, i.e. $T_{eff} \leq 5,000\,K$ and $-2.0 \leq [M/H] \leq +0.0$. We discuss ongoing efforts to incorporate molecular and grain opacities in cool stellar spectra, as well as the latest progress in deriving the effective temperature scale of M dwarfs. We especially present the latest results of the models related to the search for brown dwarfs.

1. Very low mass star models and the T_{eff} scale

Very Low Mass stars (VLMs) with masses from about 0.3 $M_{\odot}$ to the hydrogen burning minimum mass (0.075 $M_{\odot}$, Baraffe *et al.* 1995) and young substellar brown dwarfs share similar atmospheric properties. Most of their photospheric hydrogen is locked in H_2 and most of the carbon in CO, with the excess oxygen forming important molecular absorbers such as TiO, VO, and H_2O. They are subject to an efficient convective mixing often reaching the uppermost layers of their photosphere. Their energy distribution is governed by the millions of absorption lines of TiO, VO, CaH, and FeH in the optical to near-infrared, and H_2O and CO in the infrared, which leave **no** window of true continuum. But as brown dwarfs cool with age, they begin to differentiate themselves with the formation of methane (CH_4) in the infrared (Tsuji *et al.* 1995; Allard *et al.* 1996). Across the stellar-to-substellar boundary, clouds of e.g. corundum (Al_2O_3), perovskite ($CaTiO_3$), iron, enstatite ($MgSiO_3$), and forsterite (Mg_2SiO_4) may form, depleting the oxygen compounds and heavy elements and profoundly modifying the thermal structure and opacity of their photosphere (Sharp & Huebner 1990; Burrows *et al.* 1993; Fegley & Loggers 1996; Tsuji *et al.* 1996ab; Allard *et al.* 1997b).

Because these processes also occur in the stellar regime where a greater census of cool dwarfs is currently available for study, a proper quantitative understanding of VLM stars near the hydrogen burning limit is a prerequisite to an understanding of the spectroscopic properties and parameters of brown dwarfs and jovian-type planets. Model atmospheres have been constructed by several investigators over recent years with the primary goals of:

(*a*) Determining the M dwarfs effective temperature scale.

(*b*) Identifying spectroscopic signatures of substellarity i.e. gravity indicators for young brown dwarfs, and spectral features distinctive of cooler evolved brown dwarfs.

(*c*) Providing non-grey surface boundary to evolution calculations of VLMs and brown dwarfs leading to more consistent stellar models, accurate mass-luminosity relations and cooling tracks for these objects.

The computation of these atmospheres requires a careful treatment of the convective mixing and the molecular opacities. The convection must currently be handled using the mixing length formalism while a variety of approximations have been used to handle the millions of molecular and atomic transitions that define the spectral distributions of

† Present address: CRAL, ENS, 46 Allée d'Italie, Lyon Cedex 07, 69364 France

TABLE 1. Relevant Model Atmospheres

Authors	**Grid**	**T_{eff} range (K)**	**Main Opacity Treatment**
Kurucz 1992	Atlas12	3500 – ...	OS
Allard 1990	Base	2000 – 3750	SM+JOLA
Saumon *et al.* 1994	zero-metallicity	1000 – 5000	OS
Tsuji *et al.* 1995	grainless	1000 – 2800	JOLA
Brett 1995	MARCS	2400 – 4000	OS
Allard & Hauschildt 1995	Extended Base	1500 – 4500	SM
Tsuji *et al.* 1996	dusty	1000 – 2800	JOLA+Grains
Allard *et al.* 1996	NextGen	900 – 9000	OS
Allard *et al.* 1997b	NextGen-dusty	900 – 3000	OS+Grains
Marley *et al.* 1996		... – 1000	K-coefficients

VLMs and brown dwarfs. The most accurate of these methods is the so-called opacity sampling (OS) technique which consists in adding the contribution all transitions absorbing within a selected interval around each point of a pre-determined wavelength grid (typically $\approx$ 22000 points from 0.001 to 100 μm in our models). When the detail of the list of transitions is lacking for a molecule as is the case for the important absorber VO, the Just Overlapping Line Approximation (JOLA) offers an alternative by approximating the band structure based on only a few molecular rotational constants. The straight-mean (SM) and K-coefficients techniques, which consist in averaging the opacities over fixed wavelength intervals chosen smaller than the resolution of typical observations, have also been used in modeling late-type dwarf atmospheres. Their main advantage is to save computing time during the calculation of the models, often at the expense of flexibility and an accurate spectral resolution. The list of recent model atmospheres and the opacity technique they mostly rely upon is given in Table 1.

Because they mask emergent photospheric fluxes that would otherwise escape between absorption lines, the JOLA, SM and K-coefficients approximations generally lead to an excessive entrapment of heat in the atmosphere which yields systematically hotter model structures, and higher effective temperature (T_{eff}) estimates for individual stars. Allard *et al.* (1997) have reviewed the results of brown dwarfs and VLM model atmosphere calculations with respect to the effective temperature scale of M dwarfs. We reproduce in Figure 1 the $T_{eff} - (V - I)$ relation of Allard *et al.* (1997) for the models listed in Table 1.

Two double-line spectroscopic and eclipsing M dwarf binary systems, CM Draconis and YY Geminorum, offer some guidance in the sub-solar mass regime and are reported in Figure 1 according to Habets & Heintze (1981). The use of an OS treatment of the main molecular opacities, in particular for TiO, appears to yield a break-through in the agreement of T_{eff} scales with these two M dwarfs binary system. The NextGen and MARCS models yield effective temperatures that are coincidentally in good agreement with those derived empirically from the H_2O opacity profile by Jones *et al.* (1994). Note, however, that the Atlas12 OS models suffer from an inaccurate TiO absorption profile and a lack of H_2O opacities, and are therefore inadequate in the regime of VLM stars (i.e. below $T_{eff} \approx 4500$ K) where these molecular opacities dominate the stellar spectra and atmospheric structures.

Some uncertainties on the metallicity of the CM Draconis system may eventually disqualify the latter as a member of the disk main sequence (Viti *et al.* 1997). This stresses the importance of finding other low-mass eclipsing binary systems in the disk.

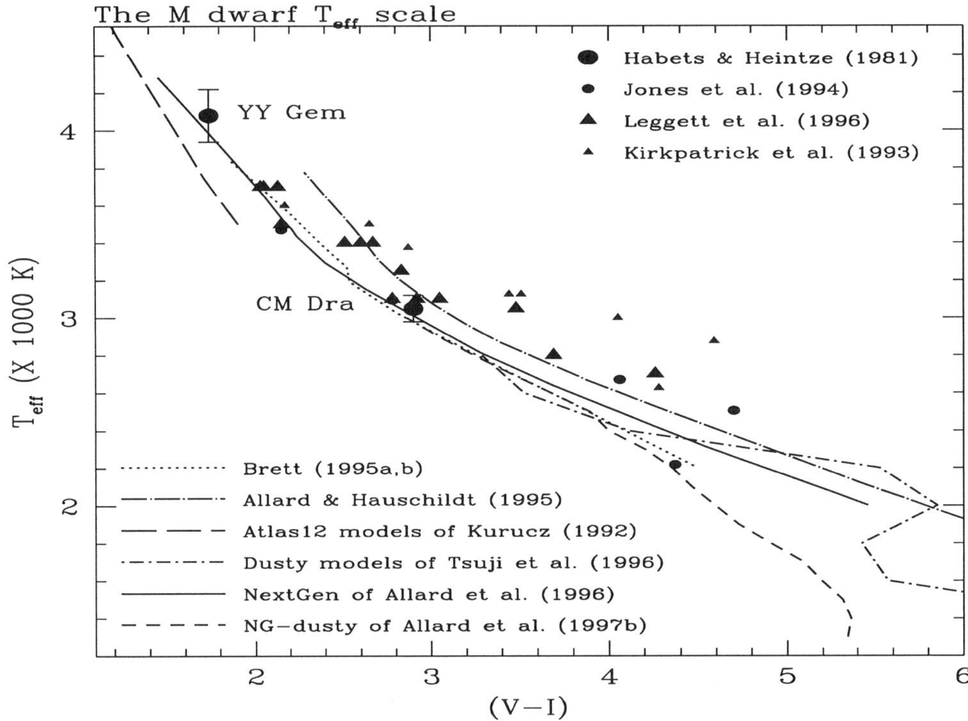

FIGURE 1. Current model-dependent effective temperature scales for cool stars down to the hydrogen burning limit. Triangles feature results from spectral synthesis of selected stars from the works of Kirkpatrick *et al.* (1993) and Leggett *et al.* (1996) as indicated. The new generation of OS models by Brett (1995a) and Allard *et al.* (1997b), as interpolated onto theoretical isochrones by Chabrier & Baraffe (1997), reproduce closely the independently-determined positions of the eclipsing M dwarf binary system CM Dra and YY Gem, and the empirical T_{eff} scale of Jones *et al.* (1994).

These are hopefully soon to be provided by the 2MASS and DENIS surveys. Much uncertainty remains, therefore, in the lowermost portion of the main sequence. The inclusion of grain formation (as discussed below) and more complete opacities of TiO (soon available from the work of Langhoff, 1997 and Schwenke, 1998) promise a better understanding of the stars and brown dwarfs in the vicinity of the hydrogen burning limit.

2. Modelling the infrared colors of brown dwarfs

The DENIS and 2MASS infrared sky surveys will soon deliver large data bases of red dwarfs, brown dwarfs and perhaps extrasolar planets, which will necessitate the best possible theoretical foundation. Brown dwarfs and giant planets emit over 65 % of their radiation in the infrared (> 1.0μm). A proper understanding of their infrared colors is essential in the search for brown dwarfs. Yet the main difficulties met by modelers in recent years has been to reproduce adequately the infrared (1.4 to 2.5μm) spectral distribution of dwarfs with spectral types later than about M6. All models listed in the central part of Table 1 underestimate the emergent flux, most as much as 0.5 mag at the K bandpass, despite the different opacity sources used by the authors. Allard *et al.* (1994) have ex-

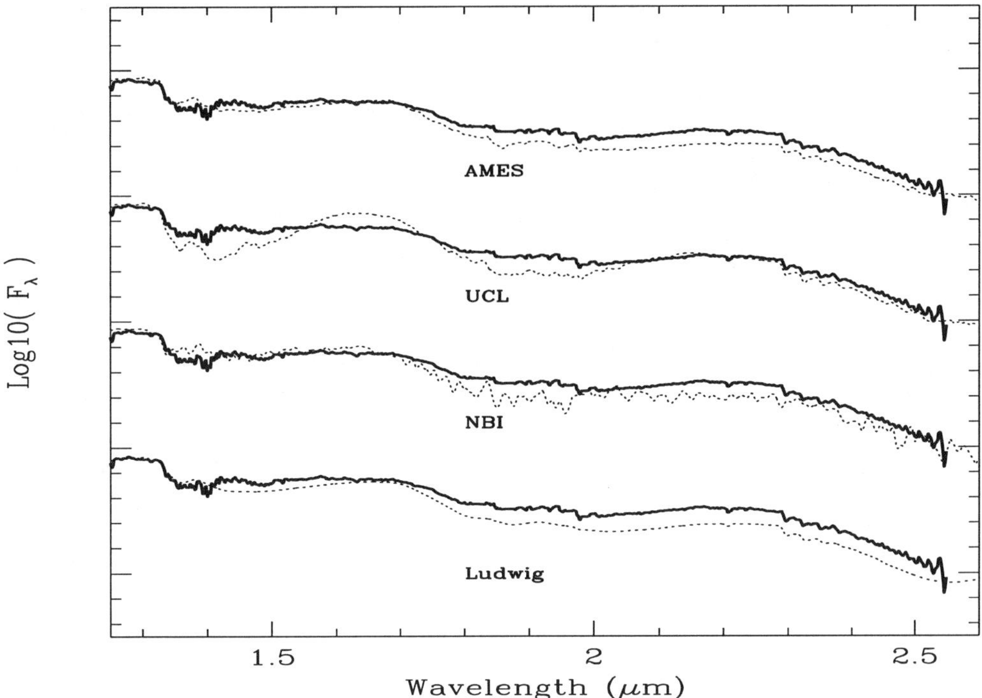

FIGURE 2. The observed infrared spectral distribution of the dM8e star VB10 as obtained at UKIRT by Jones *et al.* (1994) (bold full line) is compared to model spectra obtained using (from bottom to top): (i) the SM laboratory opacity profile of Ludwig (1971), (ii) the 20 million line list by Jørgensen (1994), (iii) the preliminary ab initio line list of 6.2 million transitions by Miller, Tennyson, *et al.* (1994), and (iv) the latest ab initio list of 300 million lines by Partridge & Schwenke (1997). The models (shown as dotted lines) are all fully converged and normalized to the observation at 1.2 μm. Their parameters were determined from a fit to the optical stellar spectra (not shown) and are nearly the same in all four cases. Note that all 300 million lines of the Partridge & Schwenke list have been included in the model construction!

plored water vapor opacity data from various sources. Figure 2 summarizes these results. The water vapor opacity profile is quite uncertain and has varied with the degree of completeness and the assumptions used in the construction of the molecular model and its potential surface. The most recent and complete line list of Partridge & Schwenke succeeds for the first time in reproducing the 1.6 μm opacity minimum, in the H bandpass, well enough for the atomic Na I resonance line to finally emerge in the synthetic spectrum, matching the observed feature. However, it fails to provide an improvement in the K bandpass where the less complete list of Miller & Tennyson still yield the best match of the models to the observed spectra. The NextGen models of Allard *et al.* (1997) are computed using the Miller & Tennyson line list and appear to be the only models to provide a match to the infrared colors of VLMs. This is shown in Figure 3 where the complete series of NextGen models — as interpolated on the Baraffe *et al.* (1997), Baraffe *et al.* (1998) isochrones for 10 Gyrs and 120 Myrs and ranging from metallicities of [M/H]= −2.0 to 0.0 — are compared to the photometric field dwarfs' samples of Leggett (1992), Tinney, Mould & Reid (1993), and Kirkpatrick, Henry & Simons (1995). Other models series including those of Brett (1995a) and the Extended grid of Allard & Hauschildt

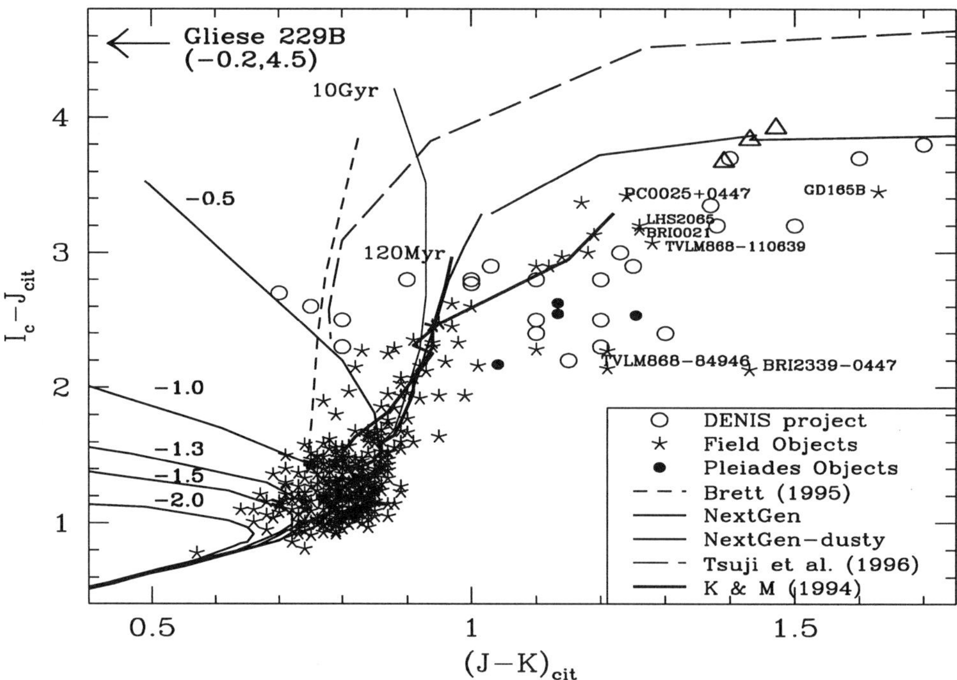

FIGURE 3. The most recent models of late type dwarfs are compared to the photometric observations of field stars and brown dwarfs, and to Pleiades objects including the brown dwarfs PPl15, Teidel and Calar3. Gravity effects are illustrated by three models (open triangles) with logg= 5.5, 5.0, and 4.5 from bottom to top. Unresolved binarity is reflected in this diagram by a red excess in $J-K$. The red dwarfs newly discovered by DENIS are also shown, although their photometry is still uncertain at this point. The field brown dwarf Gliese 229B is off the scale to the blue in $J-K$ due to strong CH_4 absorption in the K bandpass. Note that H_2 opacities depress also in K band flux and cause the NextGen models to turn to the blue at their cool ends. This diagram offers excellent diagnostics to identify brown dwarf candidates of the field (very red in either $J-K$ or $I-J$) or of the halo (very blue in both $I-J$ and $J-K$).

(1995) not shown, are distinctively bluer than the observed sequence, while the 10 Gyrs NextGen models of solar metallicity follow closely the empirical sequence† of Kirkpatrick & McCarthy (1994) down to spectral types of M6 (i.e. $J-K \approx 0.85$). Beyond this point, all grainless models fail to reproduce the bottom of the main sequence into the brown dwarf regime as defined by Gl406, VB10, BRI0021 and GD165B. They catch up with observations eventually again at the much lower T_{eff} of the evolved brown dwarf Gliese 229B, i.e. 900-1000 K (Allard *et al.* 1996, Marley *et al.* 1996).

The cause of the model discrepancies at the stellar-to-brown dwarf boundary can only be one that affects the cooler models for Gliese 229B in a far lesser obvious extent. Since the infrared spectral distribution is sensitive to the mixing length, yet without allowing for an improved fit of VLMs spectra, Brett (1995b) suggested that the problem lie in the inadequacy of the mixing length formalism for treating the convective transport in an optically thin photospheric medium. These concerns may also be augmented by

† Note that this sequence was defined by stars selected from their optical spectroscopic properties. The somewhat irregular profile of the sequence in this infrared diagram reflects uncertainties in the photometry and age of the selected stars.

uncertainties about the extent of the overshooting phenomenon in VLMs (D'Antona *et al.* 1997). The convection zone recedes gradually below the photosphere as the mass (and T_{eff}) decreases along the isochrones. Fortunately for the lithium test of substellarity (Rebolo *et al.* 1992) — which relies on the assumption that the brown dwarf is still fully convective and mixing lithium from its core to its photospheric layers after 10^8 yrs of age — the photospheric mixing breaks down only well into the brown dwarf regime i.e. for objects always cooler then about 2200 K. The presence of lithium in the spectra of a late-type (≥M10) field dwarfs, if detected, can therefore only reflect their substellar nature. The shrinking of the convection zone also allows a very good agreement between the models of Marley *et al.* (1996) (which includes adiabatic convection only for the optically thick layers of the atmosphere) and the models of Allard *et al.* (1996) (based on a somewhat more careful treatment of convection with the mixing length formalism) for the brown dwarf Gliese 229B (see Figure 5 of Allard *et al.* 1997). Yet the maximum radial extent of the convection zone occurs at around $T_{eff} = 3000$ K, while the discrepancy with the infrared observations increases steadily towards the bottom of the main sequence.

A more promising answer to the so called "infrared problem" may rather be found in the formation of dust grains in the very cool (typically $T_{layer} \approx T_{eff} - 1000$ K) upper layers of red and brown dwarf's atmospheres. Tsuji *et al.* (1996a) proposed, based on their results of including the effects of the formation and opacities of three grain species (Al_2O_3, Fe, and $MgSiO_3$) in their new "dusty" models, that the greenhouse heating of grain opacities, the resulting enhanced H_2O dissociation, and the infrared flux redistribution, can explain the infrared spectra of cool M dwarfs. The formation of perovskite dust grains at the expense of TiO may also explain the observed saturation (and disappearance in GD165B and Gliese 229B) of the TiO bands in the optical spectra of late-type red dwarfs (see also Jones & Tsuji 1997). The implications of this result is far reaching. Known field brown dwarf candidates such as BRI0021 and GD165B can be far cooler and less massive than previously suspected (see e.g. the NextGen-dusty model predictions in Figure1). If grains also form in the young Pleiades brown dwarfs PPl15, Teide1 and Calar3 ($T_{eff} \approx 3000$, 2800, and 2700 K respectively), lithium abundances derived from grainless models and synthetic spectra such as those of Pavlenko *et al.* (1995) may be overestimated, and the masses attributed to these objects possibly underestimated. Evolution models of brown dwarfs, which are sensitive to the treatment of the atmospheres (Baraffe *et al.* 1995; Chabrier & Baraffe 1997), and their predicted Mass-lithium abundance and Mass-Luminosity relations may also be affected.

And indeed, the temperatures and pressure conditions of the outer layers of red dwarfs are propice to the formation of dust grains as demonstrated years ago by Sharp & Huebner (1990). However it was not clear at the time if the inward radiation of an active chromosphere, or the efficient convective mixing from the interior, would heat up these upper photospheric layers and disable grain formation. Another concern is that, under the gravities prevailing in M dwarfs, gravitational settling may occur that would eliminate large grains and their opacities from the photospheres over relatively short time scales. These possibilities still need to be thoroughly investigated, but clearly, grain formation is a process that must be considered in the construction of M dwarf and brown dwarf model atmosphere.

In order to investigate which grains may form in the upper layers of M dwarfs, Allard *et al.* (1997b) have modified the equation of state used in the NextGen models to include the detailed calculation of some 1000 liquids and crystals, using the free Gibbs energies compiled by Sharp & Huebner. Their results show that, besides the three species considered by Tsuji *et al.* , the M dwarfs atmosphere were rich in condensates with ZrO_2, $Ca_2Al_2SiO_7$, Ca_2MgSiO_7, $MgAl_2O_4$, Ti_2O_3, Ti_4O_7, $CaTiO_3$, $CaSiO_3$ and

Ca_2SiO_4 showing up in models as hot as $T_{eff} = 2700 - 3000$ K (i.e dM8-dM6)! The preliminary NextGen-dusty models have been computed using a continuous distribution of ellipsoid shapes and interstellar grain sizes (between 0.025 and 0.25 μm) for the treatment of the opacities of the $CaSiO_3$, Ca_2SiO_4, $Ca_2Al_2SiO_7$, Ca_2MgSiO_7, $MgAl_2O_4$, Al_2O_3, Fe, $MgSiO_3$, and Mg_2SiO_4 dust grains. This contrasts with the assumption of spherical grains with 0.1 μm diameters in the dusty models of Tsuji, Ohnaka, Aoki & Nakajima (1996). Both models are shown in Figures 1 and 3. As can be seen, the dusty models of Tsuji *et al.* (elsewhere in this volume) provide the correct tendency of the coolest models to get rapidly very red (as much as $J - K = 1.65$ for GD165B) with decreasing mass for a relatively fixed $I - J$ color. Those models are however systematically too red in $I - J$ by as much as 1 mag and do not reproduce even the most massive M dwarfs while over-predicting the effects of grains in Gliese 229B type brown dwarfs (Tsuji *et al.* 1996b). The NextGen-dusty models, on the other hand, show the onset of grain formation effects for $J - K \geq 0.85$, bringing an improved agreement with the observed sequence in the region where the grainless NextGen models deviate. A similar behavior is also shown by the models in other colors such as I-J and H-K where the NextGen-dusty models provide an unprecedent match to the observed stars and brown dwarfs colors (Leggett *et al.* 1998).

Of course, much remains to be improved in the computation of models with dust grains. The size distribution of various grain species, in particular those of the perovskite $CaTiO_3$ which is responsible for the depletion of TiO from the optical spectra of late-type dwarfs and of the calcium silicates which accounts for most of the grain opacities in current models, is unknown for the conditions prevailing in M dwarfs atmospheres. It is conceivable that grains form more efficiently in M dwarfs atmospheres than in the interstellar medium, and present larger sizes are different shapes than interstellar grains do. Fortunately, the total opacity provided by a grain species is little or not affected by our assumptions on the grain sizes. Indeed, a conservation of the total number of element cores requires that the number of grains of a type be inverse proportional to the size of these grains (themselves composed of several core particulates). On the other hand, gravitational settling processes may well be accelerated for larger grains, reducing their local contribution to the total opacity compared to what is now considered in the current "static" dusty models. We may also miss a number of contributors to the opacities from e.g. species for which the scattering profile is not well-known. Further investigations including time dependent grain growth analysis will be required to determine the exact contribution of dust grains to the infrared colors of red and brown dwarfs.

In the meanwhile, diagrams like that of Figure 3 may help in distinguishing interesting brown dwarfs candidates from large data banks of detected objects, and in obtaining an appreciation of the spectral sensitivity needed to detect new brown dwarfs. Models (Tsuji *et al.* 1995; Allard *et al.* 1996; Marley *et al.* 1996) and observations of Gliese 229B have shown that methane bands at 1.7, 2.4 and 3.3 μm appear in the spectra of cool evolved brown dwarfs, and cause their $J - K$ colors to get progressively bluer with decreasing mass and as they cool over time. Yet their $I - J$ colors remain very red which allows to distinguish them from hotter low-mass stars, red shifted galaxies, red giant stars, and even from low metallicity brown dwarfs that are also blue due to pressure-induced H_2 opacities in the H–to–K bandpasses. Fortunately, uncertainties inherent to grain formation and molecular opacities are far reduced under low metallicity conditions ([M/H]< -0.5). Therefore, model atmospheres of metal-poor subdwarf stars and halo brown dwarfs are more reliable than their metal-rich counterparts at this point. This has been nicely demonstrated by Baraffe *et al.* (1997) who reproduced closely the main sequences of globular clusters ranging in metallicities from [M/H]$= -2.0$ to -1.0, as well as the sequence of the Monet *et al.* (1992) halo subdwarfs in optical color-magnitude diagrams. The colors of halo brown dwarfs as predicted by the NextGen models are therefore expected to be of quantitative quality and await **eagerly** confrontation with the infrared colors of metal-poor subdwarfs from e.g. the Luyten catalog and the US Naval Observatory surveys. Recently, Pulone *et al.* (1998) were able to obtain NICMOS photometry of the Ω Cen cluster which shows an excellent agreement with the NextGen models predictions down to the lower main sequence in these filters. This result confirms brilliantly for the first time the quality of the metal-depleted NextGen models in the infrared, and especially that of the important H_2 pressure-induced opacity as modeled by Borisow *et al.* (1997).

The sensitivity of the $I-J$ index to the chemical composition of the atmosphere (clearly illustrated by the NextGen model grid) allows to distinguish brown dwarf populations independently of an accurate knowledge of the parallaxes or distances involved. Even young brown dwarfs of lower gravity appear to form a distinct sequence at bluer $I-J$ (and redder $J-K$) values then that of their older field star counterparts as it is also evident from a comparison of the 10 Gyrs and 120 Myrs NextGen models. This gravity effect, and perhaps enhanced grain formation, may explain the scatter of spectroscopic properties observed among field dwarfs at the bottom of the main sequence, as well as the systematic differences between Pleiades brown dwarfs and older field stars of same spectral type (i.e. same VO band strengths) noted by Martìn, Rebolo & Zapatero-Osorio (1996).

3. Evolved brown dwarfs

The general spectral distributions of cool evolved brown dwarfs are well reproduced by current models despite the difference in their respective modeling techniques, and despite the uncertainties tied to grain formation and incomplete opacity data base for methane and ammonia. The models of Allard *et al.* (1996) and Marley *et al.* (1996) are compared in Figure 4 which also summarizes the predicted absolute fluxes that free-floating brown dwarfs would have at a distance of 50 pc. As can be seen, there is no clear distinction between brown dwarfs and planets; molecular bands most gradually form (dust, H_2O, CH_4 and NH_3) and recede (TiO, VO, FeH, and CO) from the stellar to the planetary regime as the atmospheres get cooler. They remain very bright in the IJK region, and become gradually redder in the near-infrared I to J bandpasses, which allows their detection from ground-based facilities. The peak of their intrinsic spectral energy distribution is located at 4.5 μm. At 5 μm, the hotter (younger or more massive) brown dwarfs and stars show strong CO bands which cause their flux to drop by nearly 0.5 dex relative to that at 4.5 μm. And between 4.5 and 10 μm, opacities of CH_4 (and H_2O in the hotter brown dwarfs) cause the flux to drop by 0.5 to 1.0 dex. Searches in the 4.5-5 μm region should therefore offer excellent possibilities of resolving free-floating brown dwarfs if space-telescope time allocations allow. On the other hand, layers of dust in their upper atmospheres may increase the albedo of extrasolar planets and cool brown dwarfs sufficiently to reflect the light of a close-by parent star, becoming therefore resolvable **in the optical** where the clouds are densest and the parent star is brightest. The detection limits of current and planned ground-based and space-based telescopes Saumon *et al.* (1996) are also indicated in Figure 4 which show that brown dwarfs within 50 pc would be easily detected by SIRTF in the 4.5-5.0 μm region. The drop in sensitivity of the various instruments redwards of 10 μm implies, however, that brown dwarfs and planets cooler than Gliese 299B have little chance to be detected in those redder bandpasses.

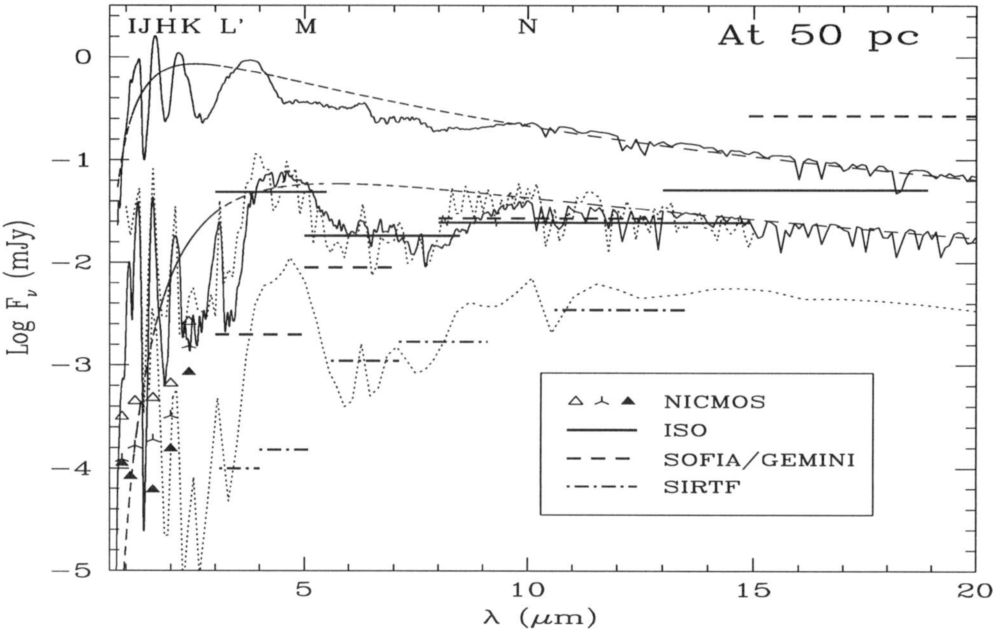

FIGURE 4. Predicted absolute fluxes of brown dwarfs at 50 pc as compared to the sensitivity of ground and space-based platforms which will be or are currently applied to the search for brown dwarfs and extrasolar planets. The latter are values reported for the 5 σ detection of a point source in 1 hr of integration, except for the three NICMOS cameras where the integration is limited to 40 minutes (Saumon *et al.* 1996). Models of both Allard *et al.* (1996) (full) and Marley *et al.* (1996) (dotted) are shown which simulate (i) a brown dwarf near the hydrogen burning limit (topmost spectrum: $T_{\rm eff}$ = 2000K), (ii) an evolved brown dwarf similar to Gliese 229B (central spectra: $T_{\rm eff}$ = 900K and 960K), and (iii) a brown dwarf closer to the deuterium burning limit (lowermost spectrum: $T_{\rm eff}$ = 500K). The corresponding black-body (dashed) are also shown for comparison.

4. Conclusions

In these exciting times where discoveries of brown dwarfs are finally breaking through, model atmospheres are also rapidly becoming up to the task of interpreting the observations and deriving new search strategies. Uniform grids of dwarf stars and brown dwarfs model atmospheres exist that extend from the tip to the toes of the main sequence – and beyond: 9000 K to 900 K, log$g$= 3.0–6.0, and [M/H]= 0.0 to −2.0 for the NextGen models. These large model grids allowed the construction of consistent interior and evolution models for VLMs that yield unprecedent agreement with globular cluster main sequences observed to 0.1 $M_\odot$ with HST. They led to the mass-luminosity relation for low mass stars, which is of primary importance for the derivation of the stellar mass function of both the halo and the disk populations. This allows now to constrain the brown dwarf density with the help of the microlensing experiments (OGLE, EROS, MACHO, etc.), and leads to the important realization that brown dwarfs cannot make up a significant fraction of the halo missing mass.

The effective temperature scale of K to M type dwarfs with spectral types earlier than M6 is now unambiguously established, with only small uncertainties remaining from a possible incompleteness of existing TiO line lists. Grain formation has been identified as an important process in M dwarfs and brown dwarfs atmospheres which could explain the long-standing difficulties of the models to reproduce the spectral distribution of dwarfs later than about M6. The results of the models indicate that it may no longer be assumed that the convection zone extends to the photosphere of late-type red dwarfs and brown dwarfs. But their photospheric lithium abundance nevertheless always reflect the substellar nature of young, hot brown dwarfs such as those found in the Pleiades cluster. Fortunately, if the lithium test cannot identify transition objects and brown dwarfs of the field, the Opacity Sampling treatment and grain formation have introduce new gravity (hence age) effects in the NextGen models that were not seen in the previous Extended models and that will potentially allow to separate younger transitional objects from field stars as readily as from their location in color-color diagrams. For this the colors of late-type red dwarfs need to be known with good accuracy i.e. better than about 0.05 magnitude, which we find is not the case of many known late M dwarfs such as Gl406, VB10, and especially LHS2924.

As cooler dwarfs are being discovered, spectral types are stretching far beyond the classical Morgan & Keenan scheme. The lack of TiO bands in the optical, and the emergence of CH_4 opacities in the infrared in GD165B and Gl229B call for an extension of the MK system beyond M9 to another spectral class. In any case, studies of the optical spectra of Gliese 229B, GD165B, the DENIS and 2MASS objects and other late-type dwarfs will soon allow to determine the stellar surface coverage of dust clouds if such are present, and to verify if intrinsic spectral-type variability afflict cool dusty dwarfs. Models will be the subject of further investigations relative to grain formation and its effect on late-type dwarfs until they can reproduce the lower main sequence and cool brown dwarfs. Finally, if brown dwarfs are not abundant in the halo, they certainly are in the galactic disk and their study remains one that shall flourish as the census of the solar neighborhood continues and the gap between planets and stars fills in.

This research is supported by a NASA LTSA NAG5-3435 and a NASA EPSCoR grant to Wichita State University, a NASA ATP grant NAG 5-3018 and an LTSA grant NAG 5-3619 to the University of Georgia. Some of the calculations presented in this paper were performed on the IBM SP2 of the UGA UCNS, at the San Diego Supercomputer Center (SDSC), the Cornell Theory Center (CTC), and at the National Center for Supercomputing Applications (NCSA), with support from the National Science Foundation.

REFERENCES

Allard, F., Alexander, D. R., Tamanai, A. & Hauschildt, P. H. 1995 Photospheric Dust Grains Formation in Brown Dwarfs, in *Brown dwarfs and extrasolar planets*, Proceedings of a Workshop held in Puerto de la Cruz, Tenerife, Spain, 17-21 March 1997, R. Rebolo, E. L. Martìn, M. R. Zapatero Osorio editors, **438**, 1998.

Allard, F. & Hauschildt, P. H. 1995 Model Atmospheres for M (Sub)Dwarfs: I. The base model grid *ApJ* **445**, 433.

Allard, F., Hauschildt, P. H., Alexander, D. R. A. & Starrfield, S. 1997 Model Atmospheres of Very Low Mass Stars and Brown dwarfs *Ann. Rev. A&A* **35**, 137.

Allard, F., Hauschildt, P. H., Baraffe, I. & Chabrier, G. 1996 Synthetic Spectra and Mass Determination of the Brown Dwarf Gl229B *ApJ* **465**, L123.

Allard, F., Hauschildt, P. H., Miller, S. & Tennyson, J. 1994 The influence of H2O line blanketing on the spectra of cool dwarf stars *ApJ* **426**, L39.

Baraffe, I., Chabrier, G., Allard, F. & Hauschildt, P. H. 1995 New evolutionary tracks for very-low-mass stars *ApJ* **446**, L35.

Baraffe, I., Chabrier, G., Allard, F. & Hauschildt, P. H. 1997 Evolutionary models for metal-poor low-mass stars. Lower main sequence of globular clusters and halo field stars

A&A **327**, 1054.

Baraffe, I., Chabrier, G., Allard, F. & Hauschildt, P. H. 1998 Evolutionary models for solar metallicity low-mass stars: mass-magnitude relationships and color-magnitude diagrams *A&A* **337**, 403.

Brett, J. M. 1995 Opacity sampling model photospheres for M dwarfs. II. The grid *A&AS* **109**, 263.

Brett, J. M. 1995 Opacity sampling model photospheres for M dwarfs. I. Computations, sensitivities and comparisons. *A&A* **295**, 736.

Burrows, A., Hubbard, W. B., Saumon, D. & Lunine, J. I. 1993 An expanded set of brown dwarf and very low mass star models *ApJ* **406**, 158.

Borisow, A., Jørgensen, U. G. & Zheng, C. 1997 Model atmospheres of cool, low-metallicity stars: the importance of collision-induced absorption. *A&A* **324**, 185.

Chabrier, G. & Baraffe, I. 1995 CM Draconis and YY Geminorum: Agreement Between Theory and Observation *ApJ* **451**, L29.

Chabrier, G. & Baraffe, I. 1997 Structure and evolution of low-mass stars *A&A* **327**, 1039.

Fegley, B. Jr. & Lodders, K. 1996 Atmospheric Chemistry of the Brown Dwarf Gliese 229B: Thermochemical Equilibrium Predictions *ApJ* **472**, L37.

Habets, G. M. H. J. & Heintze J. R. W. 1981 Empirical bolometric corrections for the main-sequence *A&AS* **46**, 193.

Jones, H. R. A., Longmore, A. J., Jameson, R. F. & Mountain, C. M. 1994 An infrared spectral sequence for M dwarfs *MNRAS* **267**, 413.

Jones, H. R. A. & Tsuji, T. 1997 Spectral Evidence for Dust in Late-Type M Dwarf *ApJ* **480**, L39.

Jørgensen, U. G. 1994 Effects of TiO in stellar atmospheres *A&A* **284**, 179.

Kirkpatrick, J. D., Henry, T. J. & Simons 1995 The solar neighborhood. 2: The first list of dwarfs with spectral types of M7 and cooler *AJ* **109**, 797.

Kirkpatrick, J. D., Kelly, D. M., Rieke, G. H., Liebert, J., Allard, F. & Wehrse, R. 1993 M dwarf spectra from 0.6 to 1.5 micron - A spectral sequence, model atmosphere fitting, and the temperature scale *ApJ* **402**, 643.

Kirkpatrick, J. D. & McCarthy, D. W. Jr. 1994 Low mass companions to nearby stars: Spectral classification and its relation to the stellar/substellar break *AJ* **107**, 333.

Kurucz, R. L. 1992 Atomic and molecular data for opacity calculations *RMxAA* **23**, 45.

Langhoff, S. R. 1997 Theoretical Study of the Spectroscopy of TiO *ApJ* **481**, 1007.

Leggett, S. K. 1992 Infrared colors of low-mass stars *ApJS* **82**, 35.

Leggett, S. K., Allard, F., Berriman, G., Dahn, C. C. & Hauschildt, P. H. 1996 Infrared Spectra of Low-Mass Stars: Toward a Temperature Scale for Red Dwarfs *ApJS* **104**, 117.

Leggett, S. K., Allard, F. & Hauschildt, P. H. 1998 Infrared Colors at the Substellar Boundary *ApJ* **in press**.

Ludwig, C. B. 1971 *Applied Optics* **10(5)**, 1057.

Marley, M. S., Saumon, D., Guillot, T., Freedman, R., Hubbard, W. B., Burrows, A. & Lunine, J. I. 1996 Atmospheric, evolutionary, and spectral models of the brown dwarf Gliese 229B *Science* **272**, 1919.

Martìn, E. L., Rebolo, R. & Zapatero-Osorio, M. T. 1996 Spectroscopy of New Substellar Candidates in the Pleiades: Toward a Spectral Sequence for Young Brown Dwarfs *ApJ* **469**, 706.

Miller, S., Tennyson, J., Jones, H. R. A. & Longmore, A. J. 1994 in Molecular Opacities in the Stellar Environment, U.G. Jorgensen editor, *Lecture Notes in Physics*, Berlin/Heidelberg: Springer-Verlag, 296.

Monet, D. G., Dahn, C. C., Vrba, F. J., Harris, H. C., Pier, J. R., Luginbuhl, C. B. & Ables, H. D. 1992 U.S. Naval Observatory CCD parallaxes of faint stars. I - Program description and first results *AJ* **103**, 638.

PARTRIDGE, H. & SCHWENKE, D. W. 1997 *J. Chem. Phys.* **106**, 4618.

PAVLENKO, Y. V., REBOLO, R., MARTÌN, E. L. & GARCIA-LOPEZ, R. J. 1995 Formation of lithium lines in very cool dwarfs *A&A* **303**, 807.

PULONE, L., DEMARCHI, G., PARESCE, F. & ALLARD, F. 1998 The Lower Main Sequence of omega Centauri from Deep Hubble Space Telescope NICMOS Near-IR Observations *ApJ* **492**, L41.

REBOLO, R., MARTÌN, E. L. & MAGAZZÚ, A. 1992 Spectroscopy of a brown dwarf candidate in the Alpha Persei open cluster *ApJ* **389**, L83.

SAUMON, D., HUBBARD, W. B., BURROWS, A., GUILLOT, T., LUNINE, J. I. & CHABRIER, G. 1996 A Theory of Extrasolar Giant Planets *ApJ* **460**, 993.

SCHWEITZER, A., HAUSCHILDT, P. H., ALLARD, F. & BASRI, G. 1996 Analysis of Keck high-resolution spectra of VB 10 *MNRAS* **283**, 821.

SCHWENKE, D. 1996 The Opacity of TiO from a Coupled Electronic State Calculation Parametrized by ab initio and Experimental Data *J. Chem. Phys.*, **in press**.

SHARP, C. M. & HUEBNER, W. F. 1990 Molecular equilibrium with condensation *ApJS* **72**, 417.

TINNEY, C. G., MOULD, J. R. & REID, I. N. 1993 The faintest stars - Infrared photometry, spectra, and bolometric magnitude *AJ* **105**, 1045.

TSUJI, T., OHNAKA, K. & AOKI, W. 1995 in Bottom of the Main Sequence — And Beyond, C. G. Tinney, Springer-Verlag, 45.

TSUJI, T., OHNAKA, K. & AOKI, W. 1996 Dust formation in stellar photospheres: a case of very low mass stars and a possible resolution on the effective temperature scale of M dwarfs *A&A* **305**, L1.

TSUJI, T., OHNAKA, K., AOKI, W. & NAKAJIMA, T. 1996 Evolution of dusty photospheres through red to brown dwarfs: how dust forms in very low mass objects. *A&A* **308**, L29.

VITI, S., JONES, H. R. A., ALLARD, F., HAUSCHILDT, P. H., TENNYSON, J., MILLER, S. & LONGMORE, A. J. 1997 The effective temperature and metallicity of CM Draconis *MNRAS* **291**, 780.

ZAPATERO OSORIO, M. R., REBOLO, R. & MARTÍN, E. L. 1997 Brown dwarfs in the Pleiades cluster: a CCD-based R, I survey. *A&A* **317**, 164.

Dust in Very Cool Dwarfs

By TAKASHI TSUJI

Institute of Astronomy, The University of Tokyo, Mitaka, Tokyo, 181-8588 Japan

Recent observations and preliminary model atmospheres of very low mass objects (VLMOs) including brown dwarfs and late M dwarfs show that these objects are mostly dusty. We first try to understand the physical reasons why dust can so easily be formed and further be sustained in the photospheres of VLMOs. By considering the thermodynamics of the grain particle formation in the photospheric environment, we found that there is a rather wide regime where dust formation can be treated within the framework of the local thermodynamical equilibrium (LTE). Actually, we consider three cases A, B, and C, which correspond to $r_{gr} = 0$, $r_{gr} < r_{cr}$, and $r_{gr} > r_{cr}$, respectively, where r_{gr} is the grain radius and r_{cr} the critical radius below which dust is unstable and can be treated by LTE. We discuss a series of model atmospheres for the cases A, B, and C, and also the cases of lower gravities representing the contracting low mass stars in young clusters and associations. Models well above the ZAMS are again dusty for $T_{\rm eff}$ $< 2,800$K. Based on these models, we discuss observable properties such as spectra and colors of VLMOs, including the gravity effect. We re-discuss the two important touchstones of the model atmospheres of VLMOs, namely GD 165B and Gl 229B.

---

1. Introduction

The idea that dust may form is stellar photospheres has been discussed already in 1960's for the case of cool giant stars. For example, Hoyle and Wickramasinghe (1962) suggested that graphite may form in the photosphere of cool carbon stars and that this should be the source of the interstellar grains. This idea has been extended by many authors, but dust formation in stellar environment has mostly been discussed in connection with outflow from the circumstellar envelope of evolved stars since then. As a result, dust formation in the envelope of cool luminous stars seems to be well understood now (e.g. Hasegawa and Kozasa 1988; Sedlmayr 1994).

On the contrary, dust formation in stellar photosphere itself has been paid little attention so far as we are aware. Here, the problem is certainly different from the case of the circumstellar envelope: For example, a difficult problem may be how to sustain the dust grains formed in stationary photosphere, even if dust could be formed. This problem has been avoided nicely in the case of the circumstellar envelope where dust formed flows out by the radiation pressure on the dust formed. Another reason why dust in stellar photosphere has been overlooked may be a lack of direct observational evidence for dust in stellar photosphere while infrared astronomy in these 30 years revealed variety of direct observational evidences for circumstellar dust.

Recent progress in observation of cool dwarfs finally revealed some serious discrepancies with the prediction of model atmospheres (e.g. Allard & Hauschild 1995; Brett 1995). We noticed that one of the reasons of this difficulty may be due to the neglect of dust in the photosphere, since the thermodynamical condition of condensation is well met in the cool and dense photospheres of cool dwarfs (Tsuji *et al.* 1996a). In our first attempt, we started from the most simple assumption that the local thermodynamical equilibrium (LTE) can be applied to the dust formation in the photospheres of cool dwarfs. This approach showed reasonable success for late M dwarfs and brown dwarf candidate GD 165B, but not for the case of the genuine brown dwarf Gl 229B. This may show that dust could no longer be sustained in the photosphere of this cool brown dwarf (Tsuji *et al.* 1996b). But this is what is normally expected, and Gl 229B is rather normal in this regard. On the other hand, further evidences for dust are found in a large number of VLMOs (e.g. Jones & Tsuji 1997; Tsuji *et al.* 1998), and it appeared that the dust is well sustained in the photosphere of most VLMOs during a long time. But, how dust could be sustained so easily in so many objects? The answer is found to be rather simple: small dust grains are unstable and this implies that their formation and destruction will be repeating (as for details, see Sect.2). This means that detailed balance between gaseous molecules and small dust particles can be realized and LTE can be applied to dust formation so long as the size of grains remain small enough. Thus, our initial assumption that LTE can be applied to dust formation found justification in observations.

Now, we believe that we understand how dust forms and how it is sustained in the stellar photosphere, at least semi-empirically. With this background, we hope to extend our model atmospheres to interpretation and analysis of observations on VLMOs including brown dwarfs. This is timely in view of the prospect of finding more brown dwarfs, as indicated by the discovery of young brown dwarfs in open clusters such as Pleiades (Rebolo *et al.* 1998) as well as in the field by the DENIS (Delfosse *et al.* 1997), by the 2MASS (Kirkpatrick *et al.* 1997) and by others (e.g. Ruiz *et al.* 1997).

2. Thermodynamics of grain formation in stellar photosphere

The thermodynamical condition of condensation is well met in the photosphere of cool dwarfs with $T_{\rm eff}$ less than about 3000 K. This is a necessary but not the sufficient condition for dust to form. Of course, thermodynamics applies strictly to any case including dust formation in stellar photosphere. However, unlike the condensation on bulk solid material as in laboratory environment for which the usual thermodynamical data are worked out, dust formed in stellar photosphere is usually a droplet which is subject to decay by the surface tension force. Thus dust may form as soon as the thermodynamical condition of condensation is met, but it will dissolve as soon as it is formed. The net effect is as if dust is not formed at all, and this is generally referred to as the super-saturation.

It is not known how far the super-saturation phase lasts, but small grains or clusters can eventually be formed under the high super-saturation ratio, which is the case of cool and dense photospheres of VLMOs. However, the cluster cannot be stable at first, since the surface tension which is proportional to the surface area ($\propto r^2$) dominates over the binding force which is proportional to the total number of monomers ($\propto r^3$) in a droplet. This applies as long as the radius of the cluster is less than a critical radius $r_{\rm cr}$, at which the change of the Gibbs free energy ΔG for the cluster formation from the monomers attains the maximum. Thus, there is a critical size below which the dust grains cannot grow. However, this case where dust is unstable is very interesting, since anyhow dust now exists in stellar photosphere. Now, it is to be remembered that any equilibrium including the chemical equilibrium of molecules is dynamical in nature and that the equilibrium is realized by the balance of formation and destruction of molecules. Thus the unstable regime of dust formation is where the the thermodynamical equilibrium of dust formation is realized. This gives a natural answer how to sustain the dust formed in stellar photosphere: dust is sustained in the photosphere because its formation and destruction are repeating so long as the grain size remains below the critical size.

Finally, dust can grow larger and the grain radius may exceed the critical radius $r_{\rm cr}$. Then, dust grains will be larger and larger, and this is the case in which dust is regarded as formed in usual sense. However, such large grains can no longer be sustained in the photosphere, and dust will segregate from the gaseous mixture. A detailed treatment of dust formation for this case is more difficult because of the onset of the non-equilibrium process, which may induce meteorological phenomena (fog, cloud, rain etc).

3. Dusty model atmospheres of very low mass objects

3.1. *Input data*

3.1.1. *Thermochemistry*

The most abundant elements H, C, N, O and S form volatile molecules such as H_2, CO, OH, H_2O, NH_3, CH_4, H_2S, and PH_3 in the cool and dense photospheres of VLMOs. The next abundant elements Fe, Si, Mg, Al, Ca etc are non-volatile and refractory compounds such as Al_2O_3 (corundum), Fe (iron), $MgSiO_3$ (enstatite), $Ca_2MgSi_2O_7$ (melitite), $CaTiO_3$ (perovskite) are formed . The other abundant elements such as He, Ne and Ar are chemically inactive and play little role in cool photosphere.

3.1.2. *Opacities*

Under such circumstances, the major opacity sources are as follow: 1) molecular bands due to volatile molecules noted above and also some non-volatile molecules such as TiO, VO, FeH, CaH etc. We apply the band model method in treating molecular opacities during the iterations for modeling. 2) The most abundant molecules – H_2 will show dipole transition induced by collision in the dense photospheres of VLMOs. Cross-sections of collision-induced absorption (CIA) are based on the Borysow's codes (Borysow *et al.* 1997). 3) Dust opacities are evaluated by the Rayleigh approximation of the Mie theory, since the grain size may be sub-micron under LTE (Sect.2). We represented the effect of dust opacities by those of corundum, iron, and enstatite.

3.1.3. *Spectroscopic database*

In evaluating the spectra after model has converged, we evaluate detailed flux by the spectral synthesis at a sampling interval of about the Doppler width. We do not use OS (opacity sampling) nor OPDF (opacity probability distribution function). Our high temperature molecular line database for this purpose now includes H_2O based on HITEMP (Rothman 1997) together with ^{12}CO,^{13}CO, OH, CN, and SiO. The HITEMP database includes $\approx 10^6$ lines of H_2O. Other polyatomics such as methane are included as a pseudo-continuum.

3.2. *Physical structure of photospheres of very low mass objects*

We have computed non-grey model atmospheres in radiative-convective equilibria by iterative method, and we first discuss the case of log g = 5.0 covering $T_{\rm eff}$ between 1,000 and 3,800 K (other parameters are $Z = Z_\odot$ and $V_{\rm micro}$ = 1km/s throughout). We consider three cases which follow the evolution of dust from $r_{\rm gr} = 0$ (case A) to $r_{\rm gr} < r_{\rm cr}$ (case B) and finally to $r_{\rm gr} > r_{\rm cr}$ (case C), where $r_{\rm gr}$ is the grain radius and $r_{\rm cr}$ the critical radius above which dust can be stabilized and grow larger (Sect. 2). In Fig. 1, the condensation lines for corundum (Al_2O_3), iron, and enstatite ($MgSiO_3$) are shown by the dotted, dashed, and long dashed lines, respectively.

3.2.1. *Case A : dust - free case*

We first consider the case in which dust never appears even if the thermodynamical condition of condensation is satisfied, and we refer this as case A. The resulting models of $T_{\rm eff}$ = 3600, 3000, 2400, 1800, and 1200 K for this case are shown by the dotted lines in Fig. 1a. This case is valid for $T_{\rm eff}$ above 3000 K (e.g. the case of $T_{\rm eff}$ = 3600 K in Fig. 1a),

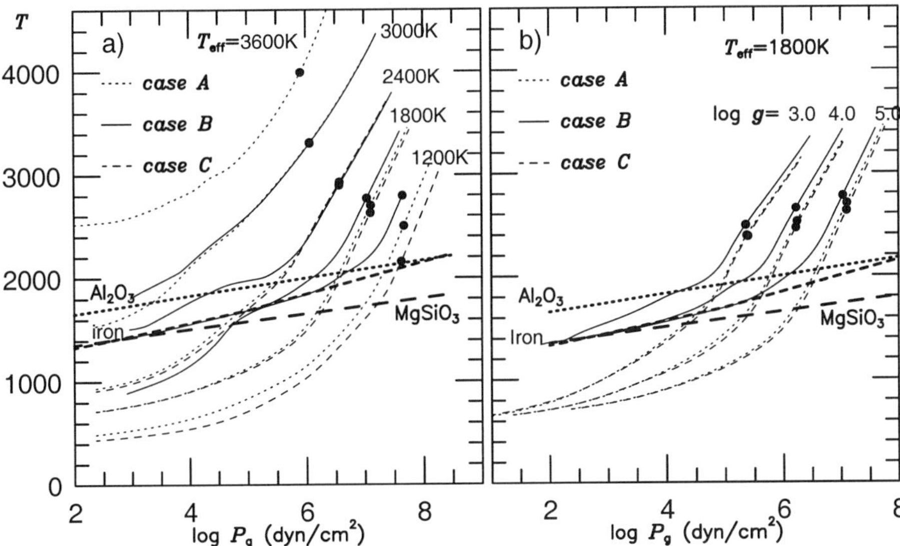

FIGURE 1. Non-grey model atmospheres of VLMOs for the cases A, B, and C are shown by dotted, solid, and dashed lines, respectively. Filled circle indicates the point where half of the total flux is carried by convection. **a)** A selected grid of models with T_{eff} = 3600, 3000, 2400, 1800, and 1200 K for log g = 5.0. **b)** The models of log g = 3.0, 4.0, and 5.0 for T_{eff} = 1800 K.

since the models never cross the dust condensation lines. Because of the highly non-grey molecular opacities, the surface temperatures are rather low in general, and the model of T_{eff} =3000 K already crosses the condensation lines. Thus the necessary condition for condensation is met, and our case A may no longer be applied to the models with T_{eff} < 3000K. However, it is possible that the case A can be applied even for $T_{\text{eff}} < 3000$K in case of super-saturation, although it is not known how far the super-saturation will last.

3.2.2. *Case B : dusty case*

The case A models with T_{eff} below 3000 K penetrate the condensation lines, and this means that the super-saturation ratio will soon be very large. Then, small dust grains (or better be called as micro clusters) will anyhow be formed. Such small grains may remain in the unstable regime in which formation and destruction of dust grains are in detailed balance as mentioned in Sect.2. By this very reason, dust formed and gaseous molecules are well mixed. Since dust opacity is by far the more effective than the molecular opacity, the thermal structures of the models of this case B shown by the solid lines in Fig. 1a are now governed by the dust opacity. It is immediately clear that the dust modifies the photospheric structure significantly. For T_{eff} = 2400 K, only corundum is formed and the surface temperature is increased by as much as 500 K by the dust opacity. For T_{eff} = 1800 K, the major opacity source is iron, which acts as a thermostat, and thermal structure converges to the iron condensation line. Finally, a large amount of enstatite is formed in cooler models and surface temperature shows a large decrease because of the infrared bands of silicate. However, temperatures of the photosphere of T_{eff} = 1200 K model of case B, for example, are still higher by as much as 1000 K as compared with the case A. It is interesting to notice that corundum works as heater, iron as thermostat, and silicate as coolant in determining the thermal structure of the dusty models..

Generally, convection zone is well below the dust forming region and convection show little interaction with dust formation. For T_{eff} near 1000 K, however, convection zone is close to the dust zone. But the large opacity due to dust does not necessarily favor the convection, since the temperature gradient is tempered by the heating of the dust. In fact, convection is pushed down in the cooler dusty models of case B (Fig. 1a).

3.2.3. *Case C : dust–segregated case*

Finally, grain size may exceed the critical size by some reason. Where and why the transition from the dust-gas detailed balance regime (case B) to dust-gas segregation phase (case C) occur are unknown. A possibility is that some mechanism to destroy dust is working until certain phase. Anyhow, the rather larger grains can no longer be sustained in the photosphere and the segregated large grains will play little role as opacity source. Instead, the opacity is again dominated by the volatile molecules and essentially the same as in case A, except that some non-volatile molecules (TiO, VO, FeH etc.) disappeared together with the dust. The resulting thermal structures of this case C are shown by the dashed lines in Fig. 1a and they differ little from those of the case A, except for very cool models with $T_{\rm eff}$ near 1000 K. However, surface temperatures are still lower in the case C than in the case A. This is because the heating by the absorption of optical photon by the non-volatile molecules no longer works.

3.3. *Effects of gravity*

In view of applications to contracting low mass stars in association and young clusters, we examined the cases of lower gravities. Since the surface temperatures at the same $T_{\rm eff}$ differ little for different gravities, the condition of condensation is satisfied as well in the objects of lower gravities. Some examples of the $p--T$ structure for log $g = 5.0$, 4.0 and 3.0 in the case of $T_{\rm eff} = 1800$ K are shown in Fig. 1b. Although the gas pressure is nearly proportional to the surface gravity g in the major part of the photosphere, the surface temperatures are nearly the same for a wide range of log g. This is because the thermal structure is largely determined by the iron which works as thermostat. The amount of dust in the photosphere is smaller in the models of the lower gravities.

4. Observables

Given model atmospheres, it is in principle possible to predict major observables such as spectra and colors. For this purpose, major problem is to have a detailed line-list, but this is still far from complete because of the lack of the necessary spectroscopic and intensity data. For this reason, our present predictions should be regarded as preliminary.

4.1. *Spectra*

4.1.1. *Optical spectra*

In optical region, we use the band model opacities and the results are mainly for showing the basic behaviors rather than for detailed quantitative analyses. We show some examples of the predicted spectra based on our models of $T_{\rm eff} = 2200$ K for the cases A, B, and C in Fig. 2. In the case A, molecular bands mostly due to TiO are very strong as was expected. This case, however, is shown only as a reference, since it is not expected that the super-saturation still lasts to $T_{\rm eff}$ as low as 2200 K. In the case B where dust forms in LTE, the molecular bands are largely masked by the strong extinction by the small dust grains which are uniformly mixed with the gas. Effect of dust may also depends on the details of dust properties (e.g. size, shape, composition etc.), and these effects should be examined further before our results can be applied to detailed analyses of observed spectra. In the case C in which dust grains grow larger and segregate from the gas, molecular opacities again dominate. But non-volatile molecules such as TiO are

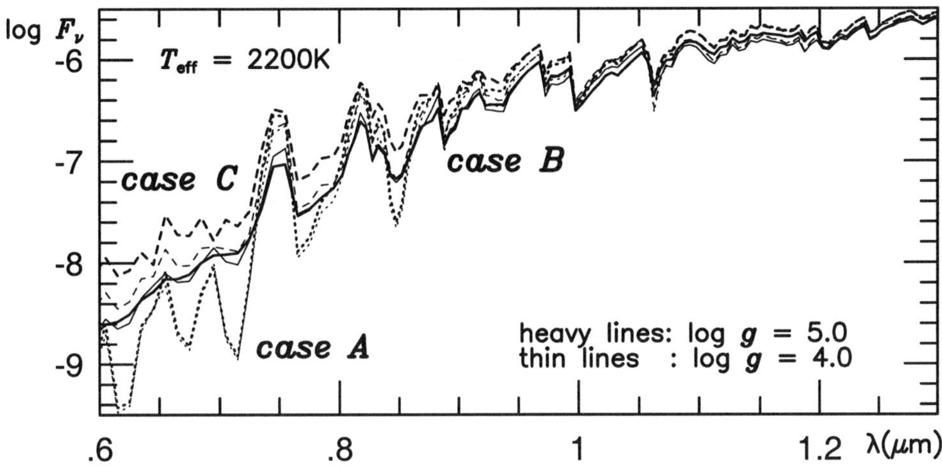

FIGURE 2. Predicted optical spectra based on our model atmospheres of cases A, B, and C are shown by the dotted, solid, and dashed lines, respectively. The heavy and thin lines are for log $g = 5.0$ and 4.0, respectively ($T_{\rm eff} = 2200$ K throughout). The unit of F_ν is erg/cm^2/sec/Hz.

depleted in dust grains, and TiO and VO bands are much weaker in the case C than in the case A. The difference between the cases A and C almost disappears in the spectral region long-ward of 1 μm where major opacity source changes from TiO to H_2O.

In Fig. 2, two cases of log $g = 5.0$ and 4.0 are shown to examine the effects of gravity, but the spectra show little change by gravity in the optical region at $T_{\rm eff} = 2200$ K. The spectra show somewhat larger fluxes at higher gravity, and this fact can be understood by the higher infrared opacity at the higher gravity (see Fig. 4).

4.1.2. *Infrared spectra*

In the infrared region, we computed the spectra based on a detailed line-list including H_2O, CO, OH, CN, and SiO, but polyatomics such as CH_4 are still treated as pseudo-continuous based on the band model opacity. Thus, we hope that our spectra offer some improvements over the previous results (e.g. Tsuji *et al.* 1998), but still preliminary until more complete line-list could be used.

Some examples of the near infrared spectra based on our models of cases B and C are shown in Fig. 3 for $T_{\rm eff} = 2400$, 2000, and 1600 K (case A may show little difference from the case C in the infrared at the $T_{\rm eff}$'s shown). The transition from stellar to substellar regimes may occur in these effective temperatures. In the model of $T_{\rm eff} = 2400$ K, which may represent late M dwarfs, the spectra of cases B and C are both dominated by the absorption bands of H_2O and CO, but these bands are already weaker in case B than in case C because of the higher surface temperature of the case B model (Fig. 1a) due to the heating by corundum. These differences are more pronounced in the model of $T_{\rm eff} = 2000$ K where iron now condensed. Since abundance of iron is larger by an order of magnitude than that of corundum and iron is more absorptive than corundum, the extinction in the shorter wavelength region is quite substantial. Finally, in the model of $T_{\rm eff} = 1600$ K, molecular bands are very weak in the case B because of the extinction by the thick dust layers. If transition to the case C from the case B occurs, however, CH_4 bands at 3.4 and 2.2 μm are now quite strong in addition to H_2O and CO bands.

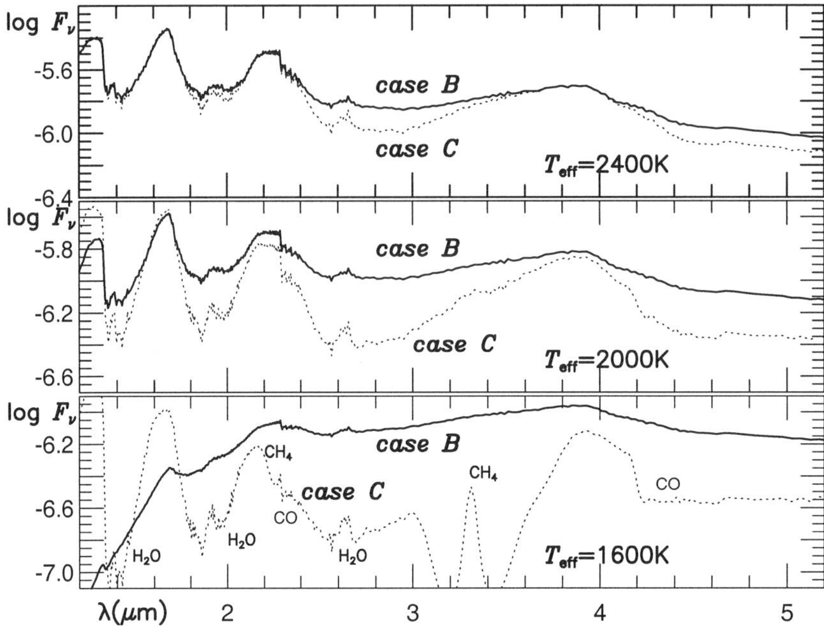

FIGURE 3. Predicted spectra based on model atmospheres of $T_{\rm eff}$ = 2400, 2000, and 1600 K (log g = 5.0 throughout). Cases B and C are shown by the solid and dashed lines, respectively.

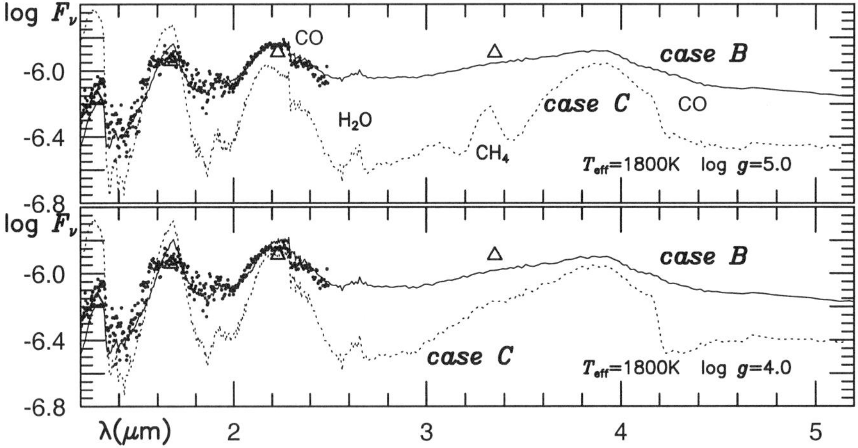

FIGURE 4. Predicted infrared spectra based on model atmospheres of log g = 5.0 (top) and 4.0 (bottom) ($T_{\rm eff}$ = 1800 K throughout). For comparison, observed spectra of GD 165B is shown by the filled circles (Jones *et al.* 1994) and by the open triangles (Tinney *et al.* 1993).

4.1.3. *GD 165B – the proto-type of the case B*

The brown dwarf candidate discovered by Becklin & Zuckerman (1988) may be remembered as a proto-type of dusty object represented by our case B. For this reason, we re-examine this object by the revised computation of the spectra using the detailed line-list and also examined the effect of gravity. The resulting spectra of the cases B and C are shown in Fig. 4 for models of $T_{\rm eff}$ = 1800 K with log g = 5.0 (top) and 4.0 (bottom). It appears that the spectra of the case B show little difference by the gravity, and this may because the thermal structures of the case B are nearly independent of the gravity by the reasons outlined in Sect.3.3. On the contrary, the spectra of the case C show considerable difference for different gravities. For example, methane band at 3.3 μm is clearly visible already at $T_{\rm eff}$ as high as 1800 K in the higher gravity model (log g = 5.0) but disappears completely in the lower gravity one (log g = 4.0). Also, the K band flux shows larger depression in the higher gravity model than in the lower gravity one, and this is due to H_2 collision-induced absorption (CIA). Thus, the effect of gravity is rather important on the infrared spectra of the case C.

We compared the observed spectrum of GD 165B by Jones *et al.* (1994) with the predicted ones in Fig. 4, and we confirm that the case B provides a reasonable fit while the case C cannot. Note that our spectrum is now based on the detailed line-list and the predicted strong H_2O bands based on our case C may no longer be an artifact of the poor H_2O data. Also, predicted H_2O bands are weaker in the lower gravity model of the case C, but still too strong to explain the observed weakness of H_2O bands. Thus, we may conclude that only the dusty models of the case B explains the spectrum of GD 165B.

4.1.4. *Gliese 229B – the proto-type of the case C*

Gl 229B discovered by Nakajima *et al.* (1995) is still the only sample of the cool brown dwarf. We have previously analyzed the SED of Gl 229B based on the narrow and broad band photometries, and found that a dust-free model of $T_{\rm eff}$ = 1000 K agreed rather well with observations. This analysis suggested that dust can no longer be sustained in the photosphere and may have segregated already (Tsuji *et al.* 1996b). However, it was based on the simple dust-free model (or case A in our present classification) which may be not physically appropriate to such a cool object. For this reason, we re-examine this object by our revised models of the case C.

Recently, a new observation of the spectrum of Gl 229B has been done by Oppenheimer *et al.* (1998), who kindly allowed to use their beautiful spectrum for comparison with our model prediction. The observed spectrum is shown by the filled circles in Fig. 5 and it is impressive to see that the methane bands are so strong. We compared this spectrum with several models of our cases A, B, and C of the different $T_{\rm eff}$, and we concluded that the model of the case C with $T_{\rm eff}$ =1200 K (log g = 5.0) provides the best fit as shown in Fig. 5 by the dotted line while the predicted spectrum for the case B is far from observation. In this very low temperature, the surface temperatures in our case C models are significantly lower than in the case A by the reason outlined in Sect.3.2.3 (Fig. 1a). For example, the temperatures of the line-forming region of the case C model of $T_{\rm eff}$ = 1200 K are nearly the same with those of the case A model of $T_{\rm eff}$ = 1000 K. For this reason, spectrum based on the case C model with $T_{\rm eff}$ = 1200 K provides the best fit to the observed one of Gl 229B, for which the case A model of $T_{\rm eff}$ = 1000 K did as well (Tsuji *et al.* 1996b). We now propose that $T_{\rm eff}$ of Gl 229B may be close to 1200 K rather than to 1000 K.

Except for the large optical flux deficiency noted by Golimowski *et al.* (1998), the predicted spectrum based on our model of the case C ($T_{\rm eff}$ = 1200 K) fits rather well to the observed one, and this fact implies that the dust actually cannot be sustained in the photosphere of Gl 229B. But, what is the fate of the segregated dust? It may be floating as clouds or may be precipitated below the visible atmosphere. It may also be possible that the dust is blown off by the radiation pressure above the photosphere. If the segregated rather large grains may form cold halo far above the photosphere, it may

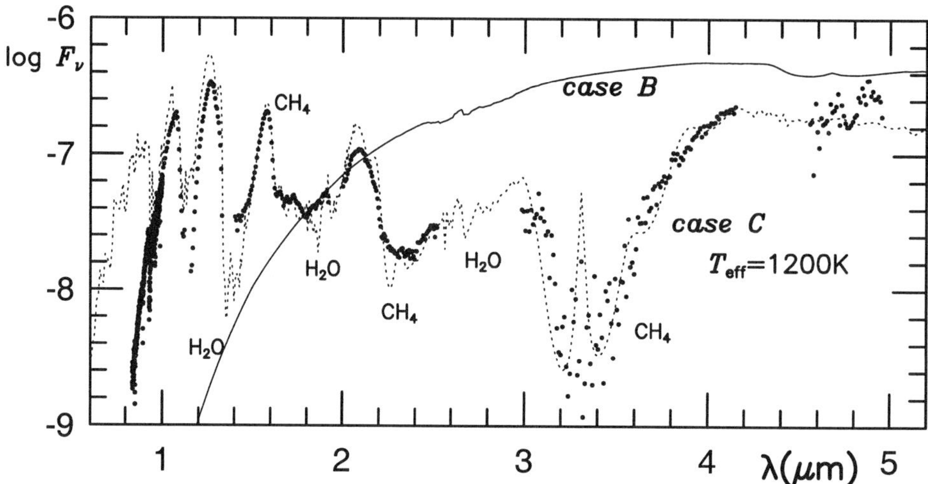

FIGURE 5. Observed spectrum of Gl 229B (Oppenheimer *et al.* 1998) shown by the filled circles is compared with the predicted ones based on our models of cases B and C shown by the solid and dotted lines, respectively ($T_{\rm eff}$ = 1200 K and log g = 5.0).

produce the extinction observed in the optical region. However, if scattering dominates, a more or less spherical halo cannot produce such a large extinction.

4.2. *Infrared colors*

Some characteristics of the spectral energy distribution (SED) can be well represented by color indexes. As examples, we calculated $J - H$ and $H - K$ from our predicted SEDs. First, the results for our case A are shown in Fig. 6a. The predicted colors for the models of $T_{\rm eff}$ from 3800 K to 2600 K appear to be confined to the rather narrow range of $H - K$ between 0.2 and 0.4 while $J - H$ remains to be nearly constant at about 0.55. These result show reasonable agreement with the mean observed colors of M2V–M6V (Legett 1992). The $H - K$, however, turns to blue at about $T_{\rm eff}$ = 2600 K, and $J - H$ at about $T_{\rm eff}$ = 1400 K in the case of log g = 5.0. These predictions never agree with observed colors of the late M dwarfs and brown dwarf candidates including GD 165B, but qualitatively consistent with the observed colors of Gl 229B.

Although our case A may be not realistic for VLMOs, we show the predicted colors for log g = 4.0 and 3.0 in Fig. 6a to see the effect of gravity on colors. In lower gravities, unlike the case of log g = 5.0, $J - H$ and/or $H - K$ show reddening even at $T_{\rm eff} < 2600$K at first, but finally turn to blue in models with $T_{\rm eff} \leq 1400$K. Yet the locus of the $T_{\rm eff}$ = 1000 K model of log g = 3.0 remains in the red regime while that of the $T_{\rm eff}$ = 1000 K model of log g = 4.0 appears in the blue regime. The colors of very cool stars should normally be red and they return to their normal colors at lower gravities in our case A. Thus, what is abnormal is the blue colors of the low temperature and high gravity models, which reflect the very peculiar SED of such models.

In fact, the SED of our case A model with $T_{\rm eff}$ = 1000 K and log g = 5.0 shows a large excess flux at around the J band in expense of a severe absorption in the K band due to H_2 CIA, H_2O and CH_4 (Tsuji *et al.* 1996b). This fact explains the blue $J - H$ and $H - K$ of very cool and dense atmosphere in the case A (this remains nearly the same in our case C shown in Fig. 6b). However, if gravity is lower, the strong absorber in the K bands, especially H_2 CIA which is highly sensitive to the density, will no longer be so effective. In other words, it is the H_2 CIA that makes the colors of high density

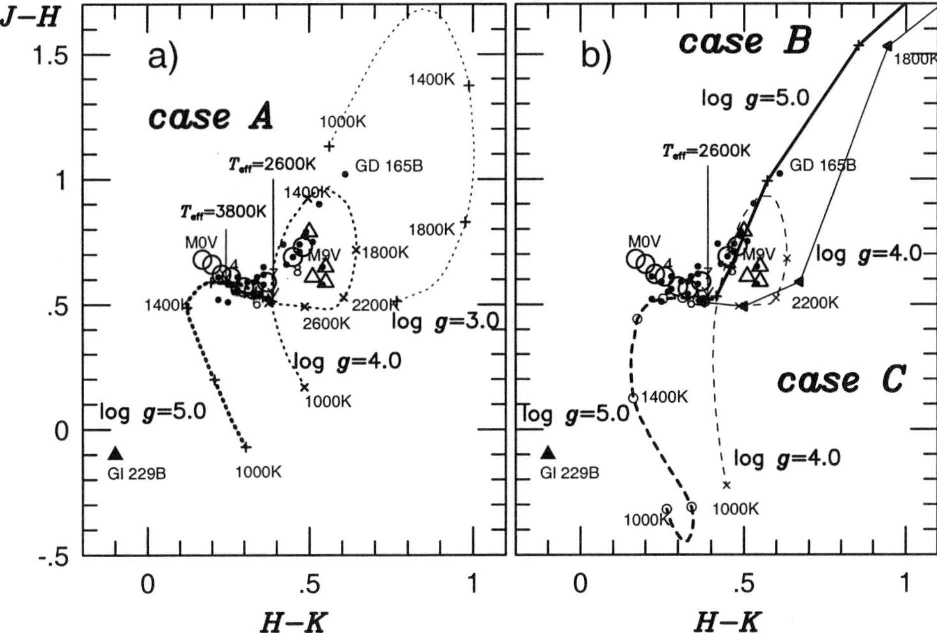

FIGURE 6. a) Predicted colors based on a grid of model atmospheres covering T_{eff} between 1000 and 3800 K (log g =5.0, 4.0, and partly 3.0). Observed colors are shown by large open circles (mean values by Legett 1992), small open circles (Tinney *et al.* 1993), open triangle (Kirkpatrick *et al.* 1997), and filled triangle (Nakajima *et al.* 1995). **a)** Case A shown by the dotted lines. **b)** Cases B and C shown by the solid and dashed lines, respectively.

stars peculiar and the colors return to the normal as soon as H_2 CIA no longer plays the major role. Now, by this return to more or less normal colors, it appears to be not impossible to explain the red colors of VLMOs including GD 165B by our case A if the gravity is somewhat lower than those normally assumed for the main sequence stars.

In Fig. 6b, we show the predicted loci for our models of the case B with $T_{\text{eff}} < 2600$K. The observed colors of VLMOs from M8-9 dwarfs to the reddest GD 165B can be well explained by the predicted ones for the case B. The effect of gravity is rather small since the infrared spectra of our case B are rather insensitive to the gravity (Fig. 4). We also show the predicted loci for our case C in Fig. 6b. The results for log g = 5.0 again show the peculiar blue colors as in the case A, but these peculiar colors may not be an artifact of modeling, since they explain the basic characteristics of the observed colors of Gl 229B rather well. The predicted loci of the case C for log g = 4.0 also show the peculiar loop as in the case A, and the reason for this is the same as for the case A.

Now, evidence for dust formation by case B in the photosphere can be observed in a larger sample of VLMOs on $(J-H, H-K)$ diagram, but only if the objects are near the main sequence (log $g \approx 5.0$). Also, similar result can be shown on $(I-J, J-K)$diagram (Tsuji *et al.* 1998), but this also applies to objects of high gravity. At lower gravities (log$g$ $\approx$ 4.0), however, reddening by the reduced IR opacity dominated by H_2 CIA produces the similar effects on the IR colors as the reddening by the dust opacity. For this reason, evidence for dust is more difficult to see on colors for stars above the ZAMS.

5. Concluding remarks

Thermodynamical condition (necessary condition) for condensation is well met for $T_{\rm eff}$ $< 3,000$ K, not only near the main sequence (log g = 5.0) but also in objects above the ZAMS. The problem is if dust actually forms in the photospheric environment and, if so, how it forms and how it can be sustained in the photosphere. Usually, dust may be regarded as formed when it is large enough to be stabilized in the regime where grain radius r_{gr} is larger than the critical radius r_{cr}, since otherwise the dust is still unstable and resolve again. However, if dust forms in this sense, it means that dust can no longer be sustained in the stationary photosphere and that dust may no longer play the major role in determining the atmospheric structure. On the other hand, if dust is still not stabilized and remains in the regime where $r_{gr} < r_{cr}$, small dust grains can be sustained in the photosphere, since formation and destruction of such small dust grains will be repeating as long as the thermodynamical condition of the condensation is fulfilled.

Thus, somewhat paradoxically, sufficient condition for a survival of dust in stellar photosphere is that it is destroyed before it will be too large. In this case, small dust grains and gaseous molecules are in detailed balance, and dust in stellar photospheres can be treated as if it is a giant molecule. This finding that dust can be treated rather easily by our case B based on LTE in the photospheric environment will have a wide application in many problems. For example, we have extended this idea to the dust formation in carbon-rich photospheres of dwarfs as well as of AGB stars (Tsuji 1996).

So far, however, evidences for dust in VLMOs are all indirect in that dust itself has never been observed. However, dust formation in the photospheres of VLMOs gives rather prominent effects on observables such as spectra and colors. Especially, the spectra as well as colors of the brown dwarf candidate GD 165B and the cool brown dwarf Gl 229B show marked contrast (compare Figs. 4 and 5), which can be understood only by considering the different stages of evolution of the dusty photospheres from our case B to C. In fact, our models are largely motivated by the observations of these two objects, GD 165B and Gl 229B, and these two actually served as the laboratories of dust formation in nature.

The photometric properties of VLMOs are reasonably well understood by our models. Some colors depend critically on surface gravity (Sect.4.2), and color alone could not provide a unique interpretation for the objects in the contracting phase unless the gravity can be estimated by other ways. We can probably assume that most of VLMOs included in Fig. 6 are near the main sequence (log $g \approx 5.0$), and then only our dusty models of the case B can explain the observed colors of VLMOs in Fig. 6. We conclude that dust forms in the photospheres of M dwarfs later than M7V and of most brown dwarfs. This fact that dust exists in the photospheres of so many objects implies that the dust survives for a long time. This is possible only in our case B where dust formation and destruction are repeating forever. An interesting implication of this result is that the validity of the assumption of LTE, which is sometimes controversial, is easily demonstrated for the case of dust formation in stellar photosphere by observational evidence itself for dust.

The progress in observations of such faint objects as VLMOs is now quite rapid (Sect.1), and more sophistication in model atmospheres should be needed for detailed interpretation and analyses of the new observations on VLMOs including brown dwarfs. Even within the classical approach, we should examine more details of dust properties such as size, shape, optical properties (e.g. dirty vs. clean silicates), core-mantle structure etc. Also, more data on molecular lines (e.g. CH_4, NH_3, VO, FeH etc) should be indispensable to complete our line-list. Further, more radical analysis should be needed to understand the transitions from the case A to B, and further to C based on the basic physics. For this purpose, one possibility may be to extend the method of meteorology and/or of planetary science to the substellar objects.

I thank Ben Oppenheimer for allowing the use of the invaluable spectrum of Gl 229B for comparison with our models. I thank Wako Aoki, Hugh Jones, and Keiichi Ohnaka for continued collaborations on various topics covered by this review. I am indebted to Prof. Rafael Rebolo for his kind support to my participation in the conference.

REFERENCES

ALLARD, F. & HAUSCHILD, P. H. 1995 Model atmospheres for M (sub)dwarf stars I. The basic model grid. *Ap. J.* **445**, 433–450.

BECKLIN, E. E. & ZUCKERMAN, B. 1988 A low-temperature companion to a white dwarf star. *Nature* **336**, 656–658.

BORYSOW, A., JØRGENSEN, U. G. & ZHENG, C. 1997 Model atmospheres of cool, low-metallicity stars: the importance of collision-induced absorption. *A. A.* **324**, 185–195.

BRETT, J. M. 1995 Opacity sampling model photospheres for M dwarfs I. computations, sensitivities and comparisons. *A. A.* **295**, 736–754.

DELFOSSE, X. *et al.* 1997 Field brown dwarfs found by DENIS. *A. A.* **327**, L25–L28.

GOLIMOWSKI, D. A., BURROWS, C. J., KULKARNI, S. R., OPPENHEIMER, B. & BRUKARDT, R. A. 1997 WFPC2 observations of brown dwarf Gliese 229B: Optical colors and orbital motion. *A. J.* **115** , 2579–2586.

HASEGAWA, H. & KOZASA, T. 1988 Condensation of dust particles. *Prog. Theor. Phys. Suppl.* **No. 96**, 107–120.

HOYLE, F. & WICKRAMASINGHE, N. C. 1962 On graphite particles as interstellar grains. *M. N.* **124**, 417–433.

JONES, H. R. A., LONGMORE, A. J., JAMESON, R. F. & MOUNTAIN, C. M. 1994 An infrared spectral sequence for M dwarfs. *M. N.* **267**, 413–423.

JONES, H. R. A. & TSUJI, T. 1997 Spectral evidence for dust in late-type M dwarfs. *Ap. J. Lett.* **480**, L39–L41.

KIRKPATRICK, J. D., BEICHMAN, C. A., & SKRUTSKIE, M. F. 1997 The coolest M dwarfs and other 2MASS discoveries. *Ap. J.* **476**, 311–318.

LEGGETT, S. K. 1992 Infrared colors of low-mass stars. *Ap. J. Suppl.* **82**, 351–394.

NAKAJIMA, T., OPPENHEIMER, B. R., KULKARNI, S. R., GOLIMOWSKI, D. A., MATTHEWS, K. & DURRANCE, S. T. 1995 Discovery of a cool brown dwarf. *Nature* **378**, 463–465.

OPPENHEIMER, B. R., KULKARNI, S. R., MATTHEWS, K. & VAN KERKWIJK, M. H. 1998 The spectrum of the brown dwarf Gliese 229B. *Ap. J.* **502**, 932–943.

REBOLO, R., MARTIN, E. L., BARSI, G., MARCY, G. W.& ZAPATERO-OSORIO, M. R. 1996 Brown dwarfs in the Pleiades cluster confirmed by the Lithium test. *Ap. J. Lett.* **469**, L53–L56.

Rothman, L. *et al.* 1997 The high-temperature molecular spectroscopy database (HITEMP). *J. Quant. Spectros. Rad. Trans.*, in press.

RUIZ, M. T., LEGGETT, S. K.& ALLARD, F. 1997 KELU-1: A free-floating brown dwarf in the solar neighborhood. *Ap. J. Lett.* **491**, L107–L110.

Sedlmayr, E. 1994 From molecules to grains. In *Molecules in Stellar Environment* (ed. U. G. Jørgensen). pp. 163–185, Springer.

TINNEY, C. G., MOULD, J. R., & REID, I. N. 1993 The faintest stars: Infrared photometry, spectra, and bolometric magnitudes. *A. J.* **105**, 1045–1059.

TSUJI, T. 1996 Dust formation in stellar photospheres: the case of carbon stars from dwarf to AGB. IAU Symp. 177 on *Carbon Star Phenomenon* (Antalya, Turkey; May 1996), The Univ. Tokyo Astron. Preprints No. 96–11.

TSUJI, T., OHNAKA, K., & AOKI, W. 1996a Dust formation in stellar phtospheres: a case of very low mass stars and a possible resolution on the effective temperature scale of M dwarfs. *A. A.* **305**, L1–L4.

TSUJI, T., OHNAKA, K., AOKI, W. & NAKAJIMA, T. 1996b Evolution of dusty photospheres through red to brown dwarfs: how dust forms in very low mass objects. *A. A.* **308**, L29–L32.

TSUJI, T., OHNAKA, K., AOKI, W. & JONES, H. R. A. 1998 Spectra of M dwarfs: dimmed by dust. In *IAU Joint Discussion on Low Luminosity Stars* (ed. J. Binney). Highlight of Astronomy, in press.

On the Interpretation of the Optical Spectra of Very Cool Dwarfs

By Y. V. PAVLENKO[1], M. R. ZAPATERO OSORIO[2] AND R. REBOLO[2,3]

[1]Main Astronomical Observatory of the Ukrainian Academy of Sciences, Golosiiv Woods, Kyiv-22, 252650, Ukraine

[2]Instituto de Astrofísica de Canarias, E–38200 La Laguna, Tenerife, Spain

[3]Consejo Superior de Investigaciones Científicas, CSIC, Spain

We present synthetic spectra in the range 640–930 nm for a sample of very cool dwarfs. The computations were performed using the recent "dusty" model atmospheres by Tsuji (this volume) and by Allard (1999), and a synthesis code (Pavlenko et al. 1995) working under LTE conditions. The absorptions of oxides (TiO and VO) and hydrides (CrH, FeH and CaH) are an important source of opacity for the coolest M-dwarfs and early L-dwarfs. We show that the densities of TiO and VO molecules and the shape and strength of their bands are sensitive to the formation of atmospheric dust. The absence of TiO and VO molecular bands in mid and late L-type dwarfs can be explained by a process of depletion of Ti and V atom into grains. The alkali elements, Li, Na and K present strong lines in the red and far-red spectrum of these objects, with Na and K contributing significantly to absorb the emerging radiation.

In order to reproduce the global shape of the optical spectra, an additional opacity is required in the computations. We have modelled it with a simple law of the form $a_\circ\,(\nu/\nu_\circ)^N$, with $N = 4$, and have found that this provides a sufficiently good fit to the data. This additional opacity could be due to molecular/dust absorption or to dust scattering. The equivalent widths and intensities of the alkali lines are significantly affected by this opacity. The lithium resonance line at 670.8 nm is more affected by the additional opacity than by the natural depletion of neutral lithium atoms into molecular species as we move to lower effective temperatures. Our theoretical spectra predict a detectable lithium resonance feature even at very cool effective temperatures (~1000 K). Changes in the physical conditions governing dust formation in L-type objects will cause variability of the alkali lines, particularly of those at shorter wavelengths.

1. Introduction

The spectral characteristics of recently discovered very cool dwarfs are drastically different from those of the well known M-dwarfs. This has prompted the use of a new spectral classification, the so-called L types (Martín et al. 1997, 1999; Kirkpatrick et al. 1999). The main molecular absorbers at optical wavelengths in early- to mid-M dwarfs (TiO and VO) are likely to be depleted at the lower temperatures of L-dwarfs because of the incorporation of their atoms into dust grains. Here we follow a semi-empirical approach to understand the relevance of different processes on the resulting spectral energy distributions in the optical for L dwarfs and Gl 229B. We have computed synthetic spectra using the latest models by Tsuji (this volume) and Allard (1999). We have taken into account the depletion of some relevant molecules associated with the formation of dust and we have investigated the effects of dust scattering and/or absorption on the formation of the optical spectra. Remarkably strong alkali lines are present in the spectra and dominate its shape in the 600–900 nm region, providing major constraints to the theoretical modelling.

2. Input data for the spectral synthesis

We used the LTE spectral synthesis program WITA5, which is a modified version of the program employed by Pavlenko et al. (1995), incorporating "dusty effects" which can affect the chemical equilibrium and radiative transfer processes. We have used the B- and C-sets of Tsuji's (this volume) "dusty" LTE model atmospheres. The B-models account for the case of equilibrium of the "dust-gas" phase transition ($r_{dust} = r_{crit}$), while the C-models correspond to the case of segregation of dust–gas phases ($r_{dust} > r_{crit}$, where r_{dust} is the size of dust particles and r_{crit} is critical size corresponding to the gas–dust detailed equilibrium state). We have also used a grid of the NextGen "dusty" model atmospheres computed recently by Allard (1999). The temperature-pressure stratification of Allard's models lie between those of the C-type and B-type models of Tsuji (this volume).

Chemical equilibrium was computed for the mix of $\approx$100 molecular species. Alkali-contained species formation processes were considered in detail because of the important role of the neutral alkali atoms in the formation of the spectra. Constants for chemical equilibrium computations were taken mainly from Tsuji (1973). In the high pressure conditions of the atmospheres of L-dwarfs some molecules can be oversaturated (Tsuji et al. 1996); and can undergo condensation. To take into account this effect we reduced the abundances of those molecular species down to the equilibrium values and imposed that molecular densities did not exceed the saturation values: $n_i \leq n_i^{sat}$. The data for n_i^{sat} were taken from Gurwitz *et al.* (1982).

We used the set of continuum opacity sources listed in Table 2 of Pavlenko et al. (2000). Opacities due to molecular band absorption were treated using the Just Overlapping Line Approximation (JOLA). Synthetic spectra of late M-dwarfs have been already discussed in Pavlenko (1997). A complete description of the molecular opacities (TiO, VO, CaH, CrH) used in our calculations can be found in Pavlenko et al. (2000). The alkali line data were taken from the VALD database (Piskunov et al. 1995). At the low temperatures of our objects we deal with saturated absorption lines of alkalis and pressure broadened profiles. At every depth in the model atmosphere the profile of the aborption lines was described by a Voigt function $H(a, v)$, where damping constants a were computed as in Pavlenko et al. (1995).

3. Atomic and molecular features

The strength of the Li, Na and K resonance doublets increases dramatically when decreasing the effective temperature ($T_{\rm eff}$). In Fig. 1 we plot synthetic spectra showing how alkalis change as a function of $T_{\rm eff}$ and gravity. These computations do not include any molecular or grain opacity in order to show more clearly how alkali absorptions change with these parameters. The overall shape of the coolest L-dwarf spectra is governed by the resonance absorptions of Na I and K I and the less abundant alkali (i.e. Li, Rb, Cs) also produce strong lines. We remark the rise in intensity of K I and Na I lines with decreasing $T_{\rm eff}$ and with increasing atmospheric gravity. From our computations we confirm that the large equivalent widths (EWs) of the K I and Na I resonance doublets, reaching several thousand Angstroms, are mainly caused by the high pressure broadening in the atmospheres of the coolest dwarfs. The subordinate lines of Na I at 819.5 nm become weaker as $T_{\rm eff}$ decreases. It is also interesting to note that the subordinate Li I line at 812.6 nm is detectable in early/mid L-dwarfs with equivalent widths not exceeding 1Å, and that the triplet at 610.3 nm appears completely embedded by the wings of the K I and Na I resonance lines.

In Fig. 2 (upper panel) we compare the observed spectrum of the early L-type brown

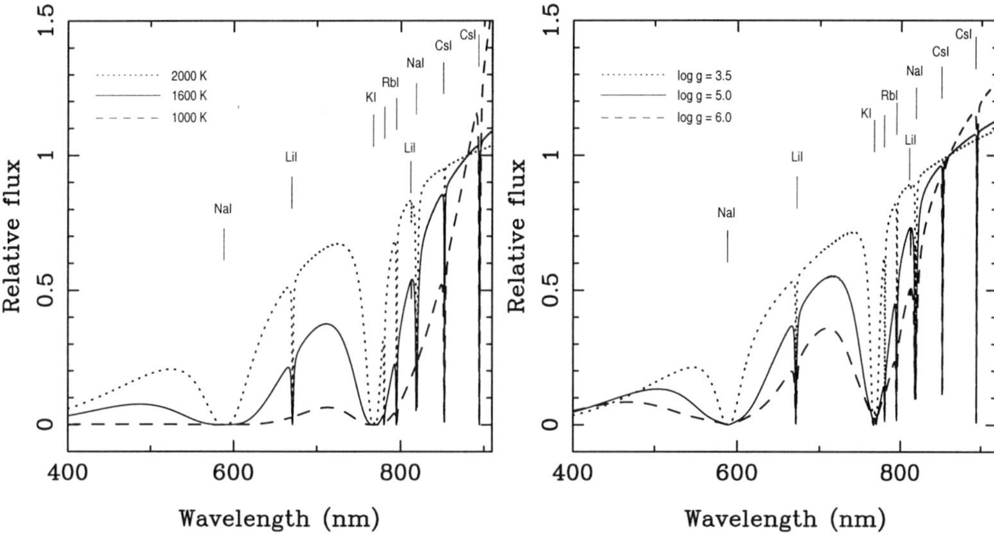

FIGURE 1. Theoretical optical spectra for the alkali elements computed with different values of $T_{\rm eff}$ and gravity. The left panel shows the dependence with temperature (computations performed using Tsuji's C-type models and $\log g = 5.0$. Spectra are normalized at 880 nm). The right panel displays the dependence with gravity at $T_{\rm eff} = 1600$ K (Allard's 1999 "dusty" models. Spectra are normalized at 860 nm). Identifications of the atomic lines are given in the top.

dwarf Kelu 1 with a synthetic spectrum obtained using the Tsuji C-type model for $T_{\rm eff} = 2000$ K and $\log g = 5$. The discrepancies of the theoretical computations with respect to the observed spectrum are notably reduced (Fig. 2, lower panel) when we impose a depletion of CaH, CrH, TiO and VO molecules accounting for the condensation of Ca, Cr, Ti and V atoms into dust grains. We implemented this "extra" depletion simply by introducing a factor R which describes the reduction of molecular densities of the relevant species over the whole atmosphere. The reduction factor for TiO ranges from 0 (complete depletion) to 1 (i.e. non depletion). Total or almost total depletion of Ti and V into the dust grains is required to explain the spectrum of the mid-type L-dwarfs DenisP J1228–1547 and DenisP J0205–1159.

4. The need for an additional opacity and its effects on the alkalis

Since the formation of dust in the atmospheres of cool dwarfs can produce additional opacity (AdO) we decided to investigate whether a simple description of this phenomenon could improve the comparison between observed and computed spectra. We adopted as law for AdO the following expression: $a_\nu = a_o * (\nu/\nu_0)^N$, where $N = 0$ corresponds to white scattering and $N = 4$ stands for the case of pure Rayleigh scattering. This AdO can be due to absorption or scattering processes, and the parameters N and a_0 should be determined from the comparison with observations. Here, we will adopt $N = 4$ which would be the most simple from the physical point of view. This new spectral synthesis provides a much better description of the data; in particular, we can explain the blue wing of the K I resonance doublet as it can be seen in Fig. 3 for two examples of L-type dwarfs. The same approach using the B-type models requires unrealistic values for N. Using C-type atmospheres from Tsuji and an index $N = 4$, we need the highest value of

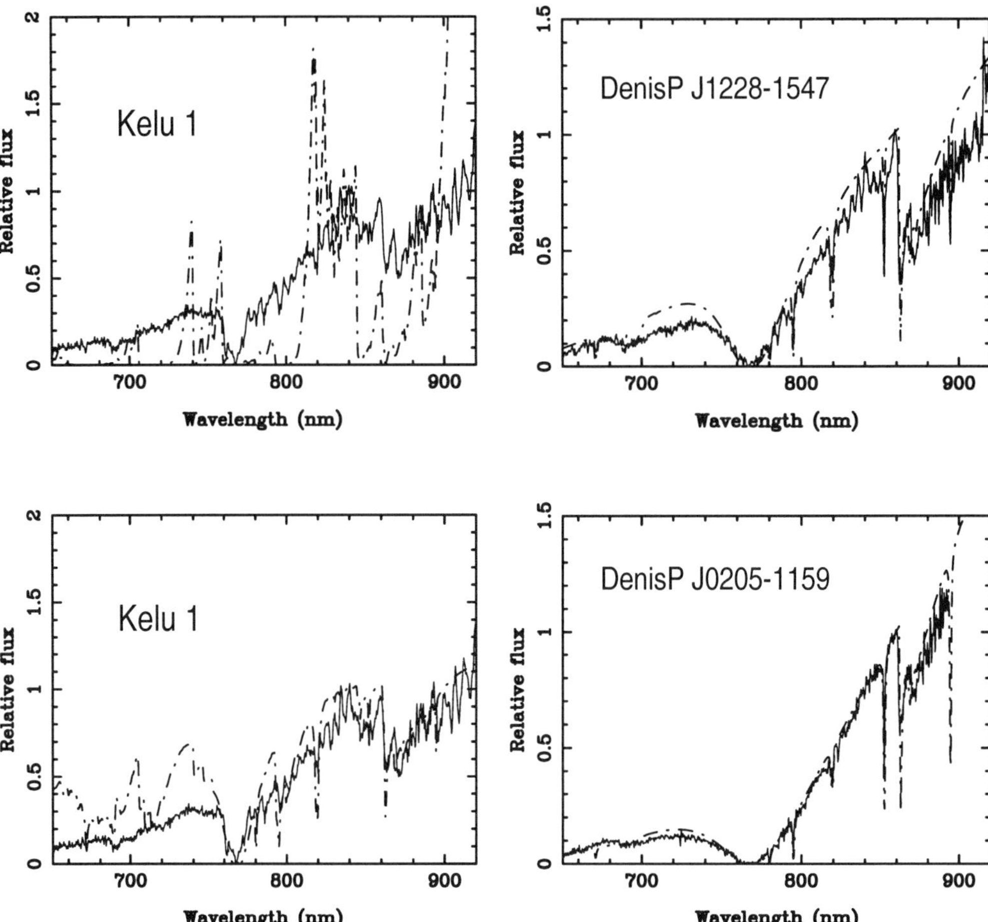

FIGURE 2. The observed spectrum of Kelu 1 (L2) and the predicted spectra (Tsuji's C-type models, 2000 K, $\log g = 5$) are shown with the full and dash-dotted lines, respectively, at a resolution of 10Å. Upper panel displays the computed spectrum considering only the chemical equilibrium of molecules. Lower panel depicts the same computations taking into account an "extra" depletion of TiO, VO, CaH and CrH.

FIGURE 3. DenisP J1228–1547 (L4.5) and DenisP J0205–1159's (L5) observed data (full line) compared to the best fits (dash-dotted line) obtained using Tsuji's C-type models (upper panel: $T_{\rm eff} = 1600$ K and $\log g = 5$; lower panel: $T_{\rm eff} = 1200$ K and $\log g = 5$). Both predicted spectra with a resolution of 10Å have been computed considering the AdO law described in the text with $a_{\circ} = 0.006$ and $N = 4$.

a_0 for understanding Gl 229B's spectrum ($a_0 = 0.1$); this can be interpreted as evidence for a very dusty atmosphere.

The strength of the resonance lines of the alkalis are considerably reduced by the presence of this AdO but they are still detectable in intermediate-resolution spectra of very cool dwarfs ($T_{\rm eff} \leq 1400$ K). The subordinate lines are, however, very much affected and may become undetectable for very high dust opacities. Lithium plays a major role as a discriminator for the substellar nature of brown dwarf candidates. Therefore, the effects of the AdO on the formation of the Li resonance line deserves detailed consideration. In the absence of the AdO the Li resonance line could present EWs of several tens of Angstroms in atmospheres as cool as 1200 K. This strength is considerably more affected by the amount of AdO needed to explain the overall spectral energy distribution than by the depletion of neutral Li into molecules. Our computations indicate that objects like DenisP J0205–1159 ($T_{\rm eff} \sim 1200$ K) and cooler objects with moderate dust opacities should show the Li I resonance doublet if they had preserved this element from nuclear burning, and consequently, the lithium test can still be applied at such low temperatures. Furthermore, even in very dusty cool atmospheres like that of Gl 229B ($T_{\rm eff} \sim 1000$ K), the lithium resonance line could be detected with an EW of several hundred mÅ. Figure 4 displays the sensitivity of the Li line to different amounts of AdO in a 2000 K-atmosphere. The abundance of lithium in Kelu 1 is consistent with complete preservation of its initial content.

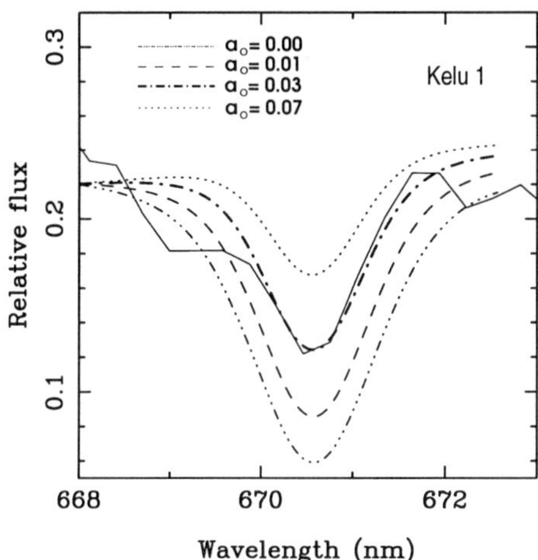

FIGURE 4. Fitting of the Li I resonace line of Kelu 1 (full line) using the Tsuji's C-type model atmosphere for $T_{\rm eff} = 2000$ K, $\log g = 5$. Computations have been performed for a lithium abundace of $\log N(\mathrm{Li}) = 3.0$ and considering different amounts of dust opacity. The resolution of all theoretical spectra is the same than the one of the observed data. The predicted model that better fits the observations coincides with the one that also nicely reproduces the overall shape of the optical spectrum ($a_\circ = 0.03$).

5. Conclusions

Strong alkali lines and oxide (TiO, VO) and hydride (FeH, CaH, CrH) molecular bands are contributing significantly to the overall shape of early L-type dwarfs red and far-red spectra, with an increasing impact of alkalis as we move to later spectral types. Our attempt to model the observed data with an LTE spectral synthesis code requires a larger molecular depletion than predicted by the chemical equilibrium. This is consistent with Ti and V being condensed into dust grains. We also need to incorporate an additional opacity which we described with the form $a_\circ\ (\nu/\nu_\circ)^N$. The consideration of this simple law, adopting $N = 4$, provides a sufficiently good fit to the observed data. This additional opacity can be ascribed to molecular/dust absorption or to dust scattering. The strength of the alkali resonance and subordinate lines in the optical spectra of L-type objects are very much affected by the additional opacity. The resonance line of lithium (widely used as a substellarity discriminator), is more sensitive to the formation of dust in the atmosphere than to the natural depletion of neutral Li atoms into molecules. In those very cool atmospheres where the additional opacity is low the lithium test for establishing the brown dwarf status of substellar candidates can still be applied. Changes in the dust concentration in the atmospheres can lead to variability of the Li resonance doublet.

REFERENCES

Allard, F. 1999, private communication

Allard, F., Hauschildt, P. H., Alexander, D. R., Starrfield, S. 1997, ARA&A, 35, 137

Gurvitz, L. V., Weitz, I. V., Medvedev, V. A. 1982, Thermodynamic properties of individual substances. Moscow. Science

Kirkpatrick, J. D., Reid, I. N., Liebert, J., et al. 1999, ApJ, 519, 802

Martín, E. L., Basri, G., Delfosse, X., Forveille, T. 1997, A&A, 327, L29

Martín, E. L., Delfosse, X., Basri, G., Goldman, B., Forveille, T., Zapatero Osorio, M. R. 1999, AJ, 118, 2466

Pavlenko, Ya. V, 1997, Astron. Reports, 41, 537

Pavlenko, Ya. V., Rebolo, R., Martín, E. L, García López, G. 1995, A&A, 303, 807.

Pavlenko, Ya. V., Zapatero Osorio, M. R., Rebolo, R., 2000, A&A, in press

Piskunov, N. E., Kupka, F., Ryabchikova, T. A., Weiss, V. V., Jeffery, C.S. 1995, A&AS , 112, 525

Tsuji, T. 1973, A&A, 23, 411

Tsuji, T., Ohnaka, K., Aoki, W. 1996, A&A, 305, 1

Absolute Dimensions for M Type Dwarfs

By ÁLVARO GIMÉNEZ

Instituto de Astrofísica de Andalucía, CSIC, Apartado 3.004, 18080 Granada, Spain, and

Laboratorio de Astrofísica Espacial y Física Fundamental, INTA, Apartado 50.727, 28080 Madrid, Spain

M-type stars define the lowest end of the main sequence and the connection between normal stars and brown dwarfs. For these reasons, the determination of accurate absolute dimensions in very low mass stars is a fundamental astrophysical problem. Moreover, they are a numerous population in our galaxy, provide limit conditions for core hydrogen burning, and allow the test of different treatments of convective energy transport.

Absolute dimensions for the comparison of empirical data with theoretical models are generally reduced to mass, radius, and temperature. The estimation of each of them as a function of colour indices, by means of either direct determinations or the use of calibration curves, is reviewed together with the available information derived from the study of well-detached double-lined eclipsing binaries.

1. Introduction

Red dwarfs are among the least massive stellar objects in the Universe. Among them, M-type stars are in the mass range from around 0.1 to 0.5 solar masses. At lower values we only find brown dwarfs or non-stellar planetary bodies. Nevertheless, low-mass stars are probably the most common type in our Galaxy, with a high potential influence in the definition of its mass function. They also provide the connection between objects with radiation generated through nuclear reactions and those which are not able to do so because of insuficiently high internal temperatures. From the structural point of view, M-type stars are very important because they are dominated by convective energy transport, and the treatment of convection is still one of the least known parts of the theory of stellar structure. Furthermore, convective motions and rotation are behind the generation of significant magnetic fields through stellar dynamos. The interaction between magnetic fields and stellar matter in establishing the final structure is also not well known. From the observational point of view, they present a challenge because of their low luminosities and, in particular, for the determination of accurate effective temperatures because of the presence of many absorption features in their spectra.

In order to really benefit of the possibilities offered by the study of very low mass stars for the understanding of stellar structure we have to carry out a detailed comparison of empirical evidences with available theoretical models. For this purpose, we should collect as much as possible accurate information about their most representative parameters or absolute dimensions: luminosity (L), radius (R), mass (m) and effective temperature (T_{eff}). On the theoretical side, important advances have been recently achieved with the work by Allard *et al.* (1997), Baraffe *et al.* (1998), and others, leading to accurate predictions for the main observational parameters of very low mass stars as a function of the initial metal content and mixing-length parameter.

In this paper, currently adopted calibrations for stellar dimensions as a function of colour indices are discussed with the help of independently derived information from double-lined eclipsing binaries and distance determinations by ESA's astrometric satellite Hipparcos. Furthermore, a brief comparison with most modern stellar structure models is included.

Star clusters and deep photometric surveys are obviously an important source of information for the study of fundamental parameters, as discussed in this workshop, but in order to get accurate values for masses and radii, double-lined eclipsing binaries are needed. In fact, the work sumarized in this paper started with the wish to extend the research carried out during the last two decades, on the accurate determination of stellar absolute dimensions by means of binary stars, to the low-mass end of the main sequence. Unfortunately, the number of accurately measured eclipsing binaries in the mass domain below 0.6 solar masses is very limited. Two systems are known since long time ago (YY Gem and CM Dra) and a new one (GJ2107A) has only very recently been discovered (Delfosse *et al.* 1998, these proceedings). Previous absolute dimension studies from binary stars have been published by Popper (1980), Harmanec (1988) and Andersen (1991). This later author, in order to keep a homogeneous temperature calibration and overall quality throughout the studied sample did not include enough systems with masses below that of the Sun. On the other hand, Popper (1980) adopted temperatures which were later found to be overestimated for some spectral types, while Harmanec (1988) found disagreements with predictions based on theoretical models precisely in the temperature range of our interest.

2. The magnitude-colour diagram

Until a good determination of the effective temperature of the stars is available, as discussed in section 4, we will consider the absolute magnitude versus colour plane as the best available approximation to the Hertzprung-Russel diagram. For this purpose, the determination of stellar distances through trigonometric parallaxes is the generally adopted procedure although the use of open clusters also helps. Examples of what can be done in these directions are given by Monet *et al.* (1992) using the survey of CCD parallaxes of faint stars from the US Naval Observatory, or by Prosser (1992) in the case of the young open cluster α Per and Hamilton & Stauffer (1993) for the Pleiades. Local studies mix stars with different ages and metal content but the luminosity determinations are independent of main sequence calibrations. On the other hand, cluster members provide samples with homogenous age and chemical composition but are generally too far for accurate trigometric distances to be determined.

We have used the data from the Hipparcos catalogue to search for late-type stars in the solar neighbourhood. A subset of all stars with colour index $B - V \geq 1.1$, a trigonometric parallax larger than 5 mas, a distance determination better than 10%, and accurate values of the colour index $V - I$ (better than 0.03), led to a list of 1428 objects out of a catalogue with 118218 entries. The evolved members were then eliminated, by imposing the absolute V magnitudes to be larger than 5, as well as those with duplicity or variability flags, allowing the compilation of a catalogue of 678 "normal" main-sequence late-type stars. For all of them, we have extracted from the Hipparcos catalogue the homogeneous values of Johnson's V magnitude, B-V colour index, and Coussin's V-I index.

A comparison of the HR diagram defined by these stars in the $M_V - (B - V)$ and $M_V - (V - I)$ planes, shows that the second is clearly better for late-type stars because of the covered range and their linear behaviour. The absolute magnitudes were computed from the parallaxes derived within the Hipparcos mission and the quoted apparent V magnitudes. No correction for reddening was applied. This may affect the more distant stars but is not expected to be important because all objects in our sample are between 3 and 60 pc from the Sun and the M-type stars ($B - V > 1.3$) are all closer than 45 pc. An analysis by Upgren *et al.* (1997) of the kinematics and M_V calibration of K and M dwarf stars using the same Hipparcos data has shown a separation between "young" and "old" disk components. The old component is found to be 0.6 magnitude fainter in M_V at the same colour index.

Generally adopted calibrations of M_V as a function of $V-I$ can then be easily tested. The linear approximation by Reid & Majewski (1993), for K and M dwarfs ($V-I > 0.92$), takes the form:

$$M_V = 2.89 + 3.37(V - I)$$

while the non-linear expression by Leggett (1992), for old-disk stars in the range $1.4 < V - I < 4.8$, is:

$$M_V = 1.763 + 4.289(V - I) - 0.1461(V - I)^2$$

Both expressions agree equally well with the Hipparcos data. A linear least-squares fit provided $M_V = 3.47 + 3.05(V - I)$, but it is clearly dominated by the more abundant earlier-type stars. As an example, restricting the least squares fit to $B - V \geq 1.5$ (and thus only 52 stars out of the original sample), we obtained $M_V = 2.9 + 3.3(V - I)$. These expressions provide a well-known method for the determination of photometric parallaxes and another approach is the comparison of the predicted distances with the observed Hipparcos values. Of course, again both expressions above for M_V provided a good agreement with the observational parallaxes but the best fit to the data allowed the obtention of $M_V = 3.26 + 3.17(V - I)$.

In order to compare the observational data with theoretical models, we extracted the zero-age main sequence for low mass stars from the models by Baraffe *et al.* (1998), for a solar initial metal abundance, by taking those with maximum $\log g$, at the end of the contraction phase. The comparison indicate a good agreement for values of $V - I < 2.0$ but cooler objects are found to be systematically above the ZAMS. The same result was already observed by other authors and can not be explained simply in terms of metal content though the fit to globular clusters isochrones and halo stars with the same models showed a perfect agreement. Baraffe *et al.* (1995) discussed further this matter. An effect of age in the sample may be the reason as pointed also out by Leggett (1992). While a 0.6 $M_\odot$ star takes $1.6x10^8y$ to reach the ZAMS, a 0.1 $M_\odot$ takes $2.5x10^9y$ and one in the stellar mass limit around 0.08 $M_\odot$ may not even have enough time within the age of the local disk.

3. The mass-luminosity relation

The mass-luminosity relation is mainly dependent on the energy generation and absorption mechanisms and should therefore be given in different forms for the different mass ranges. Because of the numerical importance of low mass stars in our Galaxy, the study of objects below 0.6 solar masses is badly needed. Empirical data can be obtained from the analysis of close visual components which provide individual masses as a function of M_V but a real mass-luminosity relation obviously requires the knowledge of the bolometric corrections for these stars. We will come later to this point.

Determinations of stellar masses at the very end of the main sequence are still not abundant though speckle interferometry has helped during the last years to increase the available sample. Systematic searches in the solar neighbourhood (Leinert *et al.* 1997) indicate that binarity in very late-type stars might be less common than in solar-type main sequence stars. New binary star masses will be derived from the analysis of the Hipparcos data, both because of new parallaxes and accurate new points in the orbits. Söderhjelm *et al.* (1997) give some information about these results. Very recently, it has also been shown that the combination of VLBI and Hipparcos measurements for some stars may provide accurate orbital solutions for very close "visual" binaries with submilliarcsecond astrometry. This is the case of the low mass companion to AB Dor studied by Guirado *et al.* (1997). They found that the mass of the discovered body should be in the range from 0.08 to 0.11 $M_\odot$, very close thus to the limit for brown dwarfs. Unfortunately no calibrated photometry is yet available leading to a new point in the $M_V - \log m$ plane. This will be very difficult if, as expected, $m_V \approx 16$ and is very close to its brighter mate ($m_V = 6.95$). VLBI measurements should nevertheless be the tool for the future since spectroscopic binaries will become visual from the point of view of spatial resolution.

Spectroscopic binaries are not very well suited for accurate mass determinations unless they show eclipses, because of the unknown value of the orbital inclination. Nevertheless, Harlow (1996) studied the case of Gl 372, with an orbital period of 47.7 days, a large orbital eccentricity ($e = 0.53$) and classified as M2V. The interesting thing about Gl 372, apart of being the only M-type spectroscopic binary with no flares or other activity indicators, is that the measured trigonometric parallax gives a distance of 16.7 pc and values of $M_V = 9.8$ with a minimum mass of 0.59 for component A and $M_V = 10.8$ with a minimum mass of 0.45 for component B. The actual masses should be very close to these values according to the calibration given below and, therefore, the system could be a good candidate to search for eclipses. The large orbital period though implies that the expected relative radii of the component stars are close to 0.01. On the other hand, the observed $V - I = 2.26$ colour for the system predicts an average value of $M_V = 10.5$ confirming the linear expressions given in Section 2. New mass determinations, better than 2%, are being obtained by Delfosse *et al.* 1998, these proceedings), for example in the cases of Gl 570B and the 2.77 days period M3.5 GJ 2069A, or Basri & Martin (1998, this workshop) in the case of the 5.83 days period binary in the Pleiades, PPL 15.

Böhm (1989) studied the available data to estimate the relation between mass and absolute magnitude in the form:

$$\log m = -0.0931 M_V + 0.4657,$$

but the stars with lower masses (below 0.2 $M_\odot$) showed brighter M_V values. A more detailed and accurate study, using new data, was carried out by Henry & McCarthy (1993). They obtained the relation:

$$\log m = 0.002456 {M_V}^2 - 0.09711 M_V + 0.4365,$$

for $M_V \leq 10.25$, and

$$\log m = -0.1681 M_V + 1.4217,$$

for larger values of M_V.

A linear least-squares fit to the data of Henry & McCarthy (1993) within the mass range of our direct interest led to an expression of the form, $\log m = -0.0896 M_V + 0.43$. Moreover, using the above given relations between M_V and $V - I$, we can have a reasonable approach to $\log m$ in terms of the colour index.

The same models adopted before from Baraffe *et al.* (1998) for the ZAMS provided an excellent fit to the observational data indicating that the problem found in the $M_V - (V - I)$ plane for later type stars is probably not present in the sample by Henry & McCarthy (1993), or in the $m - M_V$ plane. A linear fit to the ZAMS models led to an expression in the form $\log m = -0.0818 M_V + 0.42$, in good agreement with the observational data.

4. The calibration of effective temperatures and bolometric corrections

The determination of effective temperatures for M-type dwarfs is a difficult task. Their spectra are dominated by numerous molecular bands in the optical and near-infrared regions and the stellar continuum is thus never really observed. On the other hand they are intrinsically faint and very few are close enough for detailed high resolution studies. The effective temperature is defined as a measure of the stellar bolometric flux and thus should be determined through observations directly related to the bolometric flux. In other words, the determination of angular radii allows the use of the observed flux at Earth, F_E, and the Stefan-Boltzman law $\sigma T_{eff} = F$ to write, in the absence of interstellar absorption:

$$T_{eff} = \frac{4}{\sigma \theta^2} F_E$$

Angular sizes, θ, can be determined from speckle interferometry, lunar occultations or long baseline optical interferometry. Unfortunately there is no yet possibility to measure directly the radius of M-type dwarfs using interferometric techniques although the future may be different. VLBI measurements of the size of dMe stars has been attempted at 1.6 GHz by Alef *et al.* (1997) with the eclipsing system YY Gem, leading to a value of the radius (of the active region) of 2.1 photospheric radii. Though VLBI has only been used up to now for the imaging of the coronal structures (gyrosynchrotron) in late-type stars (another case is UV Cet), the time will come for photospheric observations, when optical interferometry becomes a normal technique. Until then, some kind of spectral energy distribution (SED) approach has to be accepted, like the IR flux method (IRFM) which uses the comparison of observational data with theoretical energy distributions. In fact, the IRFM has been the most commonly used method to determine temperatures in low-mass stars. This method has to be nevertheless considered with much care in the case when models do not take into account the real composition of the stars. The obtained temperature scale might be strongly metallicity dependent or completely wrong if stars are so cold that dust starts to be formed.

The results by Berriman & Reid (1987), later confirmed by Berriman *et al.* (1992), can be roughly approximated, for effective temperatures as a function of the colour index $V - I$, as:

$$T_{eff} \approx 4400 - 450(V - I)$$

for the (V-I) range between 1.5 and 4. This temperature calibration provides lower values than previous studies but it is more realistic for low mass stars because of a proper analysis of the effects of steam absorption and backwarming. Nevertheless, for the coolest range, the formation of dust in stellar atmospheres may seriously affect the determination of the effective temperature scale (Tsuji *et al.* 1996a).

Coming back again to the ZAMS models by Baraffe *et al.* (1998), a least-squares fit to the interpolated values showed a linear relation $T_{eff} = 4700 - 540(V - I)$, not very different from that derived with the empirical data of Berriman & Reid (1987). Models incorporating the effects of some grain formation have been computed by Tsuji *et al.* (1996b) claiming the resolution on this basis of the problem with the infrared spectral energy distribution. But they showed that grain formation starts to be strong enough only for temperatures below 2600 K, thus outside our range of interest.

Because of the bolometric significance of the effective temperatures, they should be linked to the determination of bolometric corrections. The bolometric corrections for main-sequence stars were analyzed in detail by Flower (1996) as a function of temperature. However he pointed out that the relation obtained for stars with temperatures below 5000 K is uncertain because of the low intrinsic accuracies of the effective temperature determinations in these stars. Greenstein (1989) combined spectrophotometric observations with broad band near-infrared measurements and trigonometric parallaxes to yield M_{bol} values for late-type dwarfs by including allowances for blanketing both from metallic lines and water. From these analyses, a least squares parabolic fit produced the average relation:

$$M_{bol} - M_I = 0.03 + 0.536(V - I) - 0.145(V - I)^2$$

which can be expressed also in terms of M_V, or BC, and $V - I$. On the other hand, Hamilton & Stauffer (1993) used data from Berriman & Reid (1987) to obtain:

$$M_{bol} = 1.805 + 0.835 M_I$$

which, using the relation between M_V and $V - I$ by Reid & Majewski (1993), can be expressed as $BC = 1.33 - 1.39(V - I)$. It can be seen that both the temperature scale and the bolometric corrections are different from those given by Flower (1996).

Models by Baraffe *et al.* (1998) also indicate larger bolometric corrections, for stars cooler than around 3500 K. A linear least-squares fit to the ZAMS data extracted from the models led to the expression: $M_{bol} - M_V = 1.89 - 1.685(V - I)$. Using this later relation between BC and $V - I$, we can transform the values for empirically determined masses in the list by Henry & McCarthy (1993), to the $\log L - \log m$ plane. A least-squares fit leads thus to $\log m = 0.90 - 0.156 M_{bol}$ (to be compared to $\log m = 0.91 - 0.151 M_{bol}$ directly derived from the ZAMS models), and the standard mass-luminosity relation for very low mass stars:

$$\log L = 2.56 \log m - 0.46$$

which can be compared with the result obtained by Giménez & Zamorano (1986) from the data by Popper (1980) for masses below 0.5 solar masses and different photometric calibrations ($\log L = 2.32 \log m - 0.61$). The determination of $\log R$ as a function of the previous parameters is then straightforward by means of the Stefan-Boltzmann law in its form:

$$\log R = 8.452 - 2 \log T_{eff} - 0.2 M_{bol}$$

though highly dependent on the calibrations adopted. The surface flux in V is given by $F_V = \log T_{eff} + 0.1 BC$ which can be approximated, with the above expressions, to $F_V \approx 3.5 - 0.2(V - I)$

5. Average calibrations

Results discussed in previous sections can be tabulated so as to provide an easier use of them. In Table 1, the values of M_V derived from the Hipparcos data are given for average binned values of the colour indices $V - I$ and $B - V$. Of course, numbers extracted from a very poor statistics (less than about 20 individual stars) should be taken with great caution. The number of stars, N, in each colour bin is given in column 2. The M_V values computed by means of the calibration by Reid & Majewski (1993) are included in column 5. Bolometric magnitudes are those derived from the bolometric corrections adopted from Hamilton & Stauffer (1993) which are consistent with the temperatures from the linear approximation to the data by Berriman & Reid (1987). Masses correspond to the calibration in terms of M_V by Henry & McCarthy (1993). Stellar dimensions included in Table 1 are thus all derived from empirical calibrations with no use of any theoretical model.

It is straightforward to check that these values agree reasonably well with the results

Table 1.

$V-I$	N	$B-V$	M_V	$M_V(c)$	M_b	BC	T_{eff}	m	R	F_V
1.50	183	1.30	8.0	7.9	7.2	-0.8	3725	0.66	0.75	3.50
1.58	151	1.35	8.3	8.3	7.3	-0.9	3690	0.64	0.71	3.48
1.69	105	1.40	8.6	8.6	7.6	-1.0	3640	0.61	0.65	3.46
1.78	73	1.43	9.0	8.9	7.7	-1.2	3600	0.58	0.62	3.44
1.90	36	1.46	9.5	9.3	8.0	-1.3	3545	0.56	0.57	3.42
2.02	39	1.48	9.8	9.7	8.2	-1.5	3495	0.53	0.53	3.40
2.07	36	1.49	9.9	9.9	8.3	-1.6	3465	0.52	0.51	3.39
2.19	27	1.50	10.1	10.3	8.5	-1.7	3415	0.50	0.47	3.36
2.28	25	1.51	10.4	10.6	8.7	-1.9	3375	0.44	0.45	3.34
2.38	21	1.52	10.7	10.9	8.9	-2.0	3330	0.39	0.42	3.32
2.47	17	1.53	10.8	11.2	9.1	-2.1	3285	0.34	0.39	3.31
2.60	16	1.58	11.2	11.7	9.4	-2.3	3230	0.29	0.36	3.28
2.67	11	1.61	11.4	11.9	9.5	-2.4	3200	0.27	0.35	3.27
2.75	3	1.63	11.9	12.1	9.7	-2.5	3165	0.24	0.33	3.25
2.89	2	1.70	13.0	12.6	9.9	-2.7	3095	0.20	0.30	3.22
2.99	2	1.75	13.2	13.0	10.1	-2.8	3055	0.17	0.29	3.20

obtained by Bessel (1990), who in addition provides spectral types, from the study of BVRI photometry of late-type dwarfs in the Gliese catalogue of nearby stars. A similar overall agreement is found with other available tabulations. A detailed comparison with stellar models (e.g. Baraffe *et al.* 1998) is not simple due to the mixture of chemical compositions and ages. Nevertheless, as expected from the adopted temperature scale, there is a tendency in Table 1 for lower values of T_{eff} than given by the models for the same masses and solar metallicities.

When the stellar radius is known, an empirical form of the mass-radius relation can be obtained (Giménez and Zamorano, 1985). Using the observational data from Popper (1980), it could be seen that there is a change of slope in the $\log R - \log m$ plane around 1.8 $M_\odot$ (thus related to the change in energy transport mechanism for the star envelopes) and a good linear fit provided by:

$$\log R = 0.053 + 0.977 \log m$$

which agrees reasonably well with data in Table 1 despite the use of different calibrations to transform M_V values of visual binaries into stellar radii.

6. Eclipsing binaries

The adoption of a temperature scale as a function of photometric colour indices is crucial for the comparison of absolute dimensions with stellar models and the derivation of average empirical relations. This is, as already mentioned, a difficult task in very low mass stars mainly due to the fact that their spectra do not always represent the actual flux of the bolometric energy distribution in the optical. A check of the obtained average relations is necessary by means of other sources and double-lined eclipsing binaries represent the best approach. An accurate and independent determination of m and R, as well as M_V, provides a test of the predicted values for the relations betwen the main stellar parameters.

Unfortunately eclipsing binaries in the low mass end of the main sequence are very difficult to find due to selection effects expected for small stars and low luminosities. We have to face the situation that well-detached systems, with relatively long periods, have a very low probability to show eclipses while the closer cases, because of their deep convective envelopes, have a rotationally induced high activity levels producing severe distortions of the light curves. If we have a binary system with two identical stars, the probability of having eclipses is proportional to $2R/a$. In the case of normal detached eclipsing binaries, the relative radii are around 0.1, i.e. a probability to find 1 case out of 5 such binaries to show eclipses. Nevertheless, this also implies that, for a normal late-type dwarf with a radius of 0.5 solar radii, the separation would be 5 solar radii. Assuming also an average mass of 0.4 solar masses, we then expect an orbital period of only 1.4 days. But these binaries, with strong tidal interaction, might be very short lived. Moreover, the component stars should according to synchronization effects be rotating with a surface velocity of around 20 Km/s. The situation is that after many years we still only have two good cases within the spectral domain of the M-type stars, namely, YY Gem and CM Dra.

The search for late-type double-lined eclipsing binaries is still going on. The objetive is obviously the determination of accurate masses and radii for stars below 1 $M_{\odot}$. Spectroscopic surveys are in progress (Popper 1996; Mayor *et al.* at Geneva observatory; Latham *et al.* at CfA, Boston, etc.), but first-rate photometry is not generally available. Clausen *et al.* (1997) therefore started a program to improve the situation, using $uvby\beta$ photometry, by the observation and analysis of known eclipsing binaries and a systematic search for new low mass eclipsing binaries. We mentioned already some of the discoveries, by means of deep surveys within some open clusters, presented in this workshop. In addition, the new efforts to discover planets around solar-type stars has open the possibility to devote large amounts of observing time to spectroscopic or/and photometric measurements of low mass stars (e.g. CM Dra). The possible candidate V4066 Sgr pointed out by its spectral classification in the General Catalogue of Variable Stars to be M5 with $V = 11$ and a period of 2.1 days was discarded by Popper (1996). V4066 Sgr has a much redder visual component while the eclipsing system shows double lines but an F spectral type. Promising candidates for the more massive part of the range of interest may be the secondary components of HP Aur and HS Aur (Popper 1997). Accurate new light curves using four colour photometry are currently being obtained. Recently Maceroni and Rucinski studied a very short period eclipsing binary, BW3 V38, with $P = 0.198$ days. Unfortunately, although the components belong to the main sequence, they are so close that reach contact configuration and no accurate absolute dimensions can be determined. Nevertheless, the $V - I = 2.3$ suggest components between those of YY Gem and CM Dra in temperature. The system was discovered during the OGLE monitoring for microlensing events and a revitalization of this field is expected with these new and efficient techniques for the identification of unknown eclipsing binaries.

Absolute parameters for these binaries can be directly determined through the accurate measurement of both m and R. Assuming the value of the bolometric correction, the observed values provide the effective temperatures through

$$T_{eff} = (10\pi R)^{1/2} 10^{-0.1(V_0 + BC - M_{bol\odot})}$$

where both R and T_{eff} are given in solar units and π denotes the trigonometric parallax. If the effective temperature is adopted from a reliable calibration in terms of colour indices, then the bolometric corrections can be derived.

The active close binary YY Gem, with an orbital period of 0.81 days and belonging to the Castor multiple system, has been studied in many opportunities but the most recent works, leading to the best available absolute dimensions, are those by Bopp (1974) and Leung & Schneider (1978). These dimensions, when compared with the models by Pols *et al.* (1995) indicate that the component stars are too big and cool to fit the models for any reasonable value of age and initial chemical composition. New models by Chabrier and Baraffe (1995) have nevertheless permitted a good fit to the data suggesting a position in the late pre-main-sequence contraction phase. Moreover, the average relations given in the previous sections as a function of colour index can be tested. The values for YY Gem of $V = 8.99$ and $(V - I) = 1.89$ could be extracted from Leung & Schneider (1978) using the transformations from Johnson's to Coussin's system given by Leggett (1992). Therefore, we expect the average of the two components of YY Gem to have $M_V = 9.26$, a mass of $m = 0.56 M_\odot$, an effective temperature of $T_{eff} = 3550K$, and a bolometric correction of $BC = -1.30$. Thus, $M_{bol} = 7.96$, $\log(L/L_\odot) = -1.33$, and $R/R_\odot = 0.58$. The observed values are in fact, $M_V = 9.03$, if the astrometric parallax is $\pi = 0.072$, and from the analysis of the light and radial velocity curves, $m = 0.59$, and $R = 0.62$. The agreement is quite fair but he system has been suggested to be composed by young disk stars, even pre-main sequence, according to its membership to the multiple system Castor. The position of YY Gem in the $M_V - (V - I)$ plane discussed in Section 2 is clearly in the upper region with respect to the rest of the stars for the same colour index.

With an orbital period of 1.268 days and a large space velocity (163 km/s), CM Dra was analyzed by Metcalfe *et al.* (1996) and is up-to-date known as the less massive main sequence eclipsing binary with double-lined spectra. This work superseded the spectroscopic study by Lacy (1977) though his photometric analysis still remains the most accurate. The photometric values by Leggett (1992), $V = 12.91$, $V - I = 2.92$ can again be combined with the previous average relations to provide the following estimations: $M_V = 12.73$, a mass of $m = 0.19 M_\odot$, an effective temperature of $T_{eff} = 3090K$, and a bolometric correction of $BC = -2.73$. Thus, $M_{bol} = 10.00$, $\log(L/L_\odot) = -2.14$, $R/R_\odot = 0.30$. The observed values, from the analysis of the light and radial velocity curves are: $M_V = 12.86$, if the astrometric parallax is $\pi = 0.0692$, $m = 0.223$, and $R = 0.245$. The agreement is not very good but we should consider the involved uncertainties in the average relations. Moreover, on the basis of the large space velocity of CM Dra, it has been suggested that it might be a metal defficient Pop II halo member. Nevertheless, the good agreement with average relations for disk stars in the solar neighbourhood, in the $M_V - (V - I)$ plane, does not add support to it. The system appears to have (when compared with models) an unusually high helium abundance $Y = 0.31$. The effective temperature and metallicity was recently analyzed by Vitti *et al.* (1997) comparing observations from 0.40 to 2.41 μm with synthetic spectra. They found discrepancies between results in the optical and the infrared. The optical spectral energy distribution indicates a temperature of 3000 K and metal-rich content while the infrared leads to low metallicity (in agreement with the Pop II nature of the system) and a temperature of 3200 K. The available absolute dimensions and the trigonometric parallax of 0.0692 ± 0.0025 arcsec, provided accurate values of the effective temperatures close to $T_{eff} = 3075K$. This later value is in excellent agreemeent with that derived from the temperature scale by Berriman & Reid (1987).

In summary, CM Dra seems to be a Population II system and YY Gem is probably still in the contracting pre main sequence phase so that we badly need new candidate eclipsing binaries in the very low mass range of the main sequence. A detailed comparison of the observations with available theoretical models was performed by Chabrier and Baraffe (1995). They found a reasonable agreement between observational data and their models for very low mass stars. For the case of CM Dra, a small metal deficiency is found but not enough to explain the large space velocity of the system. A solar-like metallicity is favoured for the case of YY Gem. Concerning the helium abundance a value close to 0.25 is found for CM Dra and to 0.30 for YY Gem.

TABLE 2.

$V-I$	M_V	M_b	BC	T_{eff}	m	R	F_V
1.7	8.3	7.3	-1.0	3635	0.72	0.74	3.46
1.8	8.7	7.6	-1.1	3590	0.65	0.67	3.44
1.9	9.1	7.8	-1.2	3545	0.59	0.61	3.43
2.0	9.4	8.1	-1.4	3500	0.54	0.56	3.41
2.1	9.8	8.3	-1.5	3455	0.49	0.51	3.39
2.2	10.2	8.6	-1.6	3410	0.45	0.47	3.37
2.3	10.6	8.8	-1.7	3365	0.41	0.43	3.36
2.4	10.9	9.1	-1.8	3320	0.37	0.39	3.34
2.5	11.3	9.4	-1.9	3275	0.34	0.36	3.32
2.6	11.7	9.6	-2.1	3230	0.31	0.33	3.30
2.7	12.0	9.9	-2.2	3185	0.28	0.30	3.29
2.8	12.4	10.1	-2.3	3140	0.25	0.27	3.27
2.9	12.8	10.4	-2.4	3095	0.23	0.25	3.25
3.0	13.2	10.6	-2.5	3050	0.21	0.23	3.23

We are tempted to use the data directly derived from the two available eclipsing binaries with absolute dimensions (YY Gem and CM Dra) to re-calibrate the adopted average relations. This should of course be made with great caution because of the possible anomalous nature of one of them, or even both, to evaluate the dimensions of normal stars in the galactic disk near to the Sun. Anyhow, we have obtained the following relations, keeping the same temperature scale as in section 4:

$$M_V = 2.00 + 3.72(V - I)$$

$$BC = 1.00 - 1.18(V - I)$$

$$\log m = -0.110 M_V + 0.77$$

These relations fit by definition the binary star data and are not found to disagree with the general relations for normal late-type dwarfs given in previouy sections. Predicted parameters for different values of the colour index $V - I$ are shown in Table 2.

REFERENCES

ALEF, W., BENZ, A. O., & GÜDEL, M. *Astron. Astrophys.* **317**, 707

ALLARD, F., HAUSCHILDT, P. H., ALEXANDER, D. R. & STARRFIELD, S. 1997. *Ann. Rev. Astron. Astrophys.* **35**, 137

ANDERSEN, J. 1991. *Astron. Astrophys. Rev.* **3**, 91

BARAFFE, I., CHABRIER, G., ALLARD, F. & HAUSCHILD, P. 1995. *Astrophys. J.* **446**, L35

BARAFFE, I., CHABRIER, G., ALLARD, F. & HAUSCHILD, P. 1998. *Astron. Astrophys.* **337**, 403

BERRIMAN, G. B. & REID, I. N. 1987. *MNRAS* **227**, 315

BERRIMAN, G. B., REID, I. N. & LEGGETT, S. K. 1992. *Astrophys. J.* **392**, L31

BESSEL, M. S. 1990. *Astron. Astrophys. S.S.* **83**, 357

BÖHM, C. *Astrophys. & Sp. Sci.* **155**, 241

BOPP, B. W. 1974. *Astrophys. J.* **193**, 389

CHABRIER, G. & BARAFFE, I. 1995. *Astrophys. J.* **451**, L29

CLAUSEN, J. V., HELT, B. E., OLSEN, E. H. & GARCIA, J. M. 1997. *In "Fundamental Stellar Properties: the Interaction between Observations and Theory"*. IAU Symp. 189 (ed. T. R. Bedding), p. 56

FLOWER, P. J. 1996. *Astrophys. J.* **469**, 355

GIMÉNEZ, A. & ZAMORANO, J. 1985. *Astrophys. & Sp. Sci.* **114**, 259

GIMÉNEZ, A. & ZAMORANO, J. 1986. *Anales de Fisica* **82**, 121

GREENSTEIN, J. L. 1989. *PASP* **101**, 787

GUIRADO, J. C., REYNOLDS, J. E., LESTRADE, J. F., PRESTON, R. A., JAUNCEY, D. L., JONES, D. L., TZIOUMIS, A. K., FERRIS, R. H., KING, E. A., LOVELL, J. E. J., MC-CULLOCH, P. M., JOHNSTON, K. J., KINGHAM, K. A., MARTIN, J. O., WHITE, G. L., JONES, P. A., ARENOU, F., FROESCHLÉ, M., KOVALEVSKI, J., MARTIN, C., LINDEGREN, L., & SÖDERHJELM, S. 1997. *Astrophys. J.* **490**, 835

HAMILTON, D. & STAUFFER, J. R. 1993. *Astron. J.* **105**, 1855

HARLOW, J. J. B. 1996. *Astron. J.* **112**, 2222

HARMANEC, P. 1988. *Bull. Astron. Inst. Czech.* **39**, 329

HENRY, T. D. & MCCARTHY, D. W. 1993. *Astron. J.* **106**, 773

LACY, C. H. 1977. ASTROPHYS. J. **218**, 444

LEGGETT, S. K. 1992. *Astrophys. J. S.S.* **82**, 351

LEINERT, CH., HENRY, T., GLINDEMANN, A. & MCCARTHY, D. W. 1997. *Astron. & Astrophys.* **325**, 159

LEUNG, K.C. & SCHNEIDER, D. 1978. *Astron. J.* **83**, 618

MACERONI, C. & RUCINSKI, S. 1997. *PASP* **109**, 782

METCALFE, T. S., MATHIEU, R. D., LATHAM, D. W., & TORRES, G. 1996. *Astrophys. J.* **456**, 356

MONET, D. G., DAHN, C. C., VRBA, F. J., HARRIS, H. C., PIER, J. R., LUGINBUHL, C. B., & ABLES, H. D. 1992. *Astron. J.* **103**, 638

POLS, O. R., TOUT, CH. A., SCHRÖDER, K. P., EGGLETON, P. P., & MANNERS, J. 1995. *MNRAS* **274**, 964

POPPER, D. M. 1980. *Ann. Rev. Astron. Astrophys.* **18**, 115

POPPER, D. M. 1996. *Astrophys. J.* **106**, 133

POPPER, D. M. 1997. *Astron. J.* **114**, 1195

PROSSER, C. F. 1992. *Astron. J.* **103**, 488

REID, I. N. & MAJEWSKI, S. R. 1993. *Astrophys. J.* **409**, 635

SÖDERHJELM, S., LINDEGREN, L., & PERRYMAN, M. A. C. 1997. *ESA SP-402*, p. 251

TSUJI, T., OHNAKA, K., & AOKI, W. 1996a. *Astron. Astrophys.* **305**, L1

TSUJI, T., OHNAKA, K., AOKI, W., & NAKAJIMA, T. 1996b. *Astron. Astrophys.* **308**, 29

UPGREN, A. R., RATNATUNGA, K. U., CASERTANO, S., & WEIS, E. 1997. *ESA SP-402*, p. 583

VITTI, S., JONES, H. R. A., SCHWEITZER, A., ALLARD, F., HAUSCHILDT, P. H., TENNYSON, J., MILLER, S., & LONGMORE, A. J. 1997. *MNRAS* **291**, 780

Theory of Low Mass Stars and Brown Dwarfs: Success and Remaining Uncertainties

By ISABELLE BARAFFE AND GILLES CHABRIER

Ecole Normale Supérieure de Lyon - CRAL - 46, allée d'Italie, 69364 Lyon, France

Important progress has been made within the past few years regarding the theory of low mass stars ($m < 1M_\odot$) and brown dwarfs. The main improvements concern the equation of state of dense plasmas and the modelling of cool and dense atmospheres, necessary for a correct description of such objects. These theoretical efforts now yield a better understanding of these objects and good agreement with observations regarding color-magnitude diagrams of globular clusters, mass-magnitude relationships and near-IR color-magnitude diagrams for young open clusters. However uncertainties still remain regarding synthetic optical colors and the complex problem of dust formation in the coolest atmosphere models.

1. Improvement of the theory

Very low mass (VLM) stars and brown dwarfs (BD) are dense and cool objects, with typical central densities of the order of 100–1000 $gr.cm^{-3}$ and central temperatures lower than 10^7 K. Under such conditions, a correct equation of state (EOS) for the description of their inner structure must take into account strong correlations between particles, resulting in important departures from a perfect gas EOS (cf. Chabrier & Baraffe (1997)). Important progress has been made in this field, in particular by Saumon, Chabrier & Van Horn (1995) who developed an EOS specially devoted to VLM stars, BD and giant planets. Since the EOS determines essentially the mechanical structure of these objects, and thus the mass-radius relationship, it can be tested against observations of eclipsing binary systems. Unfortunately, we only know two systems which can offer such a test (cf. Fig. 1 and Chabrier & Baraffe (1995)). Fig. 1 displays also the data from the white dwarf- M dwarf binary system GD 448 (Maxted *et al.* (1998). Although the radius determination of the M-dwarf is model dependent, it provides an interesting case to test the very bottom of the Main Sequence. Morover, this EOS has been successfully confronted to recent laser-driven shock wave experiments realized at Livermore, probing the complex regime of pressure-dissociation and -ionization (cf. Saumon et al. (1998)).

Another essential physical ingredient entering the theory of VLM stars and BD concerns the atmosphere modelling. VLM stars or M-dwarfs are characterized by effective temperatures from $\sim$ 5000 K down to 2000 K and surface gravities $log\, g \approx 3.5 - 5.5$, whereas BDs can cover a much cooler temperature regime, down to some 100 K. Such low effective temperatures allow the presence of stable molecules (H_2, H_2O, TiO, VO, CH_4, NH_3,...), whose bands constitute the main source of absorption along the characteristic frequency domain. Such particular conditions are responsible for strong non-grey effects and significant departure of the spectral energy distribution from a black body emission.

Tremendous progress has been made within the past years to derive accurate non-grey atmosphere models by several groups over a wide range of temperatures and metallicities (Plez, Brett & Nordlund (1992); Saumon *et al.* (1994); Brett (1995); Allard & Hauschildt (1995); Hauschildt, Allard & Baron (1998)). A detailed review of the recent progress in the field is given in Allard *et al.* (1997).

Another difficulty inherent to cool dwarf atmospheres is due to the presence of con-

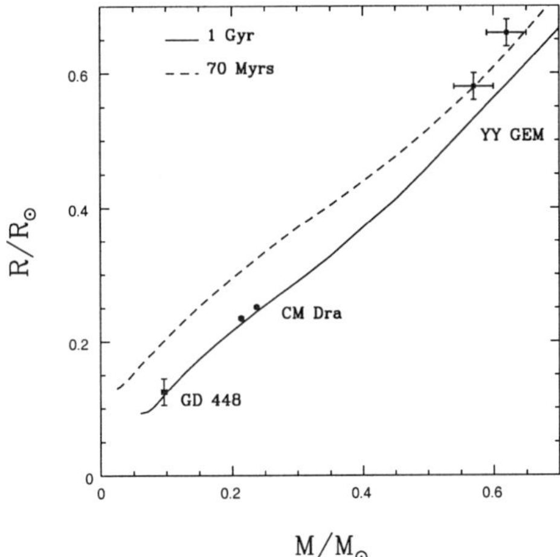

FIGURE 1. Mass-radius relationship for VLM stars down to the BD regime. The solid curves are models from Chabrier & Baraffe (1997) for solar metallicity [M/H]=0 and different ages. The two eclipsing binary systems CM Dra (Metcalfe *et al.* (1996)) and YY Gem (Leung & Schneider (1978)) are indicated. The VLM star component of the binary system GD 448 (but not eclipsing) is also shown (Maxted *et al.* (1998))

vection in the optically thin layers. This particularity is due to the molecular hydrogen recombination ($H+H \rightarrow H_2$) which lowers the adiabatic gradient and favors the onset of convective instability. Since radiative equilibrium is no longer satisfied in such atmospheres, the usual procedure based on $T(\tau)$ relationships to construct grey atmosphere models and to impose an outer boundary condition for the evolutionary models is basically incorrect (cf. Chabrier & Baraffe (1997)). An accurate surface boundary condition based on *non-grey atmosphere models* is therefore required for evolutionary models. As shown in Chabrier & Baraffe (1997), the use of *grey* outer boundary conditions, like the well known Eddington approximation, overestimates the effective temperature for a given mass and yields higher hydrogen burning minimum mass (HBMM).

2. Comparison with observations

Evolutionary calculations based on such improved physics have led to a much better representation of the observed properties of M-dwarfs. The accuracy of the theory is demonstrated by several facts. For metal-poor stars, models based on the interior models of Chabrier & Baraffe (1997) and the atmosphere models of Hauschildt, Allard & Baron (1998) reproduce *accurately* the observed main sequences of different globular clusters observed with the HST in the optical (WFPC2 camera) *and* the near IR (NICMOS camera), which correspond to metallicities ranging from $[M/H] = -2.0$ to -1.0, down to the bottom of the MS (Baraffe *et al.* (1997) (cf. Fig. 2).

The models are also in excellent agreement with the *observationally-determined* mass-magnitude relationship (Henry & McCarthy (1993)) both in the infrared and in the optical (Baraffe *et al.* (1998)) and reproduce accurately color-magnitude diagrams (CMD) in the near IR colors of disk field stars (Baraffe *et al.* (1998); Burrows *et al.* (1997)) and open clusters (Zapatero Osorio, this volume; Jameson *et al.* this volume).

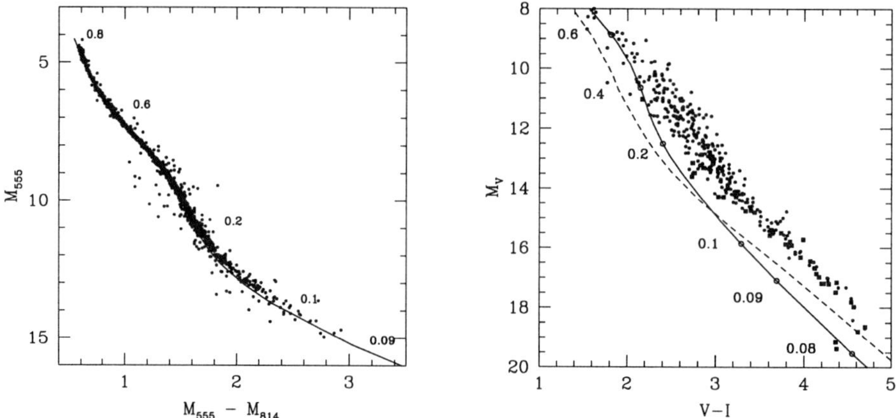

FIGURE 2. (LEFT PANEL). CMD for NGC6397 in the WFPC2 optical filters observed by King *et al.* (1998). The isochrone is from Baraffe *et al.* (1997) for a metallicity [M/H]=-1.5. See Baraffe *et al.* (1997) for distance modulus and reddening correction.

FIGURE 3. (RIGHT PANEL). CMD for disk field stars: the data are from Monet *et al.* (1992) (full squares) and Dahn *et al.* (1995) (full circles). The isochrones are from Baraffe *et al.* (1998) for metallicities [M/H]=-0.5 (dashed line) and [M/H]=0 (solid line) for an age of 5 Gyrs. Masses are indicated by open circles in the solid curve

3. Remaining uncertainties and next challenges

Problems still remain in the optical colors for solar metallicity models, which are significantly too blue for objects fainter than $M_V \sim 10$ (cf. Fig. 3). The source of this discrepancy is analysed in Baraffe *et al.* (1998) and may stem from a missing source of opacity below 1 μm, due to possible shortcomings in the TiO line list. New calculations of this line list are under progress and will hopefully solve this problem.

Moreover, below $T_{\rm eff} \sim 2800K$, grain formation starts to affect the outer layers of the atmospheres, and could possibly be responsible for the discrepancy found at magnitudes fainter than $M_V \sim 16$ (cf. Fig. 3; Allard, this volume; Tsuji, this volume). More direct evidence for the presence of grains can be found by analysing spectra of very-late type objects and their IR colors. Indeed, the first results obtained by the DENIS survey (cf. Delfosse, this volume) reveal several BD candidates showing extremely red (J-K) colors. Observed values of (J-K) > 1 cannot be reproduced by current grainless atmosphere models. Preliminary atmosphere models taking into account the effect of grain condensation can reproduce such a trend of very red IR colors (cf. Allard, this volume; Tsuji, this volume).

Since condensation affects the spectra *and* the structure of atmosphere models, it may as well alters their inner entropy and thus the evolutionary models, as shown in Fig. 4. The present evolutionary models are based on grainless atmosphere models, labeled NEG (Baraffe *et al.* (1998); Allard *et al.* (1996)), atmosphere models including grain formation but with no contribution to the opacity, labeled COND, and dusty models including the dust contribution to the opacity, labeled DUSTY (cf. Allard, this volume). Fig. 4 shows that the effect of grains on the evolutionary models is rather small. They may slightly decrease the HBMM, since the DUSTY 0.07 $M_\odot$ model achieves thermal equilibrium whereas the NEG and COND models for the same mass do not and enter the BD regime. However, models near the stellar/substellar transition are extremely sensitive to input physics and the question whether objects with masses between the

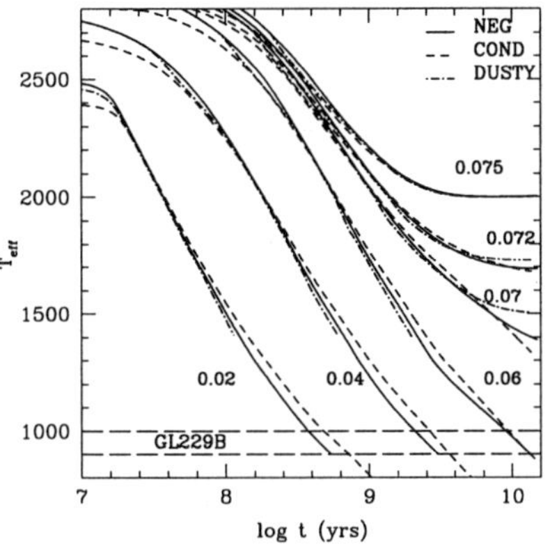

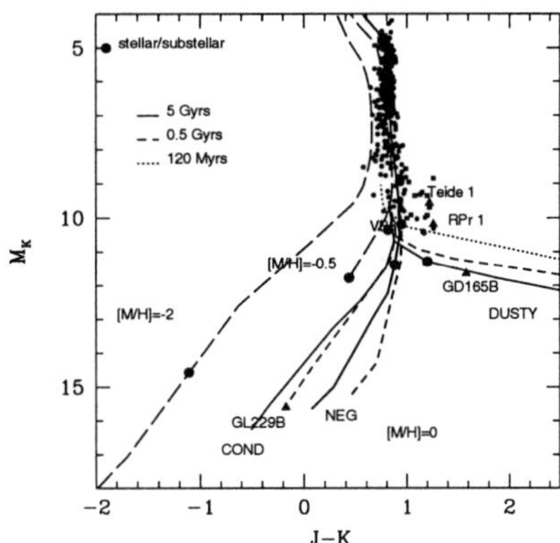

FIGURE 4. (LEFT PANEL). Evolution of the effective temperature as a function of time for VLM stars and BD models of different masses, as indicated in the figure. The outer boundary conditions are based on Allard's (this volume) models: dust-free models (NEG, solid lines); models including the formation of grain only in the EOS (COND, dashed lines); models including grains in the EOS and the opacities (DUSTY, dash-dotted lines). The range of effective temperatures of GL 229B is indicated by the long dashed lines.

FIGURE 5. (RIGHT PANEL). $(J-K)$ - M_K diagram. Observations are from Leggett (1992) (thick dots) for disk stars, Bouvier (this volume, full squares) for the Pleiades. A few BDs are as well indicated by their name (full triangles). Low metallicity isochrones are from Baraffe *et al.* (1997) (long dashed curves). Solar metallicity models are shown for different ages and based on different atmosphere models (see text): COND (solid and dashed curves on the left side), NEG (solid and dashed curves in the middle); DUSTY (solid, dashed and dotted curves on the right side). The full circles on the curves indicate the stellar/substellar transition.

current HBMM value given by dust-free models (0.075 $M_\odot$) and 0.07 $M_\odot$ are stars or BD is only a problem of semantic.

Finally, comparison of such models with observations in a $(J-K)$ - M_K CMD confirms the necessity to take into account grain formation and their opacity in order to explain the reddest objects now observed (*e.g.*, Teide 1, RPr1, GD165B, etc..). Interestingly enough, GL229B is better explained by the COND models, which omit grain opacity. This very likely reflects gravitational settling of the grain species below the photosphere.

Although encouraging, these preliminary results show the complexity of including grain formation in the atmosphere models. This problem represents certainly the next challenge for the theory to complete our understanding of cool BDs.

REFERENCES

ALLARD, F. & HAUSCHILDT, P.H 1995 *ApJ* **445**, 433

ALLARD, F., HAUSCHILDT, P.H., BARAFFE, I. & CHABRIER, G. 1996 *ApJ* **465**, L123

ALLARD, F., HAUSCHILDT, P.H., ALEXANDER, D.R. & STARRFIELD, S. 1997 *ARAA* **35**, 137

BARAFFE, I., CHABRIER, G., ALLARD, F. & HAUSCHILDT, P. 1997 *A&A* **327**, 1054

BARAFFE, I., CHABRIER, G., ALLARD, F. & HAUSCHILDT, P. 1998 *A&A* **337**, 403

BRETT, J.M 1995 *A&A* **295**, 736

BURROWS, A. ET AL. 1997 *ApJ* **491**, 856

CHABRIER, G. & BARAFFE, I. 1995 *ApJ* **451**, L29

CHABRIER, G. & BARAFFE, I. 1997 *A&A* **327**, 1039

Dahn, C.C., Liebert, J., Harris, H.C. & Guetter, H.H. 1995 In *The bottom of the main-sequence and below* (ed. C. Tinney).

Hauschildt, P.H., Allard, F. & Baron, E. 1998 *ApJ*, in press

Henry, T. & McCarthy, D.W.Jr. 1993 *AJ* **106**, 773

King, I.R., Anderson, J., Cool, A.M. & Piotto, G. 1998 *ApJ* **492**, L37

Leggett, S.K. 1992 *ApJS* **82**, 351

Leung, K.C. & Schneider, D. 1978 *AJ* **83**, 618

Maxted, P.F.L., Marsh, T.R., Moran, C., Dhillon, V.S. & Hilditch, R.W. 1998 *MN-RAS*, in press

Metcalfe, T.S., Mathieu, R.T., Latham, D.W. & Torres, G. 1996 *ApJ* **456**, 356.

Monet, D.G. et al. 1992 *AJ* **103**, 638

Plez, B., Brett, J.M., Nordlund, A. 1992 *A&A* **256**, 551

Saumon, D., Chabrier, G., Wagner, D.J., Xie, X., 1998 *Phys. Rev.*, submitted

Saumon, D., Chabrier, G. & Van Horn, H.M. 1995 *ApJS* **99**, 713

Saumon, D., Bergeron, P., Lunine, L.I., Hubbard, W.B. &Burrows, A. 1994 *ApJ* **424**, 333

III. CONVECTION, ROTATION AND ACTIVITY

Convection in Low Mass Stars

By FRANCESCA DANTONA

Osservatorio Astronomico di Roma, Monteporzio, Italy

I review the two main aspects of convection modeling important for the stellar structure: *i)* the determination of the temperature gradient in the stellar interior, and in particular in the superadiabatic part of the envelope, which plays a key role for the determination of the stellar $T_{\rm eff}$; *ii)* the description of chemical mixing -mainly in the presence of nuclear burning. I discuss these two aspects in general, and their importance for low and very low mass stars and brown dwarfs structures.

In particular, I discuss the uncertainty (of $\sim$ 200K) in the $T_{\rm eff}$ of masses $M \lesssim 0.1 M_{\odot}$in the phase of Deuterium burning, and the role of mixing in the problem of Lithium burning in very low masses. For this latter problem, I show that the relation between the age of a cluster and the luminosity of the lithium depletion edge is not only independent of the mixing timescale, but also independent of the metallicity, in the population I range, so that the Lithium test can be safely used as an age indicator.

1. Modeling convection: MLT and FST models

A convection model for general use in stellar structure relies on the computation of two main quantities: the *fluxes* and the *scale length* Λ. As the models presently available are all *local*, one can not expect that the description be unique, so there will be also one or more *tuning* parameters. There are two main models presently available: the Mixing Length Theory (MLT) (Böhm-Vitense 1958), and the Full Spectrum of Turbulence (FST) models by Canuto and Mazzitelli (1991). Other attempts to model convection more in detail (e.g. in two or three dimensions, with consideration of rotation and/or magnetic fields) are limited to the solar case, while the more widely used Large Eddy Simulations (LES) are not yet able to deal with more than a few tens of the billions of scales of turbulence present in a star (see Canuto and Christensen Dalsgaard 1997).

The first main input of a convection model is the *convective flux*: the MLT is a "one eddy" theory: it adopts a phenomenological description of convection based on the hypothesis that the convective energy is carried by convective elements of a unique size, that in the end dissolve into their surroundings. The FST employs the results of the computation of the whole energy distribution of the convective eddies. The MLT fluxes are *much smaller* than the experimental ones (Castaing *et al.* 1989), while the FST parameters are based on the results of laboratory experiments extrapolated to the stellar conditions. Thus the FST fluxes come out from a better description than the MLT ones, and, furthermore, they *can be improved*: we can use today the analytic computations by Canuto and Mazzitelli (1991, CM) which take into account the whole eddies distribution, and the more recent results by Canuto Goldman and Mazzitelli (1996, CGM) which also account for the non linear interactions in the growth rate; further results may come out in the future from other kind of modeling of the Navier Stokes equations, or from the Large Eddy Simulations (LES).

The second input in the model is the *scale length.* The MLT scale length is assumed to be proportional to the pressure scale heigth, $\Lambda = \alpha H_p$), while the FST approach adopts the distance $\Lambda = z$ of the convective layer from the boundary of convection (or a function of this distance: —e.g. Ventura *et al.* (1998) adopt the harmonic mean between the distance from the top and the distance from the bottom—). Also in this respect the FST approach is then more physically sound than the MLT approach, as the values of α

which fit the solar models and most stellar models are so large (from ~ 1.5 to ~ 2.3) that the very basis of the theory is unplausible: we are assuming that a convection element carries on without interacting with the environment for a length over which the pressure has varied by a factor 5 or more. That the geometrical distance from the convective boundary should be taken as scale length is based on experimental grounds: even in the MLT framework the first models have attempted to use this parameter, but the solar $T_{\rm eff}$ is $\sim 10\%$ off with this choice, and this approach was abandoned, until Canuto and Mazzitelli (1991) showed that it allowed a good fit of their solar model, thanks to the larger FST convective fluxes.

Both models are local and need a *tuning* parameter: The tuning of MLT is made by calibrating α (till a few years ago, only on the solar model; today there is a tendency to calibrate it for each star, each evolutionary phase and metallicity, and even through the interior of a star. It is evident that the more we complicate our free parameter, the smaller is the predictive power). The FST tuning is made by adding a small parameter to Λ, namely we define:

$$\Lambda = z + \beta H_p^{top} \tag{1.1}$$

where β is ~ 0.1 (CGM fluxes) or ~ 0.2 (CM fluxes) to fit precisely the solar radius. Although still parametric, then, the FST model is less adjustable than the MLT: the range of variation allowed for the parameters does not allow, in principle, to fit *whatever* observation, so that this model is more easily falsifiable than the MLT. Notice in addition that we can consider among the 'parameters' also the Kolmogorov constant, which today is thought to be in the range 1.7±0.1; its value could be better defined by future experiments, lowering even more the degree of freedom.

2. Tests of the FST model

The FST model has been tested on almost all evolutionary phases for which the external convection is important. It seems then a good time to make a list of its performances. Being a model less tunable than the MLT, its success is not to be undervaluated: it really seems that the FST provides a reasonable description of turbulence in stars in many different evolutionary phases.

1. **Tests on white dwarfs** show that the FST model correctly describes the location of the DB instability strip (Mazzitelli and D'Antona (1991) and the sharp blue edge of the ZZCeti strip (Althaus and Benvenuto 1996, Mazzitelli 1995);

2. **Massive red giants** $T_{\rm eff}$'s are well described with FST models, while they require different α's for different mass if the MLT is employed (Stothers and Chin 1995). In fact, $\Lambda = z$ is verified from 1 to $20M_\odot$ by Stothers and Chin 1997.

3. **Open and Globular Clusters HR diagrams** can be reasonably fit within the FST framework (D'Antona *et al.* 1992; D'Antona *et al.* 1997);

4. **Hot bottom burning** and lithium production are naturally achieved in massive asymptotic giant branch stars using FST (D'Antona and Mazzitelli 1996, Mazzitelli *et al.* 1999);

5. **Atmospheric convection** in the FST framework has been studied by Smalley and Kupka (1997) who show that it provides better *uvby* colors for A, F and G stars. Another interesting hint comes from the works of Fuhrmann *et al.* (1993) and van't Veer & Mégessier (1996): using MLT atmospheric models for normal dwarfs, they find that consistency between the temperature derived from the continuum, and that derived from the absorption line profiles can only be achieved if the value of α is ~ 0.5: in other words, in the outer superadiabatic layers convection must be much more *inefficient* than

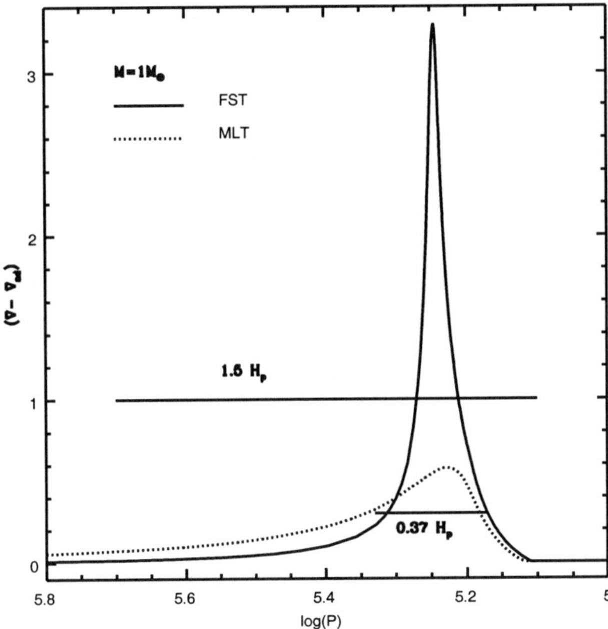

FIGURE 1. The superadiabatic gradient $\nabla - \nabla_{\rm ad}$ in the external layers of the solar model versus the logarithm of the pressure. The FST gradient is much more peaked than the MLT gradient. Notice that the extension in pressure of this latter peak is $\sim 0.4H_p$. We show for an easy comparison the extent of the 1.5 pressure scale height which is the value of α needed to fit the solar radius in the MLT model. It is clear that the MLT model adopts a schematization which is totally inadequate to describe the superadiabatic part of the solar envelope.

implied by the general use of values of α much larger than one to fit the solar radius. This resembles the FST modeling, in which convection is very inefficient in the superadiabatic layers. In fact, Bernkopf (1998) shows that the FST can provide at the same time the fit of the Balmer line profiles in the solar model (which would require $\alpha \simeq 0.5$) and of the solar radius (which would require $\alpha \simeq 1.7$ in the MLT framework).

6. Pre–Main Sequence evolution for masses below $\sim 1M_\odot$ is coherently described: models based on the CM fluxes (D'Antona and Mazzitelli 1994, DM94) provide an overall description of the TTauri loci more satisfying than the MLT description. This problem is discussed in some detail in D'Antona and Mazzitelli 1997 (DM97), and I devote the next section of this review to clarify this point.

7. The solar oscillation spectrum: the residual $\delta\nu$ between the theoretical and observed high frequency solar oscillation are much closer to zero in the FST solar model than in the MLT model (Paterno' *et al.* 1993, Basu and Anthia 1994, Monteiro *et al.* 1996: see for a review Canuto and Christensen Dalsgaard 1998). This problem is still one of the most discussed: the frequency is proportional to the sound velocity and this latter is proportional to the square root of the temperature. The exact temperature profile depends on the overadiabaticity $\nabla - \nabla_{\rm ad}$, which is noticeably different from zero only at the border of the convective zone, where convection becomes "inefficient". Since the MLT is structurally a "high efficiency" model, it is incapable of dealing accurately with regions of low efficiency. This is easily understood when we look at Figure 1, showing the outer solar superadiabaticity versus pressure: the peak of the superadiabatic gradient $\nabla - \nabla_{\rm ad}$

in the MLT spans $\sim 0.4H_p$, but we are describing it with a model having $\alpha = 1.5H_p$: so how can the model provide an accurate description of the outer solar structure? The FST superadiabaticity peak is *much higher*, and this is the characteristic which provides the temperature stratification in better agreement with the high frequency oscillation spectrum. Some numerical simulations of the solar convection (e.g. by by Kim *et al.* 1996) also indicate that the MLT overestimates the efficiency of convection in the superadiabatic layer. Other recent numerical simulations carried out by Nordlund and coworkers find a behaviour of $\nabla - \nabla_{\rm ad}$ more similar to that found in the MLT description and shown in Figure 1. Although this is a model versus model comparison, it is taken very seriously among workers in this field. However, the LES simulations largely depend on the Prandtl number (usually ~ 10 orders of magnitude larger than in stars) and on the subgrid model. The MLT stratification is based on a very poor modeling, as remarked above and shown in Figure 1, so the fact that both LES and MLT provide the same kind of result can not be taken as a proof of their reliability.

In the following we try to understand the role of convection for the modeling of low mass stars and brown dwarfs.

3. Convection in low mass stars

Convection determines the $T_{\rm eff}$ of the stellar model by the evaluation of the *superadiabatic* gradient $\nabla - \nabla_{\rm ad}$. The smaller is the overadiabaticity (e.g. main sequence (MS) stars versus giants (RGs); low mass stars versus more massive ones) the less relevant is the exact treatment of convection. The above statement is (qualitatively) independent of the convection model adopted: the (pre–MS and post–MS) RGs location in the HR diagram varies by several hundreds of degrees if we change Λ/H_p in the MLT, while the MS location is less sensitive –see, e.g., the location of pre–MS and MS solar models for varying convection assumptions in DM94. Below $\sim 0.5M_\odot$ the MS does not shift significantly by changing Λ/H_p, unless we go to extremely small values of the ratio.

This is also true for FST based models: the MS location of low mass models is not much sensitive to the convection treatment. As shown by Baraffe *et al.* 1995 (BCAH95) and Chabrier and Baraffe 1997 (CB97), for the MS of low masses the atmospheric boundary conditions are the most important input for the $T_{\rm eff}$ determination.

Nevertheless, convection in VLMs ($M \lesssim 0.3M_\odot$) presents another complication: it extends into the *optical* atmosphere. In this case, the convection scale definition provided in equation 1.1 should become:

$$\Lambda = z_{phot} + z_{atm} + \beta H_p^{top} \tag{3.2}$$

where z_{phot} is the distance to the photosphere and z_{atm} is the convective atmospheric thickness. If we *do not* consider z_{atm}, that is, we take $z_{atm} = 0$, even the dense structures of low mass preserve a *residual* overadiabaticity and the $T_{\rm eff}$ is $\sim 100-200$K smaller than with the larger Λ. We can justify the choice not to consider z_{atm} on the basis that we expect a more inefficient convection when dealing with the radiative atmospheric losses.

This choice in modeling the scale length is at the basis of the differences in $T_{\rm eff}$ between the low mass model location of the DM94 tracks (adopting the scale in 3.2) and the more recent DM97 release (assuming $z_{atm} = 0$; see also D'Antona and Mazzitelli 1998). Figure 2 shows the comparison between 4 sets of MS locations: the BCAH95 versus CB97, the DM94 versus DM97 models. We see that the $T_{\rm eff}$ location of the MS depends both on the *atmospheric integration and opacities* (e.g. the difference between CB97 and BCAH95 models) and on the detailed treatment of the convection model (difference between the DM94 and DM97 models). We obviously know and acknowledge that this approach is

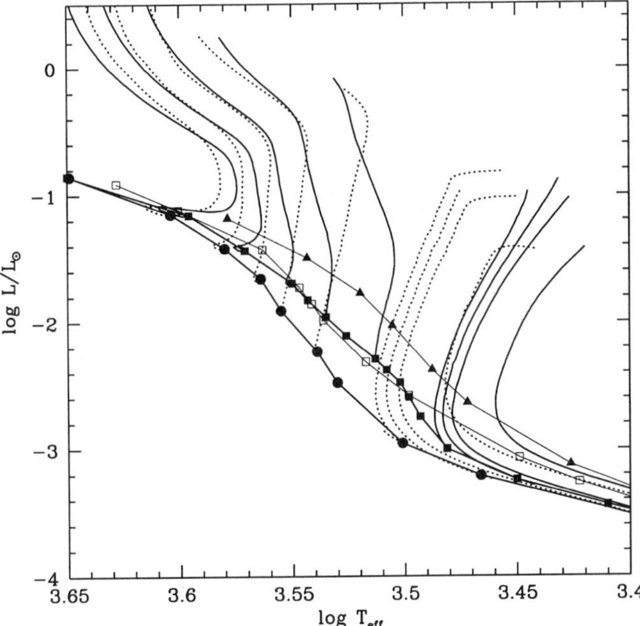

FIGURE 2. Tracks of 0.6, 0.5, 0.4, 0.3, 0.2, 0.1, 0.09, 0.08, and 0.05 $M_\odot$. The continuous lines are the DM 1997 tracks, the dashed tracks are from DM94. The main sequences of BCAH95 (triangles) and from CB97 (open squares) are also shown, both having non grey atmospheric integration and different opacity treatment. The improvement in the opacity treatment from BCAH95 to CB97 *increases* the $T_{\rm eff}$ of the models with non–grey atmosphers by 100 − 200K. The difference discussed in the text in the assumed scale length of convection *decreases* the $T_{\rm eff}$ of the DM94 grey-atmosphere models by 100 − 200K. The $T_{\rm eff}$'s of CB97 and DM97 result to be in close agreement. Notice, however, that this agreement is totally *fortuitous* and that we would need non–grey atmosphere boundary conditions computed with FST before we can meaningfully compare the results.

still very preliminary: if we are reasonable convinced that the FST model does a good job in describing convection in many evolutionary phases, to advance in this field we badly need to attempt to overcome the MLT model *in the non–grey model atmospheres* for VLMs, and join them to FST interior models.

Looking at Figure 3 we see that the transition between stars and BDs, which in the figure is exemplified by the $0.1M_\odot$, is characterized by a sharp change of slope -or drop in luminosity- in the D–burning band. The D-burning phase for VLMs and BDs is so long lived (several Myr) that it is highly probable to sample many VLMs in this phase in the star forming regions. The theoretical uncertainty by $\sim$ 200K in the location of this feature can be possibly observationally checked out. A poster by Ventura *et al.* (1999) shows how study of a large sample of pre–MS low mass objects could help in choosing among the different convection models.

4. The FST model and Pre Main Sequence Evolution

The study of the stellar content of very young stellar populations attempts to shed light on important problems of star formation: how large is the age spread in star forming regions, what is the initial mass function and how it is comparable to the IMF derived

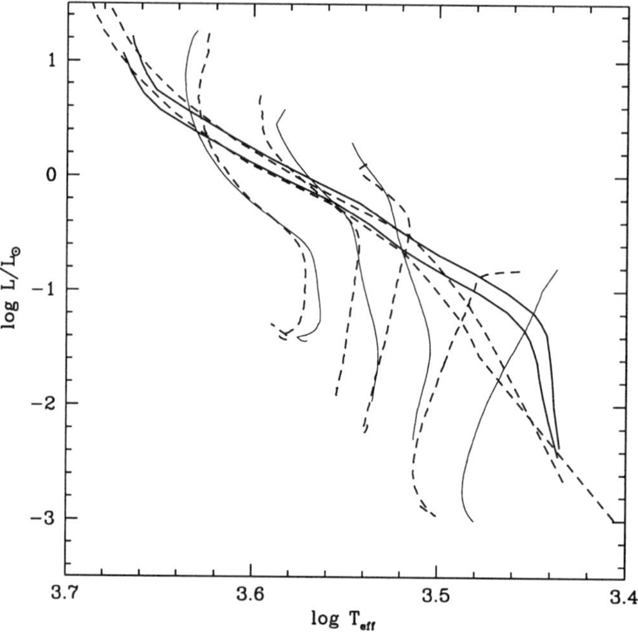

FIGURE 3. Deuterium burning bands and evolutionary tracks from DM94 (Alexander CM set, dashed) and DM97 (full lines) for 0.5, 0.3, 0.2 and $0.1 M_\odot$from left to right. The D–burning phase is very important for VLMs, for which it lasts several million years. Study of many young associations could be able to define observationally the 'turning point' of this region (Ventura *et al.* 1999), at masses $\lesssim 0.1 M_\odot$. The figure shows that there is a difference of $\sim$ 200K between the DM97 and DM94 location of this turning point, due to the different assumptions made for the contribution of atmospheric convection to the scale length.

from MS stars. Of course these question have a particular appeal for the low mass stars and brown dwarfs, which are much more luminous in the pre–MS phase than in MS. Nevertheless, the problem of derivation of IMF and age spread are strictly connected with the treatment of convection. The most used approach to derive age and mass distribution is in fact to attribute to each object a mass and an age by comparison with a given set of evolutionary tracks. Different treatments of convection provide of course different answers both for age and mass. This is exemplified in Figure 4, where the masses and ages for a given observational point are shown from the evolutionary sets which have been mostly used in recent years, by F.J. Swenson (reference in Swenson *et al.* 1994) DM97 and CB97. Notice that the difference in the mass is a factor two and the age a factor three! (This in part depends on the fact that the CB97 models employ a value of mixing length $l/H_p = 1$ which is generally not considered 'reliable' –notice that the choice does not affect the reliability of the MS location of CB97 models, however). One possibility often mentioned is that pre–MS binaries may make it possible to calibrate the convection model. *A unique* calibration in terms of mixing length ratio will not hold for the whole diagram, and a calibration "point by point" of the HR diagram not only seems very difficult but also useless in terms of predictive power. In recent years the comparisons made with the DM94 models employing the FST have given a reasonable performance both at relatively large masses ($\sim 1 M_\odot$ and at the VLM end, so that they seem to provide a fair description of the whole low mass pre–MS evolution (e.g. Lawson *et al.* 1996, Luhman & Rieke 1998). As we have mentioned, for a fair description at

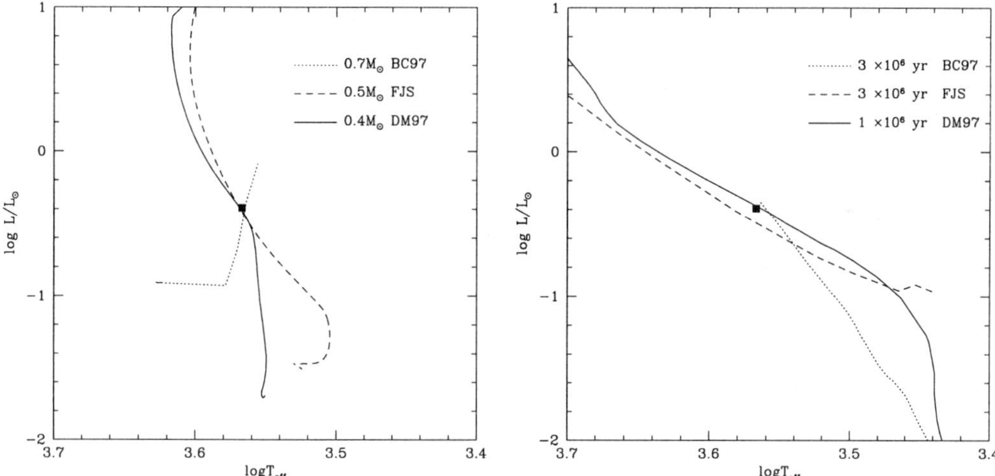

FIGURE 4. For a given observational point, exemplified by the dot, the left figure shows the mass calibration and the right figure the age calibration in terms of three different sets of tracks and isochrones.

low $T_{\rm eff}$, where non–grey boundary conditions are needed, we will need non–grey model atmospheres built including the FST convection model.

5. Mixing timescales

Another important aspect of convection is the *mixing* timescale. The convection model employed provides the average velocity of convective elements at each point, and the convective turnover timescale can be obtained by:

$$\tau_{mix} = \int_{r-bot}^{r-top} dr/v_{conv} \tag{5.3}$$

where the integral is over the whole convective region. Any model up today available for convection is *local* so that, by definition, $v_{conv} = 0$ at the boundaries, and $\tau_{mix} = \infty$. Therefore it is customary to calculate this quantity excluding the meshes close to the convection boundaries.

In the FST tretment, v_{conv} is computed by means of equations 88, 89 and 90 in CGM, and takes into account again the whole eddy distribution. There are noticeable differences between the average FST and MLT velocities and mixing timescales. These are exemplified in Figure 5 in which we show these quantities for 0.1 and $0.05M_{\odot}$. Notice that the FST timescale is a factor $\sim 2-3$ smaller than the MLT one.

Chemical mixing must be treated carefully when we have to deal with nuclear reactions occurring inside the convective region. To have a self–consistent approach, in fact, we must use a diffusion approximation to compute the time derivative of each nuclear species of abundance X_i, and to couple it to nuclear evolution, solving the set of equations;

$$(\frac{dX_i}{dt}) = (\frac{\partial X_i}{\partial t})_{nucl} + \frac{\partial}{\partial m_r}[(4\pi r^2 \rho)^2 D \frac{\partial X_i}{\partial m_r}] \tag{5.4}$$

This has been done, e.g., to deal with the hot bottom burning nuclear processing and mixing in the asymptotic giant branch stars (Sackmann and Boothroyd 1992) and has

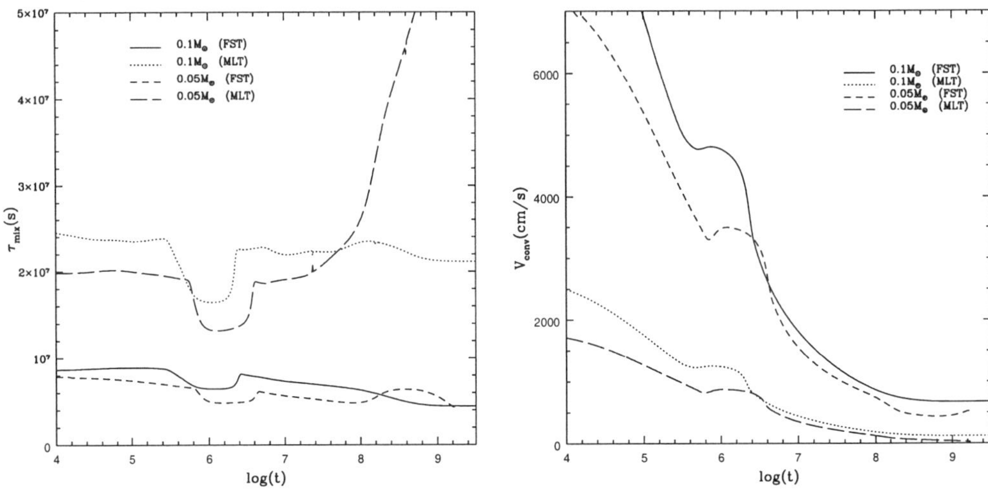

FIGURE 5. Mixing timescale and average velocities as function of age for the tracks of 0.1 and $0.05M_\odot$. The MLT and FST model are compared.

been recently introduced by Ventura *et al.* (1998) for the treatment of 14 elements evolution (22 reactions are included in the network). In particular, Deuterium and Lithium are explicitly treated. The diffusion coefficient is treated in a local approximation as

$$D = \frac{v_{conv} \times \Lambda}{3} \tag{5.5}$$

where the convective velocity and the scale length Λ are computed via the FST.

For what concerns VLMs and BDs, both in the MLT and FST treatment the mixing times are much shorter than the nuclear burning timescales for the reactions important in these objects, namely those involving proton capture by Deuterium ($T \gtrsim 1 \times 10^6$K) and by Lithium ($T \gtrsim 3 \times 10^6$K), so that the Deuterium and Lithium burning can be treated in the approximation of instantaneous mixing (see the poster by Zeppieri *et al.* 1999 and Ventura and Zeppieri 1999).

6. The Lithium test in brown dwarfs

Based on the previous result, we can confidently say that when a fully convective object shows lithium at its surface, it still preserves lithium in its interior. This renders the "lithium test" (Rebolo 1991, Magazzù *et al.* 1991: see in this book the review by Martín 1999) a very precise indicator of the central temperature of the star, and so of its possible brown dwarf nature.

The presence of Lithium in VLMs in a young cluster is also a powerful method to date it, as outlined by DM94 and applied for the first time to the Pleiades by Basri *et al.* 1996, and more recently, e.g., by Stauffer *et al.* 1998. Figure 6 shows the Lithium versus luminosity tracks and isochrones for VLMs and BDs. Lithium depletion occcurs at smaller luminosity and larger age the smaller the mass, and thus the luminosity boundary between objects having already depleted lithium and objects which still preserve it (the "lithium edge") is an age indicator. The best range of ages to date a cluster is between $\sim$ 30 and $\sim$ 400Myr: below 30Myr, depletion is scarce; above 400Myr the lithium

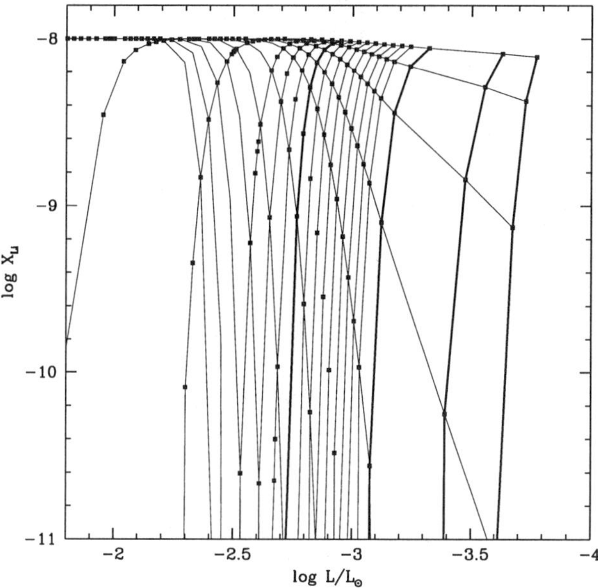

FIGURE 6. Lithium isochrones and evolutionary tracks for solar composition. From left to right the tracks of 0.1, 0.095, 0.09, 0.085, 0.08, 0.075, 0.07, 0.068, 0.065, 0.062 and 0.06$M_\odot$ are shown: smaller masses deplete lithium at lower luminosity, until the 0.06$M_\odot$ preserves the initial content. The isochrones from 3×10^7yr to 4×10^8yr are shown. The thick lines are the 10^8, 2, 3 and 4×10^8yr isochrones. The wide separation of the isochrones for ages between 30 and 400Myr shows that in this range the lithium test is very powerful.

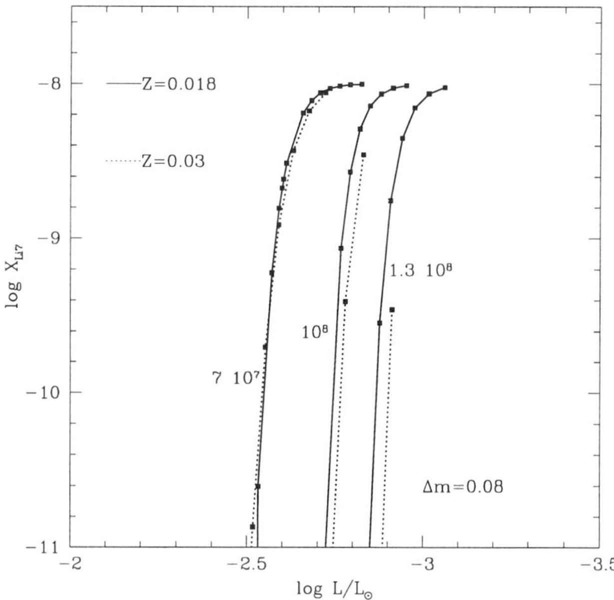

FIGURE 7. Lithium isochrones for solar (metal mass fraction Z=0.018) and larger than solar (Z=0.03) metallicity. While the masses along each of the isochrone points are different in the two cases, the "lithium edge" is not much dependent on the metallicity. This is another reason why the lithium test can be significantly used for dating purposes.

edge becomes dimmer and dimmer, posing hard detection problems in clusters, and the isochrones become much closer together, losing their dating capability.

Interestingly enough, the lithium edge isochrones *do not depend significantly on the cluster metallicity*, as it is shown in Figure 7 by comparing the case of Z=0.018 and Z=0.03. This characteristic renders the lithium test even more powerful, as it is not necessary to know precisely the cluster metallicity to date it.

I thank the Scientific Organizing Committee for the invitation to deliver a talk at this meeting, and acknowledge the role of my coworkers Italo Mazzitelli, Paolo Ventura and Anna Zeppieri in defining the issues described above.

REFERENCES

ALTHAUS, L. G. & BENVENUTO, O. 1996 *MNRAS* **278**, 981.

BARAFFE, I., CHABRIER, G., ALLARD, F., HAUSCHILDT, P.H. 1995 *ApJ* **446**, L35 (BCAH95).

BASRI, G., MARCY, G.W., GRAHAM, J.R. 1996, *ApJ*, **458**, 600.

BASU, S. & ANTHIA, H.M. 1994, *Journal of Astrophysics and Astronomy*, **15**, 143.

BERNKOPF, J. 1998, *A&A*, **332**, 127.

BÖHM-VITENSE, E. 1958, *Z. Astrophys.*, **46**, 108.

CANUTO, V.M., & MAZZITELLI, I. 1991 *ApJ* **370**, 295.

CANUTO, V.M., GOLDMAN, I. & MAZZITELLI, I. 1996 *ApJ* **473**, 550.

CANUTO, V.M., & CHRISTENSEN DALSGAARD, J. 1998 *Ann. Rev. Fluid Mech.* **30**, 167.

CASTAING, B., GUNARATNE, G., HELSOT, F., KADANOFF, L., LIBCHABER, A., *et al.* 1989, *J. Fluid Mech.* **204**, 1.

CHABRIER, G., BARAFFE, I. 1997 *A&A* **327**, 1039 (CB97).

D'ANTONA, F., MAZZITELLI, I., & GRATTON, R.G. 1992, *A&A*, **257**, 539.

D'ANTONA, F. & MAZZITELLI, I. 1994, *ApJS* **90**, 467 (DM94).

D'ANTONA, F., MAZZITELLI, I. 1996, *ApJ*, **470**, 1093.

D'ANTONA, F., CALOI, V., & MAZZITELLI, I. 1997, *ApJ*, **477**, 519.

D'ANTONA, F. & MAZZITELLI, I. 1997, in *"Cool stars in clusters and associations"*, *Mem. S.A.It.* **68**, 807 (DM97).

D'ANTONA, F. & MAZZITELLI, I. 1998, in *'Brown Dwarfs and Extrasolar Planets"*, PASP Conference Series 134, eds. R. Rebolo, E. Martín, M.R. Zapatero Osorio, p. 442.

FUHRMANN, K., AXER, M.,GEHREN, T. 1993, *A&A*, **271**, 451.

KIM, Y.-C., FOX, P.A., DEMARQUE, P. 1996, *ApJ*, **461**, 499.

LAWSON, W.A., FEIGELSON,E.D., HUENEMOERDER, D.P. 1996, *MNRAS*, **280**, 1071.

LUHMAN, K.L., RIEKE, G.H. 1998, *ApJ*, **497**, 354.

MAGAZZÙ, A., MARTÍN, E. L., REBOLO, R. 1991, *A&A*, **249**, 149.

MARTÍN, E. L. 1999, in this book.

MAZZITELLI, I. & D'ANTONA, F. 1991 *"White Dwarfs"*, eds. Vauclair, G. & Sion, E., (Kluwer Publ., 1991) 305.

MAZZITELLI, I. 1995 *White Dwarfs*, eds. D. Koester and K. Werner, Lecture Notes in Physics **443**, 58.

MAZZITELLI, I., VENTURA, P., D'ANTONA, F. 1999, in preparation.

MONTEIRO, M. J. P. F. G., CHRISTENSEN-DALSGAARD, J., THOMPSON, M. J. 1996, *A&A*, **307**, 624.

PATERNÒ, L. *et al.* 1993, *ApJ*, **402**, 733.

REBOLO, R. 1991, *in IAU Symposium* **145**, ed. G. Michaud and A. Tutukov, (Kluwer, Dordrecht), p.85.

SACKMANN, I.J. & BOOTHROYD, A.I. 1992, *ApJ*, **392**, L71.

SMALLEY, B., KUPKA, F. 1997, *A&A*, **328**, 349.

STAUFFER, J.R., SCHULZ, G. & KIRKPATRICK, J.D. 1998, *ApJ*, **499**, L199.

STOTHERS, R.B., & CHIN, C.–W. 1995, *ApJ*, **440**, 297

STOTHERS, R.B., & CHIN, C.–W. 1997, *ApJ*, **478**, L103.

SWENSON, F.J. *et al.* 1994, *ApJ*, **425**, 286.

VAN'T VEER, C. & MÉGESSIER, C. 1996, *A&A*, **309**, 879.

VENTURA, P., ZEPPIERI, A., MAZZITELLI, I. & D'ANTONA, F. 1998, *A&A*, **334**, 953.

VENTURA, P., ZEPPIERI, A. 1999, *A&A*, **340**, 77.

VENTURA, P., ZEPPIERI, A., D'ANTONA, F., MAZZITELLI, I. 1999, Proceedings of Euroconference *"Very Low-Mass Stars and Brown Dwarfs in Stellar Cluster and Associations "*, eds. R. Rebolo & M.R. Zapatero Osorio, in press.

ZEPPIERI, A., VENTURA, P., D'ANTONA, F., MAZZITELLI, I. 1999, Proceedings of Euroconference *"Very Low-Mass Stars and Brown Dwarfs in Stellar Cluster and Associations "*, eds. R. Rebolo & M.R. Zapatero Osorio, in press.

Rotation Law and Magnetic Field for M Dwarf Models

By G. RÜDIGER AND M. KÜKER

Astrophysikalisches Institut Potsdam, An der Sternwarte 16, 14482 Potsdam, Germany

In stellar convection zones and fully convective stars, the rotation profiles are determined by the balance between the Reynolds stress and the meridional circulation. Due to the Coriolis force, the Reynolds stress has a non-diffusive component called Λ-effect that drives both differential rotation and meridional motions. The solar differential rotation pattern is almost perfectly reproduced by a mixing-length model of the convection zone that takes into account the influence of the Coriolis force on the convective motions. The same model also yields the turbulent electromotive force that together with rotational shear drives the solar dynamo.

The model has recently been applied to a fully convective pre-main sequence star. We find that for a strictly spherical star without any latitudinal gradients in temperature, density and pressure the rotation is very close to the rigid-body state. We conclude that the stellar magnetic field must be generated by a mechanism quite different from that in the Sun, namely an α^2 rather than an $\alpha\Omega$-dynamo. It is thus very likely to have non-axisymmetric geometry and not to show cyclic behavior.

We study the analogous problem for M dwarfs. Like the T Tauri stars, these objects are fully convective and may hence be expected to have similar rotational profiles and magnetic field structures, respectively. As their Coriolis numbers are, however, closer to solar values than to those of pre-main sequence stars, the rotation may also be of solar-type.

1. Introduction

At present, the idea is widely accepted, that the solar magnetic field is generated in the overshoot layer at the bottom of the convection zone. This model has replaced the older one of a convection zone dynamo for three reasons. First, helioseismology has revealed that the radial rotational shear necessary to produce the toroidal field is only found in the overshoot layer. Moreover, the shear is positive and hence an $\alpha\Omega$-type dynamo with the α-effect acting in the convection zone would produce a poleward drift of the magnetic field. Second, in the convection zone the magnetic field is subject to buoyancy forces that transport the flux from the bottom to the surface within about one month and thus inhibit the generation of a large-scale field. A third problem for the convection zone dynamo is the fact that the orientation of the field that emerges from bipolar regions is almost parallel to the equatorial plane, which suggests that the origin of the field is a region that is less turbulent than the convection zone.

While the overshoot dynamo explains the solar cycle quite well, it can obviously not work in fully convective stars. We are thus expecting the M dwarf activity as substantially different from the solar activity. In opposition to the solar case fully convection stars can answer the question whether convection of rotating stars is able to produce an alpha-effect or not. That the answer is Yes in the moment we only know for the case which are simulated by the geophysicists (Glatzmaier & Roberts 1996).

As well-known the internal stellar rotation is one of the key processes for the amplification of magnetic fields. Consequently, we have to know the stellar rotation law in order to design a particular dynamo model. We therefore present here first results for a theory of differential rotation in cool M stars.

2. Theory of differential rotation

In a convective star the turbulent convective motions cause an additional stress on the mean (global) motion known as Reynolds stress. While this stress can be described as an additional viscosity in case of a non-rotating convection zone this is no longer correct as soon as the rotation period becomes comparable with the convective turnover time. In that case a non-viscous contribution, the Λ-effect, arises forcing differential rotation.

In Küker *et al.* (1993), Reynolds stress was considered the only transporter of angular momentum in the solar convection zone. This model yields a rotation pattern that agrees almost perfectly with the observations from helioseismology and thus confirms the assumption, that the meridional flow is of minor importance for the problem of solar differential rotation. The latter assumption can, however, not be correct for stars in general since a model based on Reynolds stress alone always yields a normalized differential rotation that increases with increasing rotation rate, in contradiction to the observations.

We therefore present a model of a rapidly rotating fully convective main sequence star in which Reynolds stress and meridional circulation are treated consistently, i.e. we solve the full Reynolds equation,

$$\rho\left[\frac{\partial\bar{\boldsymbol{u}}}{\partial t}+(\bar{\boldsymbol{u}}\cdot\nabla)\bar{\boldsymbol{u}}\right]=-\nabla\cdot(\rho Q)-\nabla\bar{p}+\rho\boldsymbol{g}+\nabla\cdot\pi. \tag{2.1}$$

Here

$$Q_{ij}=\langle u_i'(\boldsymbol{x},t)u_j'(\boldsymbol{x},t)\rangle \tag{2.2}$$

is the correlation tensor of the fluctuating part $\boldsymbol{u}'$ of the velocity field, $\bar{\boldsymbol{u}}$ denotes its mean velocity. The molecular stress tensor π can be neglected since it is many orders of magnitude smaller than the Reynolds stress.

We only treat the axisymmetric case. The velocity field can then be separated into a global rotation and the meridional flow:

$$\bar{\boldsymbol{u}}=r\sin\theta\,\Omega\,\boldsymbol{e}_\phi+\boldsymbol{u}^{\mathrm{m}}, \tag{2.3}$$

where $\boldsymbol{e}_\phi$ is the unit vector in the azimuthal direction. The azimuthal component of the Reynolds equation reads:

$$\frac{\partial\rho r^2\sin^2\theta\,\Omega}{\partial t}+\nabla\cdot\boldsymbol{t}=0, \tag{2.4}$$

where

$$\boldsymbol{t}=r\sin\theta\left[\rho r\sin\theta\Omega\boldsymbol{u}^{\mathrm{m}}+\rho\langle u_\phi'\boldsymbol{u}'\rangle\right]. \tag{2.5}$$

The meridional circulation can be obtained by taking the azimuthal component of the curl of (2.1):

$$\frac{\partial\omega}{\partial t}=-\left[\nabla\times\frac{1}{\rho}\nabla(\rho Q)\right]_\phi+r\sin\theta\frac{\partial\Omega^2}{\partial z}+\frac{1}{\rho^2}(\nabla\rho\times\nabla p)_\phi, \tag{2.6}$$

where $\omega=(\nabla\times\bar{\boldsymbol{u}})_\phi$. In (2.6), we have omitted all nonlinear terms except the one including Ω^2. We use the anelastic approximation

$$\nabla\cdot(\rho\bar{\boldsymbol{u}})=0, \tag{2.7}$$

i.e. the density is constant with time but varies with depth. Both density and pressure are assumed to be functions of the fractional stellar radius only and their gradients are thus aligned. As a consequence, the last term in (2.6) vanishes and the rotation pattern is determined by the balance between the meridional flow and the Reynolds stress.

In the correlation tensor Q, a viscous and a non-viscous part can be distinguished,

$$Q_{ij} = -\mathcal{N}_{ijkl}\frac{\partial \bar{u}_k}{\partial x_l} + \Lambda_{ijk}\Omega_k. \tag{2.8}$$

The viscous part is given by

$$\begin{aligned}\mathcal{N}_{ijkl} &= \nu_1(\delta_{ik}\delta_{jl} + \delta_{jk}\delta_{il}) \\ &+ \nu_2\left(\delta_{il}\frac{\Omega_j\Omega_k}{\Omega^2} + \delta_{jl}\frac{\Omega_i\Omega_k}{\Omega^2} + \delta_{ik}\frac{\Omega_j\Omega_l}{\Omega^2} + \delta_{jk}\frac{\Omega_i\Omega_l}{\Omega^2} + \delta_{kl}\frac{\Omega_i\Omega_j}{\Omega^2}\right) \\ &+ \nu_3\delta_{ij}\delta_{kl} - \nu_4\delta_{ij}\frac{\Omega_k\Omega_l}{\Omega^2} + \nu_5\frac{\Omega_i\Omega_j\Omega_k\Omega_l}{\Omega^4}\end{aligned} \tag{2.9}$$

(Kitchatinov *et al.* 1994). The viscosity coefficients,

$$\nu_n = \nu_0\phi_n(\Omega^*), \qquad n = 1\ldots 5, \tag{2.10}$$

depend on the angular velocity as well as on the convective turnover time, $\tau_{\rm corr}$, via the Coriolis number

$$\Omega^* = 2\tau_{\rm corr}\Omega. \tag{2.11}$$

In the limiting case of very slow rotation, $\Omega^* \ll 1$, the viscous stress becomes isotropic and reduces to the well-known stress-strain relation with the viscosity coefficient

$$\nu_0 = c_\nu\tau_{\rm corr}u_{\rm T}^2, \tag{2.12}$$

which we therefore use as the reference value of the turbulence viscosity. In (2.12), c_ν is a dimensionless number smaller than unity, $\tau_{\rm corr}$ the correlation time of the turbulence and $u_{\rm T} = \sqrt{\boldsymbol{u}'^2}$ the amplitude of the velocity fluctuations. In stellar convection zones $\tau_{\rm corr}$ equals the convective turnover time while $c_\nu \sim 1/3$.

The second contribution, the Λ-effect, is the source of differential rotation. In spherical polar coordinates, it is only present in the components $Q_{r\phi}$ and $Q_{\theta\phi}$, i.e.

$$Q^\Lambda_{r\phi} = \nu_0(V^{(0)} + V^{(1)}\sin^2\theta)\sin\theta\Omega, \tag{2.13}$$

$$Q^\Lambda_{\theta\phi} = \nu_0 H^{(1)}\sin^2\theta\cos\theta\Omega. \tag{2.14}$$

The functions ϕ_n have been derived in Kitchatinov *et al.* (1994) and those for $V^{(0)}$, $V^{(1)}$, and $H^{(1)}$ can be found in Kitchatinov & Rüdiger (1993). Both the diffusive and non-diffusive terms in the Reynolds stress are suppressed and deformed by rotation. From our point of view the question is how the effects combine to the resulting rotation law.

3. The alpha-puzzle

The evolution of the mean magnetic field $\bar{\boldsymbol{B}}$ is governed by the dynamo equation

$$\frac{\partial\bar{\boldsymbol{B}}}{\partial t} = \nabla\times(\bar{\boldsymbol{u}}\times\bar{\boldsymbol{B}} + \boldsymbol{\mathcal{E}}), \tag{3.15}$$

where $\boldsymbol{\mathcal{E}}$ is the turbulent electromotive force, $\boldsymbol{\mathcal{E}} = \langle\boldsymbol{u}'\times\boldsymbol{B}'\rangle$. As usual, we assume approximate scale separation and write

$$\mathcal{E}_i = \alpha_{ij}\bar{B}_j + \eta_{ijk}\bar{B}_{j,k}. \tag{3.16}$$

Basic for the theory is the knowledge of the tensors α and η. For slow rotation and weak magnetic field these tensors take the simple and well-known forms

$$\alpha_{ij} = \alpha_0\delta_{ij} \quad \text{and} \quad \eta_{ijk} = \eta_{\rm T}\epsilon_{ijk}, \tag{3.17}$$

hence

$$\boldsymbol{\mathcal{E}} = \alpha_0\bar{\boldsymbol{B}} - \eta_{\rm T}\,{\rm rot}\,\bar{\boldsymbol{B}}. \tag{3.18}$$

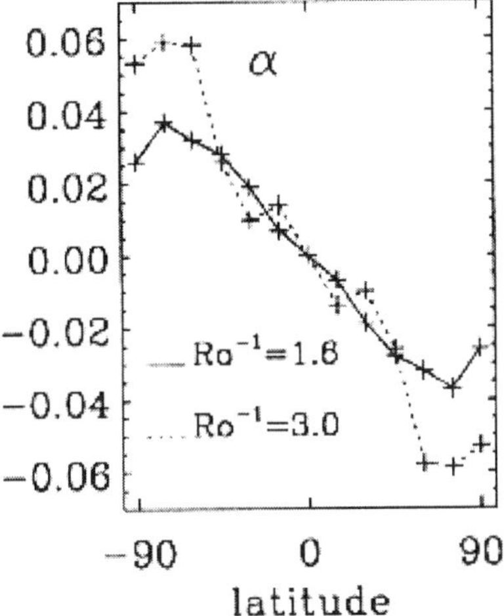

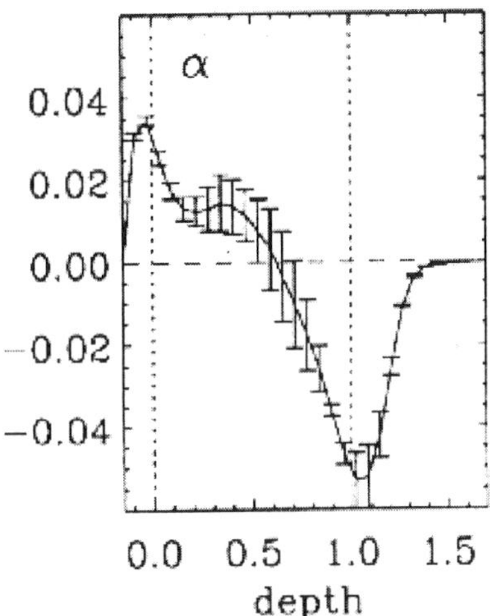

FIGURE 1. LEFT PANEL: The latitudinal dependence of α for 2 simulations. RIGHT PANEL: The depth dependence of α for the north pole. Note the negative values at the bottom of the convective domain (Brandenburg 1994)

In case of arbitrary rotation rate but weak magnetic field, both the α-effect and the magnetic diffusivity tensor become strongly anisotropic. The α-effect consists of the contributions from the stratifications of density and turbulence velocity. The contribution from density stratification reads

$$\alpha_{ij}^{\rho} = -\delta_{ij}(\boldsymbol{G\Omega})\alpha_1^{\rho} - (G_i\Omega_j + G_j\Omega_i)\alpha_2^{\rho} - (G_i\Omega_j - G_j\Omega_i)\alpha_3^{\rho} - \frac{\Omega_i\Omega_j}{\Omega^2}(\boldsymbol{G\Omega})\alpha_4^{\rho}, \quad (3.19)$$

where $\boldsymbol{G} = \nabla \log \rho$. A similar expression holds for the α-effect from the stratification of the turbulence, i.e.

$$\alpha_{ij}^{u} = -\delta_{ij}(\boldsymbol{U\Omega})\alpha_1^{u} - (U_i\Omega_j + U_j\Omega_i)\alpha_2^{u} - (U_i\Omega_j - U_j\Omega_i)\alpha_3^{u} - \frac{\Omega_i\Omega_j}{\Omega^2}(\boldsymbol{U\Omega})\alpha_4^{u}, \quad (3.20)$$

with $\boldsymbol{U} = \nabla \log \sqrt{u_T^2}$. The vectors $\boldsymbol{G}$ and $\boldsymbol{U}$ point to opposite directions. Hence, the terms (3.19) and (3.20) have different signs. While in the convectively stable overshoot layer at the bottom of the solar convection zone the total sign of α is that of α^u, the α-effect due to the density stratification dominates in stellar convection zones. A widely used approximation for the α-effect is its relation to the helicity $\mathcal{H} = \langle \boldsymbol{u}' \operatorname{rot} \boldsymbol{u}' \rangle$:

$$\alpha \propto -\tau_{\text{corr}}\mathcal{H} \quad (3.21)$$

(Krause & Rädler 1980). Models of SN explosion indeed seem to confirm this relation (Korpi *et al.* 1998). Recently, the helicity has also been derived from solar surface mesogranulation pattern observations as negative (left-handed, see Rüdiger *et al.* 1998). The expected α-value in convection zones is thus positive at the northern hemisphere.

For slow rotation α is proportional to Ω. The numerical simulations are in agreement with this finding. The α-effect results as positive at the top of the convection zone of the northern hemisphere (Fig. 1, Brandenburg 1994). This is in accordance with the well-known α-relation

$$\alpha_{\phi\phi} = -l_{\text{corr}}^2\Omega\frac{d}{dr}\log(\rho u_T)\cos\theta, \quad (3.22)$$

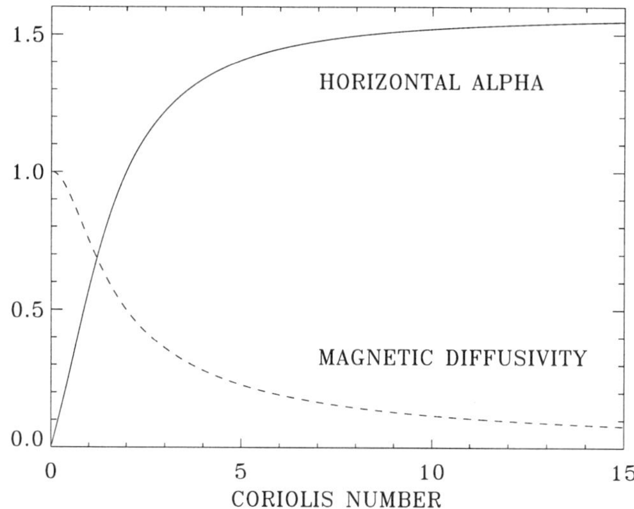

FIGURE 2. The horizontal part of the α-effect, $\alpha_{\phi\phi}$, and the magnetic diffusivity vs. the Coriolis number. Both functions are given in dimensionless units

where the helicity is expressed by the influence of rotation on stratified turbulence. In (3.22), $l_{\rm corr}$ denotes the mixing length. Two remarks are reasonable:

i) The vertical component α_{zz} can have the opposite sign to the horizontal components of the α-tensor (Brandenburg *et al.* 1990). Investigating the expansion flow caused by SN explosions in the interstellar medium Ferrière (1993) also found that α_{zz} can have the opposite sign to the horizontal components of the α-tensor. A negative α_{zz} has also been obtained by Kaisig *et al.* (1993) using an axisymmetric compressible simulation. Rüdiger & Kitchatinov (1993) found a negative value for α_{zz} for stratified turbulence and intermediate values of the inverse Rossby number.

ii) At the base of the convection zone we have a steep *decrease* of the turbulence intensity so that (3.22) turns there to negative values (at northern hemisphere) (Krivodubskij & Schultz 1993). Again, the simulation presented in Fig. 1 is in accord with this argument. Negative values for the α-effect are necessary for the operating of the solar dynamo (Rüdiger & Brandenburg 1995).

There is a new discussion with the α-effect in shear flows. A quasilinear computation (SOCA) of the influence of the differential rotation on the α-effect leads to

$$\alpha \simeq -l_{\rm corr}^2 \frac{\partial\Omega}{\partial\theta}\frac{d\log(\rho u_{\rm T})}{dr}\sin\theta \tag{3.23}$$

for a sphere or

$$\alpha \simeq -l_{\rm corr}^2 \frac{d\Omega}{ds}\frac{d\log\rho}{dz} \tag{3.24}$$

for an accretion disk (Pipin *et al.* 1999), where s is the distance from the axis of rotation. The latter relation for accretion disks with $\partial\Omega/\partial s < 0$ yields negative values in the northern hemisphere and positive values in the southern hemisphere (Brandenburg & Donner 1997). For the solar overshoot region, however, with (3.23) negative values for the northern α only result for the steep decrease of the turbulence intensity with depth. The majority of the turbulence quantities in the mean-field theory vanishes for high rotation rates (Kitchatinov *et al.* 1994). Interestingly enough it is not true for the

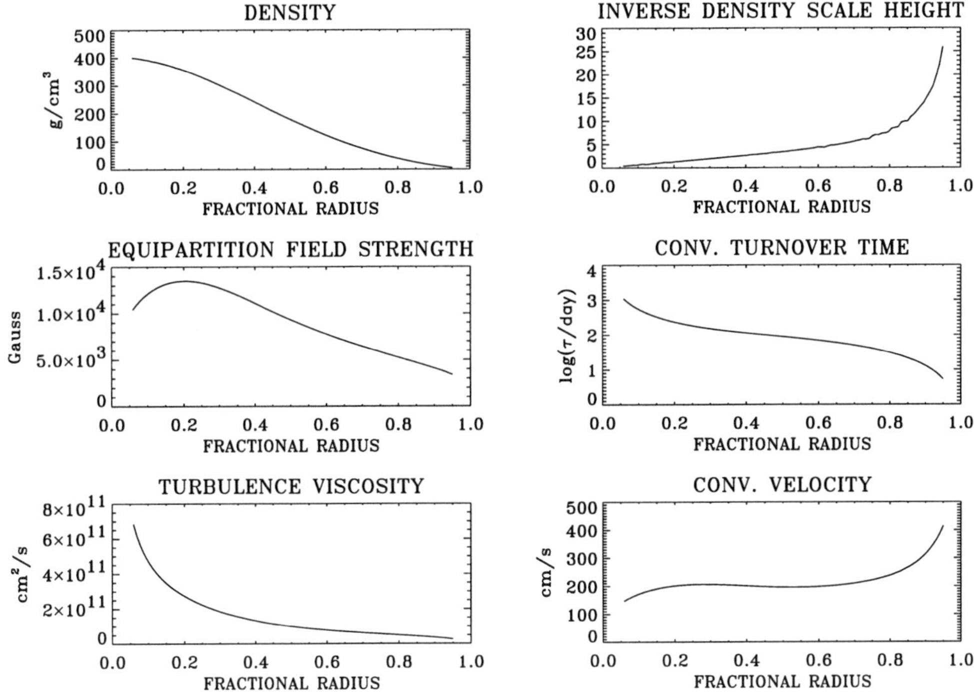

FIGURE 3. The stratification of an M dwarf with $0.1M_\odot$. All quantities are plotted vs. the fractional stellar radius.

α-effect itself. For fast rotation the α-effect saturates but becomes highly anisotropic (Rüdiger 1978; Rüdiger & Kitchatinov 1993).

In Fig. 2, the dependences of the horizontal part of the α-effect, $\alpha_{\phi\phi}$, and the magnetic diffusivity on the Coriolis number are shown. In case of rapid rotation, i.e. large Coriolis numbers, $\alpha_{\phi\phi}$ approaches a finite limit while the magnetic diffusivity decreases as $1/\Omega^*$. For sufficiently rapid rotation the dynamo will thus always become supercritical.

4. Models and their internal flows

Stellar structure models by Chabrier and Baraffe (1997) are used, where all stars with less than 0.35 solar masses were found to be fully convective. The first model describes a star with 0.3 solar masses at an age of 5.6 Gyr. This star has a radius of 0.3 solar radii and a central density of 90 g/cm^3. The second star has 0.1 solar masses, an age of 6.6 Gyr, 0.12 solar radii, and a central density of 400 g/cm^3. Fig. 3 shows its stratification. From top left to bottom right, the plots show density, amplitude of the density stratification vector, $\boldsymbol{G} = \log \nabla\rho$, equipartition field strength $B_{\rm eq}$, convective turnover time $\tau_{\rm corr}$, reference value ν_0 of the turbulence viscosity and convection velocity $u_{\rm T}$ from mixing-length theory. While the radius of this star approximately equals the depth of the convection zone of the Sun, the densities are two orders of magnitude larger. The convection velocities, on the other hand, are much smaller than those in the Sun. The convective turnover times of about three months result in Coriolis numbers larger than one, especially for young stars with rotation periods of several days or even less than one day. Hence, in the context of mean-field magnetohydrodynamics, M dwarfs must be regarded as rapid rotators. Slow rotation is only reached for M dwarfs with rotation periods exceeding 1

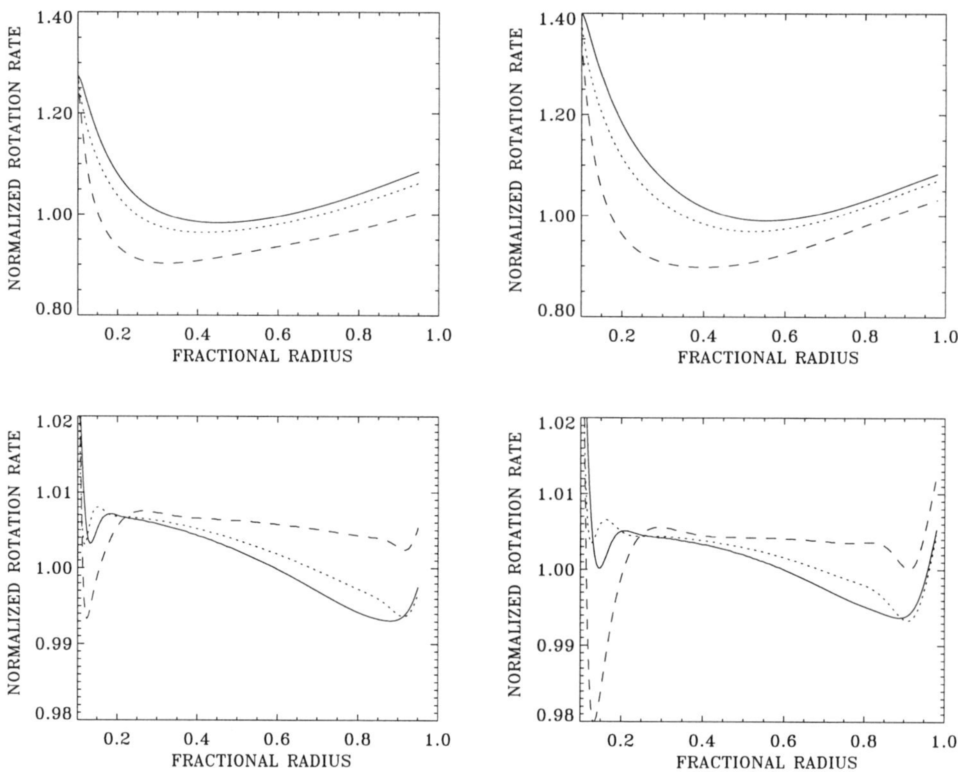

FIGURE 4. Normalized rotation rates at the equator (solid lines), 30° latitude (dotted), and 60° latitude (dashed) for M dwarfs with $M = 0.1M_\odot$ (left column) and $M = 0.3M_\odot$ (right column). The average angular velocities are $10^{-7}\mathrm{s}^{-1}$ (top) and $10^{-6}\mathrm{s}^{-1}$ (bottom).

year.† To determine the rotation profiles, we have solved the system (2.4,2.6) for the stellar models described above with an explicit time dependent finite difference code in spherical polar coordinates. The stellar surface is assumed to be stress-free,

$$Q_{r\phi} = Q_{r\theta} = 0. \tag{4.25}$$

For technical reasons, we exclude the innermost part of the star by imposing a second boundary at a fractional radius of 0.1 with the same boundary conditions.

Fig. 4 shows the resulting rotation profiles for two different angular frequencies, namely $\Omega = 10^{-7}\mathrm{s}^{-1}$ and $\Omega = 10^{-6}\mathrm{s}^{-1}$. The first value corresponds to very slow rotation with a period of 727 days while in the second case the period of 72 days is still three times longer than that of the solar rotation. In the first case (top row), the Coriolis numbers are larger than one at small radii and smaller than one close to the surface. Hence, there are different rotation patterns. While the central parts of the stars show negative radial shear, the rotation profiles are flatter and the shear is positive in the upper layers. The plots in the bottom row show a completely different picture. The rotation has become almost rigid. There is now a very small negative shear throughout the whole star except two thin layers at the boundaries and the interior of the tangent cylinder around the (artificial) inner boundary. For more rapid rotation the stars continue to rotate more and more rigidly.

† Proxima Cen: 31.5 days, Lalande 21185: 47 days

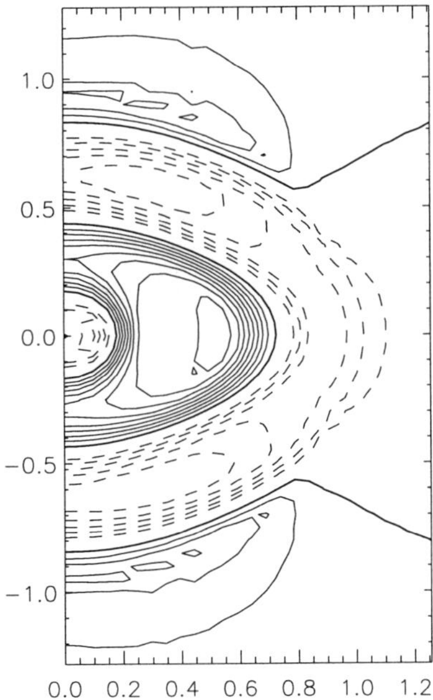

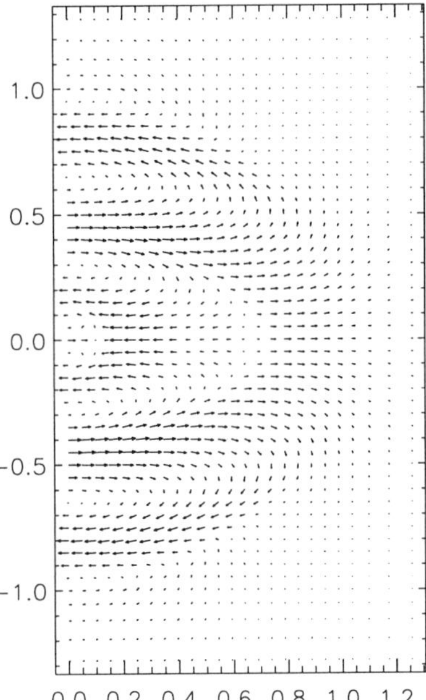

FIGURE 5. Magnetic field structure of an M dwarf with 0.1 $M_\odot$. The left picture shows the azimuthal component of the field in the $\phi = 0$ half plane. Solid contours denote positive and dashed contours negative values of B_ϕ. The right picture is a vector plot of the field components in the same half plane. The stellar radius has been used as unit length.

5. The M star dynamo model

After all, the dynamo operating in the fully convective M dwarfs should clearly be of the α^2-type. This seems as a simplification but it is not. In α^2-dynamos the anisotropy of the α-tensor plays an important role (Rüdiger & Elstner 1994). If – as it is the result of quasilinear turbulence theory – the α_{zz} has the opposite sign as the $\alpha_{\phi\phi}$ – the resulting magnetic field becomes nonaxisymmetric with massive consequences for nonlinear computations (cf. Moss & Brandenburg 1995). The anisotropy of the α-effect rather than its sign gives the main uncertainty in the theory.

44madrid The growth of the mean magnetic field is limited by its back reaction on the electromotive force. We use a simple α-quenching prescription,

$$\alpha_0 = \frac{\tilde{\alpha}_0}{1+\beta^2}, \tag{5.26}$$

where

$$\beta = 2\pi\,\frac{|B|}{B_{\rm eq}} \tag{5.27}$$

is the normalized magnetic field and

$$B_{\rm eq} = \sqrt{\mu_0 \rho u_{\rm T}^2}. \tag{5.28}$$

is the turbulence-equipartition field.

The magnetic diffusivity tensor assumes the form

$$\eta_{ijk} = \eta_{\rm T}\epsilon_{ijk} + (\eta_{\|} - \eta_{\rm T})\epsilon_{ijl}\frac{\Omega_l\Omega_k}{\Omega^2} \tag{5.29}$$

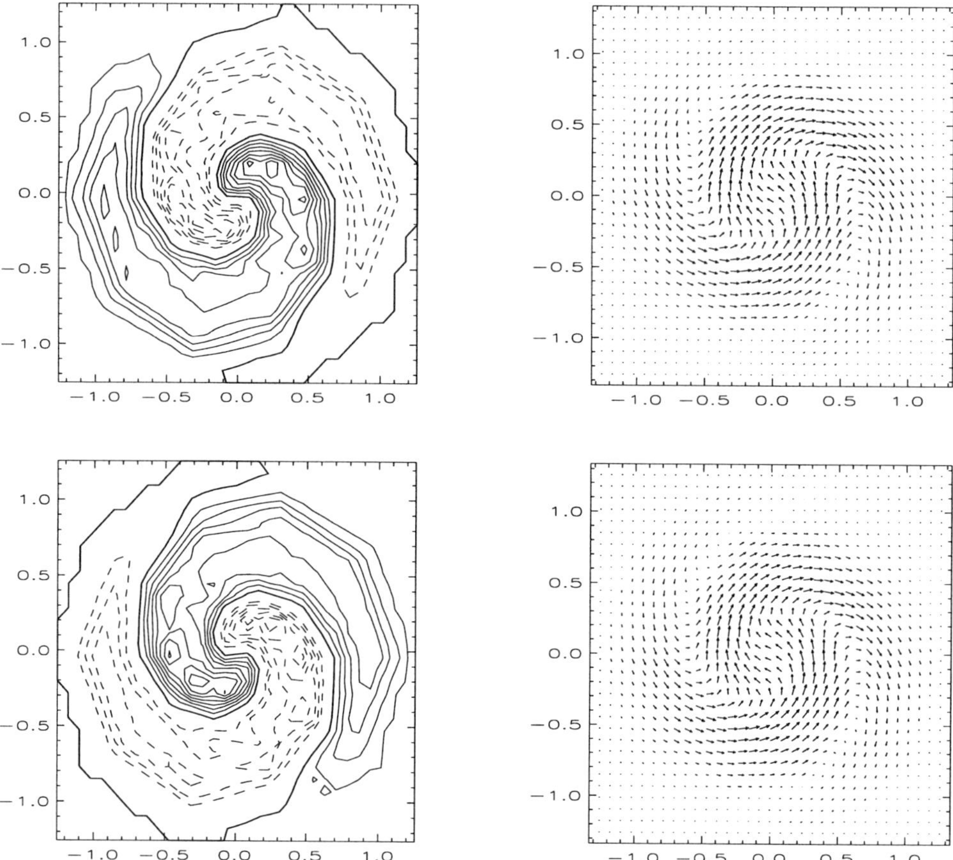

FIGURE 6. The magnetic field of an M dwarf with 0.1 $M_\odot$ in two planes parallel to the equatorial plane. The pictures in the top row show a plane with $z > 0$, while the pictures in the bottom row show a plane with $z < 0$. The pictures in the left column show contour plots of the z component, those in the right column the r and ϕ components.

with

$$\eta_{\mathrm{T}} = (\phi_1 + \phi_2)\eta_0, \qquad \eta_{\|} = 2\phi_1\eta_0 \tag{5.30}$$

and

$$\phi_1 = \frac{3}{4\Omega^{*2}}\left(-1 + \frac{\Omega^{*2}+1}{\Omega^*} \arctan \Omega^* \right), \tag{5.31}$$

$$\phi_2 = \frac{3}{2\Omega^{*2}}\left(1 - \frac{\arctan \Omega^*}{\Omega^*} \right) \tag{5.32}$$

(Kitchatiov *et al.* 1994). The reference value η_0 of the eddy diffusivity is

$$\eta_0 = c_\eta u_{\mathrm{T}}^2 \tau_{\mathrm{corr}} \tag{5.33}$$

with $c_\eta \simeq 0.30$. To solve the induction equation, we use a 3D time dependent fully explicit second order finite-difference scheme in cylindrical polar coordinates. The computational domain is a cylinder with a radius of two and a height of four stellar radii. The star is located at the center of the cylinder. The surrounding medium is considered co-rotating with the star and poorly conducting. Comparisons with Moss & Branden-

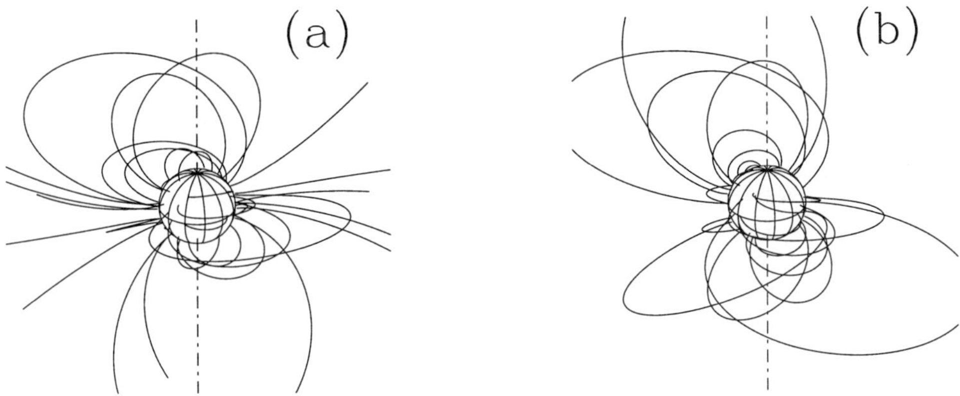

FIGURE 7. Visualization of a magnetic field configuration similar to the one we find by Moss & Brandenburg (1995).

burg (1995) show that with the same electromotive force, our model yields very similar results to their model, which assumed a star surrounded by vacuum. The assumption of small conductivity (large magnetic diffusivity) outside the star is thus a rather perfect approximation for a vacuum boundary condition.

As there is no rotational shear in our model, the stellar rotation rate enters only via the Coriolis number which we thus vary as an input parameter. There are no further free parameters. For both stellar models the critical value for dynamo excitation lies between one and two. The field geometry for a fairly supercritical case with $\Omega^* = 2$ is shown in Figs. 5 and 6. The field is completely non-axisymmetric without any axisymmetric contribution and symmetric with respect to the equatorial plane. The total field energy is constant with time but the field rotates with the star. The cross sections with constant z show a spiral-type geometry while the field distribution in the r-z half plane is rather shellular. For larger values of the Coriolis number the field resembles that of a tilted dipole with the dipole axis lying in the equatorial plane. This type of field geometry was also found by Moss & Brandenburg (1995). The similarity of the results is due to the fact that α^2-dynamos are not sensitive to the sign of α.

6. Conclusions

Very slow rotators and very fast rotators are almost rigid rotators. As M dwarfs with rotation periods shorter than 1 year must be considered as very fast rotators, their differential rotation proves to be small. There is thus no doubt that the dynamo operating in the fully convective M dwarfs is of the α^2-type. The anisotropy of the α-tensor, suggested by both quasilinear as well as nonlinear simulations, finally leads to nonaxisymmetric and non-oscillating large-scale magnetic fields (cf. Fig. 7). Hence, a distinct rotational modulation should be observable for the active M dwarfs. Our scenario, supported by nonlinear simulations of internal rotation, meridional flow and mean magnetic fields leads to clear predictions for observations. Only very slow rotators with $P_{\rm rot} > 1$ year should adopt solar-type solutions. It is an open question, however, whether the magnetic spin-down of M dwarfs is effective enough to produce such rotation times within the age of the universe.

Our warmest thanks go to Isabelle Baraffe and Gilles Chabrier for their kind cooperation concerning structure and evolution of the used M dwarf models

REFERENCES

BRANDENBURG, A., NORDLUND, Å., PULKKINEN, P., STEIN, R. F. & TUOMINEN, I. 1990 3-D simulation of turbulent cyclonic magneto-convection. *Astron. & Astrophys.* **232**, 277.

BRANDENBURG, A. 1994 Hydrodynamical simulations of the solar dynamo. In *Advances in Solar Physics* (eds. G. Belvedere & M. Rodono). Lecture Notes in Physics, vol. 432, p. 73. Springer.

BRANDENBURG, A. & DONNER, K. J. 1997 The dependence of the dynamo alpha on vorticity. *Month. Not. Roy. Astron. Soc.* **288**, L29.

CHABRIER, G. & BARAFFE, I. 1997 Structure and evolution of low-mass stars. *Astron. & Astrophys.* **327**, 1039.

FERRIÈRE, K. 1993 Magnetic diffusion due to supernova explosions and superbubbles in the galactic disk. *Astrophys. J.* **409**, 248

GLATZMAIER, G. & ROBERTS, P. H. 1996 An anelastic evolutionary geodynamo simulation driven by compositional and thermal convection. *Physica D* **97**, 81.

KAISIG, M., RÜDIGER, G. & YORKE, H. W. 1993 The α-effect by supernova explosions. *Astron. & Astrophys.* **274**, 757.

KITCHATINOV, L. L. & RÜDIGER, G. 1993 Lambda-effect and differential rotation in stellar convection zones. *Astron. & Astrophys.* **276**, 96.

KITCHATINOV, L. L., PIPIN, V. V. & RÜDIGER, G. 1994 Turbulent viscosity, magnetic diffusivity, and heat conductivity under the influence of rotation and magnetic field. *Astron. Nachr.* **315**, 157.

KORPI, M. J., BRANDENBURG, A., SHUKUROV, A. & TUOMINEN, I. 1998 Supernova driven interstellar turbulence. In *Proc. of the 2nd Guillermo Haro Conf., Interstellar turbulence* (eds. J. Franco & A. Carraminana), p. 16. Cambridge University Press.

KRAUSE, F. & RÄDLER, K.-H. 1980 Mean-field magnetohydrodynamics and dynamo theory. Akademie-Verlag Berlin.

KRIVODUBSKIJ, V. N. & SCHULTZ, M. 1993 Complete alpha-tensor for solar dynamo. In *Proc. IAU Symp. 157, The cosmic dynamo* (eds. F. Krause, K.-H. Rädler & G. Rüdiger), p. 25. Kluwer, Dordrecht.

KÜKER, M., RÜDIGER, G. & KITCHATINOV, L. L. 1993 An alpha-omega model of the solar differential rotation. *Astron. & Astrophys.* **279**, 1.

MOSS, D. & BRANDENBURG, A. 1995 The generation of non-axisymmetric magnetic fields in the giant planets. *Geophys. Astrophys. Fluid Dyn.* **80**, 229.

PIPIN, V. V., KITCHATINOV, L. L. & RÜDIGER, G. 1999 The α-effect in stratified shear flows. in preparation

RÜDIGER, G. 1978 On the alpha-effect for slow and fast rotation. *Astron. Nachr.* **299**, 217.

RÜDIGER, G. & KITCHATINOV, L. L. 1993 α-effect and α-quenching. *Astron. & Astrophys.* **269**, 581.

RÜDIGER, G. & ELSTNER, D. 1994 Non-axisymmetry vs. axisymmetry in dynamo-excited stellar magnetic fields. *Astron. & Astrophys.* **281**, 46.

RÜDIGER, G. & BRANDENBURG, A. 1995 A solar dynamo in the overshoot layer: cycle period and butterfly diagram. *Astron. & Astrophys.* **296**, 557.

RÜDIGER, G., BRANDENBURG, A. & PIPIN, V. V. 1998 A helicity proxy from horizontal flow patterns. *Solar Phys.* acc.

Doppler Imaging of Cool Dwarf Stars

By KLAUS G. STRASSMEIER

Institut für Astronomie, Universität Wien, Türkenschanzstraße 17, A-1180 Wien, Austria

Previous Doppler images of evolved RS CVn stars and of single pre-main-sequence stars reveal interesting differences in the types of magnetic activity seen in these two classes of objects and our Sun; the presence and nature of polar spots may be one of the most striking differences found. The time seems ripe now to extend the Doppler-imaging technique to the very cool end of the main sequence. There, rapidly-rotating stars are thought to be fully convective and have no convective overshoot layer as in the Sun and similar stars. Since this is exactly the region where the solar/stellar dynamo is supposed to be located , one could expect a fundamentally different field topology and thus a qualitatively different surface temperature distribution. However, recent magnetic images of main-sequence stars suggest no basic differences to active giants. In this paper I will discuss the basic principles of Doppler imaging and the observational requirements, its application to the cool dwarf stars YY Gem and LQ Hya and future applications to brown dwarfs. In principle, Doppler imaging also contains the possibility to image the transits of extra-solar planets down to the size of a few terrestrial radii.

---

1. Introduction: why would we want to resolve stellar surfaces?

Doppler imaging for stars that have spots of cooler or greater temperature on their surface, amounts to recovering the surface temperature distribution from the integral equation that relates the distribution of surface temperature to the observed line-profile and light-curve variations. Modern approaches invert the observations to recover a unique surface image within an otherwise ill-posed problem. It is the spot distribution that we believe must be a signature of the magnetic surface field and thus the underlying stellar dynamo. The long-term goal is to systematically obtain Doppler images of rapidly-rotating stars from various parts in the Hertzsprung-Russell diagram. Then, we may be able to supply the much needed observational constraints on the evolution of stellar magnetic fields and their impact on stellar evolution in general.

As a reminder, a particularly salient example of the contrary predictions of stellar physics and cosmology is the problem of the age of the oldest stars in some globular clusters, estimated with the help of stellar evolution theory, which turns out to be larger than the currently assumed age of the Universe. Either the Universe is too young or a better theory of stellar evolution is needed to solve this paradox. I guess, everybody will agree with me that the latter is true no matter how old the universe now really is and, again I am sure everybody will agree, that stellar rotation and magnetic fields play important roles, probably key roles, for the evolution of an approximately solar-mass star. For such a task, precise observational data are needed in order to constrain theoretical models. This is where Doppler imaging comes into the game. It uses a star's rotation to map its surface inhomogeneities and thus links stellar rotation with surface indicators of its magnetic field.

The magnetic fields of cool stars are believed to be generated by a dynamo seated somewhere in the convective envelope. The basis for this picture is our Sun, where buoyant magnetic fields produced by the dynamo rise through the convection zone to form bipolar surface fields and related phenomena like cool starspots. The sum of these phenomena on other stars is loosely called "stellar magnetic activity". The solar dynamo is believed to be located in the overshoot region below the convection zone, because only there is the growth rate of the Parker-instability small enough, i.e. the local temperature gradient smaller than the adiabatic gradient, to allow the fields to remain there for a long enough time to be enhanced by the dynamo action in order to reach the observed field strengths throughout the 22-year magnetic cycle (Spruit & van Ballegooijen 1982).

But dynamo action also occurs in supposedly fully convective M-dwarfs, suggesting that the existence of an interface between radiative interior and convection zone is not essential to the dynamo process. The first Doppler image of a M-dwarf became only very recently available but its interpretation is severely hampered by the complexity of the spectra of which these images are extracted. This will be discussed in some detail in Sect. 3. Other gaps in our knowledge involve stellar butterfly diagrams and the possible existence of active longitudes (e.g. Berdyugina & Tuominen 1998), polar spots (e.g. Bruls *et al.* 1998) or plages, and even the scale height of the emission from coronal loops. To answer all these questions we need approximate 3D images of the atmospheres of a set of stars well sampled over the H-R diagram. The cool dwarf stars could very well be the Rosetta stone for understanding stellar dynamos.

Table 1 is a summary of published Doppler images of dwarf stars as well as of cool pre-main-sequence objects. Here, the term "cool" is always meant with respect to the Sun (other people use Vega as the standard of everything!). Only three of these stars are of M-type, two of them in a close binary system (YY Gem A+B) and one is even a classical T Tauri star (DF Tau). Clearly, not a representative sample to draw general conclusions from. Instead, I will focus on the what-could-be-done's and what-could-be-expected's rather than present a review of the few existing literature. In the first section I outline the principles of Doppler imaging, the second section will deal with the observational requirements and some applications, and the third section outlines a possibility to map transits of extra-solar planets or brown dwarfs.

2. The principles of Doppler imaging

2.1. *Perspective*

The theoretical angular resolution of an optical telescope is proportional to λ/D, where D is the telescopes aperture diameter and λ the wavelength. The 10-m Keck telescopes have therefore a resolution capability of at most 0.01 arcsec. Optical interferometry with the Fine Guidance Sensors abord the Hubble Space Telescope achieves a resolution of 0.001 arcsec for point sources, while ESO's VLTI is hoped to achieve even 0.0005 arcsec in the visible. The resolving power of Doppler imaging is not directly comparable to these numbers since it is an *indirect* technique via spectral-line variations. The resulting effect for the stellar surface, though, is the same. Doppler imaging can resolve stars with an equivalent angular resolution of 0.000001 arcsec – as just recently achieved with the 3.6m Canada-France-Hawaii telescope for a $9^{\rm th}$-magnitude star (Strassmeier & Rice 1998b) – i.e. three orders of magnitude better than any optical interferometer to date for a star at least five magnitudes fainter. With a true angular resolution like this, one could easily read a newspaper in a Viennese Café; from the beach in La Palma, of course.

2.2. *General principles of Doppler imaging*

Doppler imaging is similar to medical tomography and is designed to invert a series of high-resolution spectral line profiles into an "image" of the stellar surface. Appropriate recent reviews are the papers by Rice (1996) and Piskunov & Rice (1993), or the dedicated IAU Symposium No. 176 on "Stellar Surface Structure". Fig. 1 illustrates how the crucial latitude dependence enters into the line profile. The late Armin Deutsch of the Palomar Observatory first applied this idea in his equivalent-width analysis of a rapidly-rotating Ap-star (Deutsch 1958). The beginnings in cool-star work, i.e. to recover the surface

TABLE 1. Main-sequence stars (upper half of the table) and pre-main-sequence (PMS) stars with published Doppler images.

Spectral type	**Star name**	$T_{\rm eff}$ **(K)**	$P_{\rm rot}$ **(days)**	**Obs. code**	**References**
G0V	ER Vul A	5900	0.7	NOT	Piskunov (1996)
G1.5V	HD 129333	5870	2.6	CFHT	Strassmeier & Rice (1998a)
G2V	YY Eri	5800	0.31	NOT	Maceroni *et al.* (1994)
G2V	ER Vul B	5750	0.7	NOT	Piskunov (1996)
G3V	He699	5720	0.61	AAT	Barnes *et al.* (1998)
G5V	He520	5620	0.49	AAT	Barnes *et al.* (1998)
K0V	AB Dor	5250	0.51	ESO	Kürster *et al.* (1994)
				AAT	Collier Cameron & Unruh (1994)
				AAT	Collier Cameron (1995)
				AAT	Donati & Collier Cameron (1997)
				AAT	Donati *et al.* (1998)
K0V	HII3163	5250	0.42	KECK	Stout-Bathala & Vogt (1996)
K2V	LQ Hya	5000	1.61	CFHT	Strassmeier *et al.* (1992)
				NSO	Saar *et al.* (1994)
				CFHT	Rice & Strassmeier (1998)
				AAT	Donati (1998)
K2V	V471 Tau	5000	0.52	McD	Ramseyer *et al.* (1995)
K4V	HII686	4500	0.40	KECK	Stout-Batalha & Vogt (1996)
M1V	YY Gem A	4000	0.82	McD	Hatzes (1995a)
M1V	YY Gem B	4000	0.82	McD	Hatzes (1995a)
			PMS stars		
G2	SU Aur	5800	3.09	NOT	Petrov *et al.* (1995)
G5	V824 Ara A	5400	1.68	ESO	Kürster *et al.* (1992)
				ESO	Dempsey *et al.* (1998)
G5	HDE283572	5300	1.55	PdM	Joncour *et al.* (1994a)
				CFHT	Strassmeier & Rice (1998b)
K0	V824 Ara B	5050	1.68	ESO	Dempsey *et al.* (1998)
K0	Par 1724	5000	5.7	KECK	Stout-Batalha (1997)
				McD	Neuhäuser *et al.* (1998)
K3	Sz68	4800	3.9	McD	Johns-Krull & Hatzes (1997)
K4	V410 Tau	4400	1.87	CFHT	Strassmeier *et al.* (1994)
				PdM	Joncour *et al.* (1994b)
				McD	Hatzes (1995b)
				CFHT	Rice & Strassmeier (1996)
M2	DF Tau	3750	8.5	WHT	Unruh *et al.* (1998)

Observatory codes: AAT - Anglo Australian Telescope, CFHT - Canada France Hawaii Telescope, ESO - European Southern Observatory (CAT), KECK - Keck Telescope, McD - McDonald Observatory (2.1 m), NOT - Nordic Optical Telescope, NSO - National Solar Observatory (McMath), PdM - Pic du Midi, WHT - William Herschel Telescope.

temperature distribution as opposed to the surface abundance distribution for Ap-star, is inevitably connected with the work of Vogt & Penrod (1983) and Vogt *et al.* (1987). They showed that cool spots on the surface of a late-type star produce distortions in the line profiles that can be modelled throughout a rotational cycle.

Doppler imaging uses the one-to-one relation between the radial velocity and the intensity of a point on the rotating stellar surface and its projected signature in the broadened

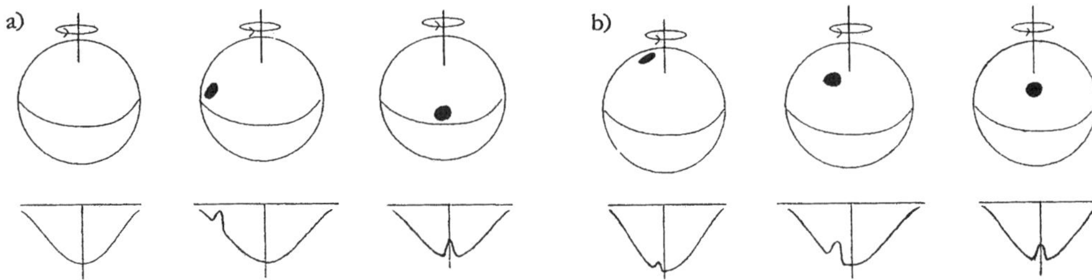

FIGURE 1. The relation between the position of a dark spot on the stellar surface and its appearance as a bump in the broadened spectral line profile. The left three rotational sequences (a) are for a low-latitude spot and the right sequences (b) are for a high-latitude spot. Adopted from Rice (1996).

line profile (Struve 1930). This relation makes up the forward integral equation

$$R_{\text{comp}}(\lambda,\varphi) = \frac{\iint I_c(M,X(M))R_{\text{loc}}(M,\theta,X(M),\lambda+\Delta\lambda_D(M,\varphi))\cos\theta\, dM}{\iint I_c(M,\theta,X(M))\cos\theta\, dM} \tag{2.1}$$

where I_c is the continuum intensity, R_{loc} the local line profile, M the position on the stellar surface, θ the foreshortening angle, $\Delta\lambda_D$ the Doppler shift of area element dM at a particular rotation phase, and $X(M)$ the unknown surface parameter that is solved for. In our case, $X(M) \equiv T_{\text{eff}}(M)$. Then, the problem is to alter the assumed initial $T_{\text{eff}}(M)$ until a satisfactory fit to the observations is found. This is done with a discrepancy function D that contains the squared differences between the computed and the observed line profiles at all phases. It must be minimized with the help of more or less clever algorithms. The discrepancy function is

$$D_\lambda(\varphi) = \frac{1}{n_\varphi n_\lambda}\sum_{\varphi=1}^{n_\varphi}\sum_{\lambda=1}^{n_\lambda} g(\varphi,\lambda)[R_{\text{obs}}(\varphi,\lambda) - R_{\text{comp}}(\varphi,\lambda)]^2 \ , \tag{2.2}$$

where g could be a pre-determined statistical weight for each pixel in each line profile if needed (by default $g = 1$). The trick is now to add additional information with the help of a penalty function P whose purpose is to increase D if surface detail appears that is not compatible with the signal-to-noise ratio of the data. The function that is minimized is called the error function E,

$$E(X(M)) = D(X(M)) + P(X(M)). \tag{2.3}$$

Usually, a technique such as conjugate gradients is used for the minimization (see Press *et al.* 1992). If the data are of infinite S/N ratio, the choice of P is irrelevant. Even for more realistic S/N ratios of, say, at least above 100:1, no significant differences between different choices of P are evident (Piskunov 1990).

Naturally, such an image reconstruction process is an ill-posed mathematical problem, meaning, that there are more solutions than the line-profile data can constrain. This led to the application of a variety of penalty functions to regularise the solution. The most commonly used is the Maximum-Entropy function, others are the Tikhonov functional, and the CLEAN algorithm. The so-called "Occamian" approach finds a solution that is determined solely from the error gradient without a regularisation functional. The latter two applications were developed and are used by Kürster (1993) and Berdyugina (1998), respectively. A detailed discussion of these techniques would exceed the page limit of this contribution and I refer to the original publications and, to some extend, also to the review by Rice (1996).

3. The better the data, the better the image

Table 1 also identifies the observatories where high-resolution high-S/N spectroscopic observations for Doppler imaging are currently being made. Without exceptions, these are the best telescopes and spectrographs available today (1998-Sept). Unfortunately, one of the most efficient instruments of its size, the ESO 1.4m Coudé Auxiliary Telescope (CAT) and Coudé Echelle Spectrograph (CES) at La Silla, was closed in 1997. Another, very efficient small telescope, the KPNO 0.9m coudé feed telescope, is awaiting the same destiny. However, new telescope generations with high spectral resolutions come online in the near future, most notably the many 8–10m giants currently under construction. The capabilities and limitations of Doppler imaging with the ESO VLT unit telescope and the UVES spectrograph were discussed by Strassmeier (1995).

3.1. *General observational constrains*

The spectral lines used for the mapping should be reasonably free of blends because, first of all, unrecognized blends artificially enlarge the equivalent width of the mapping line and cause a mismatch in the $\log g$–$T_{\rm eff}$–$[m/{\rm H}]$ plane ($[m/{\rm H}]$ being the relative abundance of the elements used in Doppler imaging). Starting out with such a mismatch can prevent the code from converging at all or, even worse, converge on a wrong solution. Unruh & Collier Cameron (1995) found that neglecting blends at considerable distances from the mapping-line center will lead to spurious banding in the reconstructed image. These artificial bands show up at higher latitudes the further a blend is away from the mapping-line center. From detailed simulations, Unruh & Collier Cameron estimate that the ratio of the wavelength displacement of the blend to the line broadening due to the $v \sin i$ of the star had to be of the order of 1/3 for artifacts to appear, with assumed equivalent widths for the blend between 10% and 50% of the mapping line.

Atomic lines of cooler stars become progressively weaker because the thermal energy is too small to populate the higher atomic levels and also because more atoms are being bound in molecules and are missing for the atomic line formation. In Fig. 2, we show a series of 6160-Å spectra of K–M dwarf stars obtained with the KPNO 0.9m coudé feed telescope. This wavelength region is commonly used in work on late-type stellar photospheres and contains many, relatively unblended lines (a complete library of F2–M8 dwarf spectra was recently presented by Montes & Martín 1998). However, if we approach the surface temperature of an M0 star ($\approx$4000 K) most of these lines are severely blended, diluted by molecular bands, or simply not excited anymore – e.g. the Fe I line at 6180 Å with a lower excitation potential of 2.73 eV – and thus can not be used for Doppler imaging of cool dwarfs. Finding a useful spectral line is crucial though. Within the wavelength region shown in Fig. 2, one line appears to be reasonably unblended even at M1, i.e. Fe I at 6141.73 Å ($\log gf = -0.50$, $\chi_{\rm low} = 3.60$ eV). Even at the latest spectral types it is not affected by the red wing of the nearby, strong Ca I line at 6122 Å nor by the TiO bands redwards of 6150 Å. Beside other spectral lines, this line was already used to obtain a Doppler image of the young K2-dwarf LQ Hya and its atomic parameters are reasonably well known (Strassmeier *et al.* 1992).

Because the spot shape on the stellar surface is determined from the shape of the tiny bump in the line profile, spectra with the highest obtainable S/N ratio must obtained. E.g., if we would like to detect a 1% spot-bump amplitude (i.e. 0.01 of the normalized continuum) at a level that is 3σ the uncertainty of a profile point, we need a S/N ratio

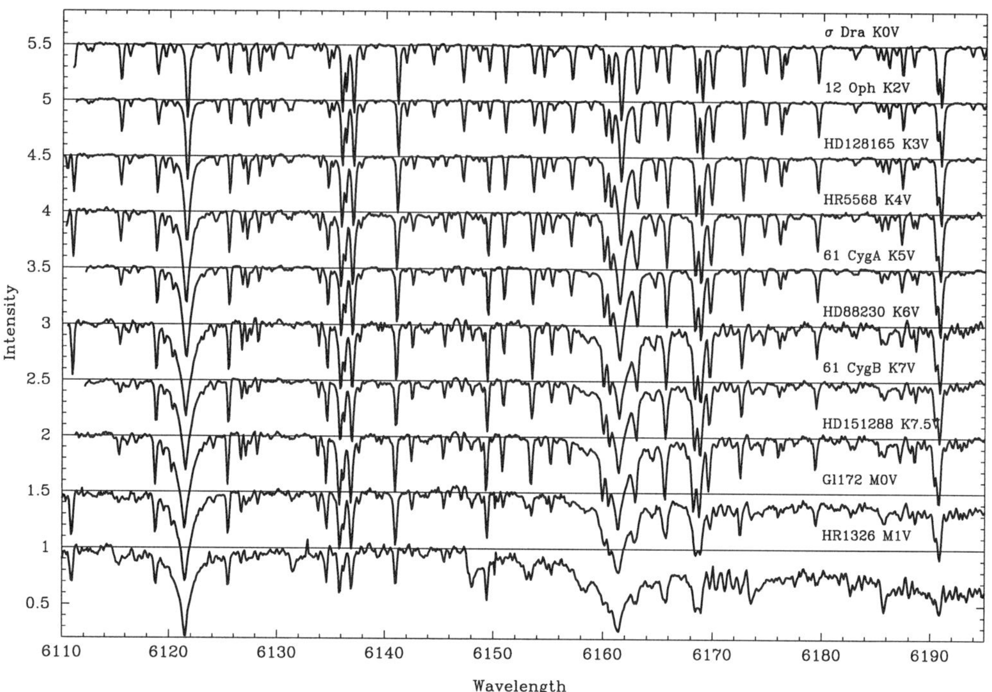

FIGURE 2. Representative spectra of K–M dwarfs. The spectra were taken in one of the frequently used Doppler-imaging wavelength regions near the Ca I triplet lines at 6122 and 6162 Å. Notice the increasing strength of the line wings of the two Ca lines and the increasing continuum deviation of the spectrum redwards of ≈6150 Å due to titanium oxide (TiO) molecular bands. For example, the often used Fe I 6180-Å line appears nearly unblended until a spectral type of K5, but nearly disappeared at a spectral type of M1.

of 300:1 per pixel. At a resolution of around 100,000, such high-quality spectra can be achieved up to a magnitude of ≈8.5 mag in V with a 3–4m class telescope, a superb spectrograph and a CCD with high quantum efficiency and low read-out noise. Of course, the higher the resolution, the shorter the wavelength range and the lesser lines one can work with. A noble technique to achieve superior S/N is to combine the many spectral lines from one spectrum into a single artificial spectral line (Donati *et al.* 1997). This is basically done by cross correlating all spectral lines in all orders of an echelle spectrum with themselves, thereby achieving S/N ratios of several thousand to one through the process of co-adding. Of course, the line-profile inversion and interpretation must then be done from an artificial spectral line profile, which is a combination of many lines of different elements, different temperature sensitivities, different limb-darkening dependences, and different strengths or, equivalently, different formation heights in the stellar photosphere. The recovered stellar surface information is thereby solely determined from the shape of the line profile and is independent of its equivalent width. As we will see later in Sect. 3.3, this is exactly the tool needed for the very cool, and thus very faint, stars.

For the bright ($V < 10^m$) and longer-period ($P > 1^d$) stars, where a sufficiently high S/N ratio can be achieved in a short series of single exposures, it is the *external* uncertainty of every spectrum that limits the stellar surface information content (unless you are one of the lucky ones who has access to the Keck telescopes or similar, see Fig. 3). The most

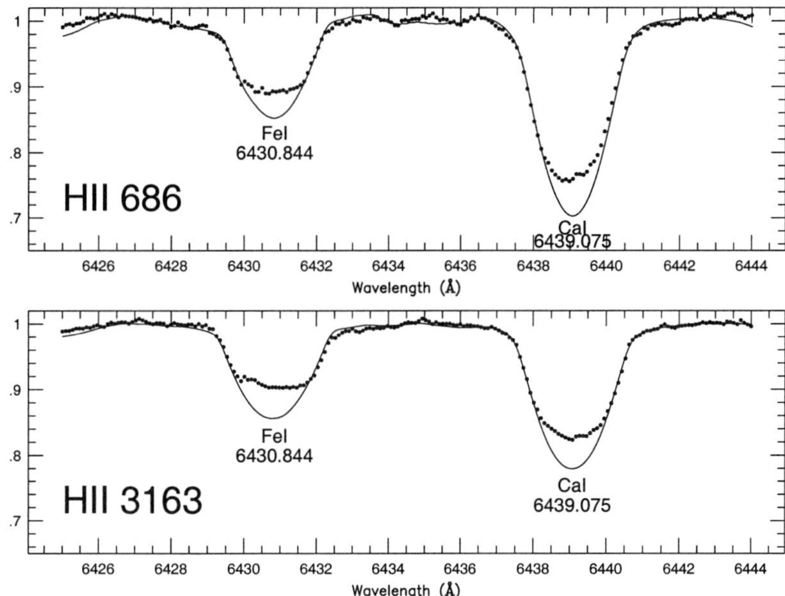

FIGURE 3. Spectra of the cool Pleiades dwarfs HII 686 (K4, V=13.5 mag) (top panel) and HII 13163 (K0, V=12.8 mag) (bottom panel). The spectra were obtained with the 10-m Keck I telescope and the HIRES spectrograph at R=67,000 with an exposure time of 30 minutes. It shows the Doppler-imaging lines Fe I 6430 Å and Ca I 6439 Å. The thin lines are spectral-synthesis fits from appropriate model atmospheres. S/N ratio is 160:1 for HII 3163 and 120:1 for HII 686. Courtesy of Steve Vogt, reproduced from Stout-Batalha & Vogt (1996).

notable sources of such errors are time-dependent, wavelength-dependent, and location-dependent variations of the CCD flat field. Furthermore, very small nonlinearities in the wavelength calibrations from night to night due to temperature changes and/or mechanical motions of the CCD and the grating or, unnoticed asymmetric illumination of the image slicer, the slit, or the fiber entrance, cause uncertainties in the line profiles that dominate over the photon noise. Additional sources are straylight, a misalignment of the spectrograph dispersion direction with the CCD pixel rows, dewar warming, variable sky transparancy and humidity, simple dirt on filters and grisms etc. Unfortunately, the true amount of the sum of these external errors remains basically unknown and continues to be a source of uncertainty in Doppler imaging.

3.2. *Systematic differences from spectral line to spectral line*

This subsection appeared in Rice & Strassmeier (1998), and I felt that it would help the reader of this paper to better judge the overall reliability of Doppler imaging.

There are a number of factors which contribute to systematic differences in the maps we obtain for different spectral lines. One is the problem of getting the value of $\log gf$ (the transition probability) correct. Generally, in temperature sensitive lines, the mapping program compensates for small errors in $\log gf$ by adjusting the average temperature of the surface so that the equivalent width of the computed line matches the equivalent width of the observed line. If the $\log gf$ is in error by more than about 0.3 or so the problem is usually quite obvious but smaller errors will have small effects that cannot be determined from the mapping output.

A second systematic difference arises in the strong lines because of the assumption of

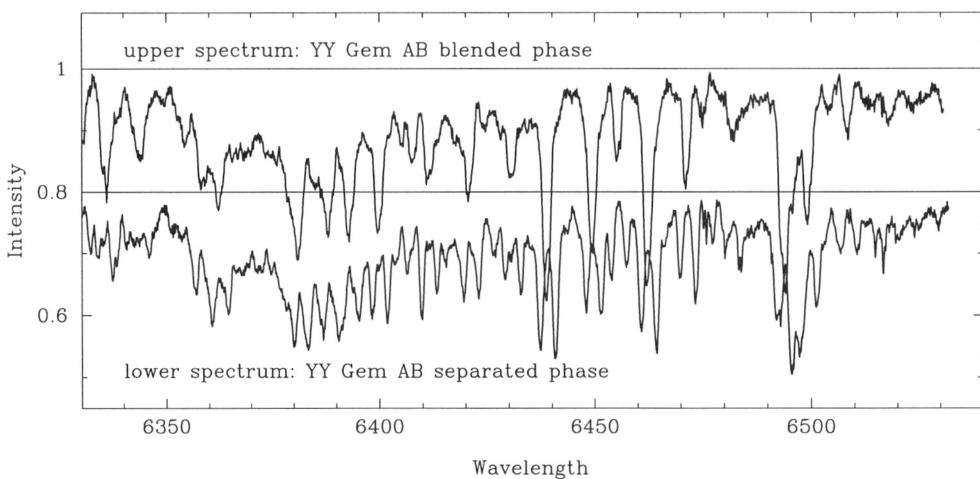

FIGURE 4. Two spectra of YY Gem demonstrate the inherent difficulties in mapping M-dwarf stars. YY Gem is a double-lined spectroscopic binary consisting of two rapidly-rotating dM1e stars in a 0.8-day synchronous orbit. The figure compares two CFHT spectra, one taken at conjunction (upper spectrum) and one at quadrature (lower spectrum; shifted by −0.2 in continuum for better visibility). The atomic lines in both spectra are severely diluted by the presence of TiO molecular bands at 6384.17 Å ($\gamma R_2(14)3$–1), 6420.61 Å ($\gamma R_3(13)4$–2), and 6447.90 Å ($\gamma R_2(13)4$–2).

LTE in calculating local line profiles (for NLTE effects see Bruls *et al.* 1998). One should expect that the core representation in the local line profiles would be poor for very strong lines. In some cases there may even be core emission present in the actual local line profiles in variable amounts over the stellar surface and we are unable to allow for that. In a relatively slowly-rotating star such as LQ Hya, where the $v \sin i$ is $\approx$27 km s^{-1}, the errors in representing the local line profiles become serious, especially in the strongest lines where we not only might expect the greatest error in the calculation of the local line profile but where, in addition, the ratio of rotational broadening to local line width is smallest.

A third systematic effect is the effect of setting the continuum accurately during data reduction. Errors in judging the correct location of the continuum affect weak lines more than strong lines and would (if the error tends to be systematic) have an effect on the average surface temperature calculated by the program. For the very coolest stars, even guessing the continuum becomes a major task.

3.3. *A closer look to two stellar examples: YY Geminorum and LQ Hydrae*

YY GEM is one of the three known eclipsing binaries that consist of two M dwarf stars. It is also the brightest and most active of the three (the others are CM Dra and GJ 2069A). The latter's discovery is most recent and was announced in the poster contribution by X. Delfosse *et al.* during this meeting! YY Gem A+B are also the only M stars with a published Doppler image (Hatzes 1995a), and it is quite worthwhile to take a closer look at the spectrum and the map of YY Gem.

Figure 4 shows two medium-resolution CFHT spectra of YY Gem. One of the spectra was taken at a phase where the lines from the two stellar components were blended, and the other spectrum shows the lines separated. Hatzes' (1995a) keen approach was now to extract the components' Fe I 6430 Å spectral line profiles from the combined normalized contiuuum, rescale them to the individual continua, and neglect the molecular bands. To first order, the latter is a reasonable assumption because the electronic oscillator strengths and even wavelengths of the many TiO-band systems can be seriously in error (Valenti *et al.* 1998) and ignoring them altogether is not such a bad approximation as it might sound. However, it was the rotational modulation of the strengths of optical molecular bands such as TiO and VO (vanadium oxide) that provided the conclusive proof that starspots are indeed regions cooler than the surrounding stellar photosphere (Vogt 1981). In the meantime, some groups even proposed to use these molecular bands for Doppler imaging (O'Neal *et al.* 1996). As we know, one man's noise is other man's data! The point though is that the varying band strengths result in an additional variation of the local continuum (despite the actual continuum changes due to the photospheric light) and must be taken into account if the mapping line equivalent width is to be preserved. The approach that Hatzes (1995a) took in his Doppler imaginary of YY Gem included the computation of the specific line profiles from pretabulated model atmospheres across the stellar disk. Clearly, not a straightforward process if molecular bands must be taken into account whose theoretical parameters are more than uncertain. At the moment (1998 Sept.), for the cool M-dwarf stars we would favour a Doppler imaging approach that uses only the line *profile* information rather than fully synthesizing the whole spectral region. Anyway, Hatzes (1995a) made the beginning, and his images of YY Gem A and B are shown in Fig. 5.

Another "cool" target is the single K2-dwarf LQ Hya (HD 82558), believed to have just arrived on the zero-age main sequence (ZAMS) and, so far, to have maintained most of its rotational momentum gained during the contraction towards the ZAMS. The star also displays extreme chromospheric activity as well as a variable photospheric spot distribution mainly at low to intermediate latitudes but not the typical large polar spot as seen in some RS CVn's. Over the years, the star was mapped independently by three different groups in Vienna and Brandon (Strassmeier & Rice), at CfA and Uppsala (Saar & Piskunov), and in Toulouse (Donati) (see Table 1). Fig. 6 shows a Doppler image of LQ Hya in spherical projection (from Rice & Strassmeier 1998), where the grey scale denotes temperature in the sense that the darker the grey scale the cooler the surface.

All four earlier images of LQ Hya from 1991–1993 show a stronger and larger polar feature than the image in March 1995. The overall tendency for large moderately cool spot regions to be concentrated to the equatorial zone is consistent among all images as is the tendency for mid-latitude regions to be unspotted or hotter. There does seem to be a tendency for the spots to be within a broad equatorial band (i.e. extending to greater latitudes in both a positive and negative sense) in the images from 1991 compared to 1993 and the 1995 image reproduced in this paper. Since the retreat of the spots to a narrower equatorial band coincides with an apparent reduction in the intensity of the polar feature, this might be an indication of some longer-term cycle in the spot distribution similar to what is seen during the 11 year solar cycle. At least, the existence of a polar feature is consistent among the images in all years.

4. Could Doppler imaging be used to image the transit of an extra-solar planet?

In principle, yes. But Doppler imaging is probably not the right tool for *discovering* transits of planets and suchlike; high-speed and wide-field CCD photometry is certainly a much more efficient technique here. However, once a planet transit can be predicted, Doppler imaging may not only provide independent confirmation but also aid to determine the planetary radius.

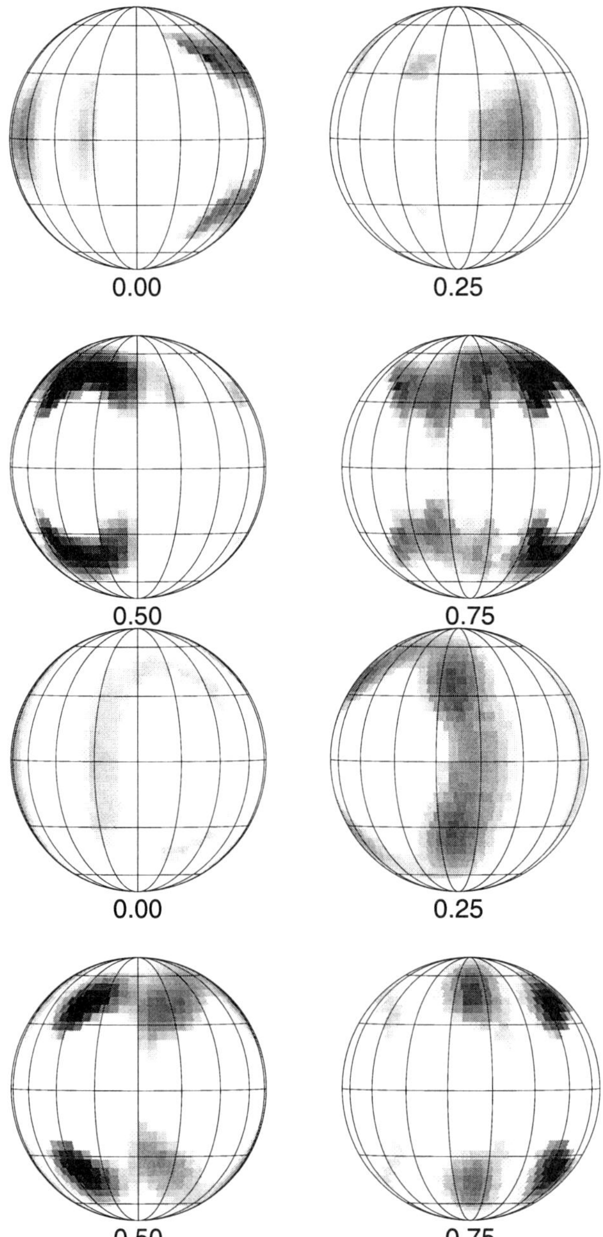

FIGURE 5. Doppler images of the two dM1e components of YY Gem. The maps are shown in a spherical projection at four rotational phases. *Left:* YY Gem A, the darkest regions are 300 K below the photospheric temperature. *Right:* YY Gem B, the darkest region are about 500 K cooler. From Hatzes (1995a), courtesy of A. Hatzes, McDonald Observatory.

The effect of a dark planet in front of a stellar disk on a spectral line profile is very similar to that of a cool starspot and is determined by the ratio of the local stellar line broadening to the stellar rotational broadening

$$r_{\text{planet}} \equiv \frac{FWHM_{\text{Doppler}}}{FWHM_{v \sin i}} \,, \tag{4.4}$$

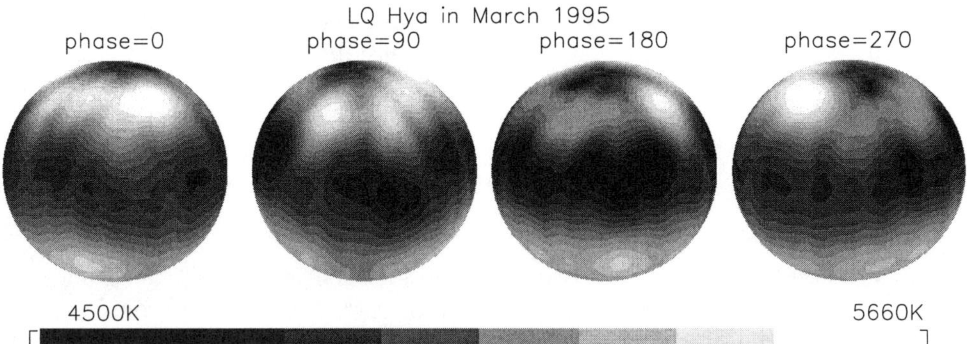

FIGURE 6. A Doppler image of LQ Hya. LQ Hydrae is a single K2 dwarf star that is thought to have just arrived on the main sequence and to be approximately the age of the Pleiades. The image in this figure is the average map from altogether 8 spectral regions and three continuum bandpasses. Despite the star's relative rapid rotation, it does not show a prominent cap-like polar spot as other, more evolved stars, but rather a broad equatorial band of spots and a comparably small and asymmetric polar feature.

where the "Doppler" term stands for the (missing) velocity signature of the area of the stellar surface that is occulted by the planetary disk. The smallest detectable surface area or, equivalently, the smallest velocity signature in the line profile is given by the ratio of the intrinsic width of the observed spectral line to the rotational broadening $v \sin i$,

$$x_{\rm line} \equiv \frac{FWHM_{\rm line}}{FWHM_{v \sin i}} \; . \tag{4.5}$$

To resolve the velocity signature of the "missing" surface element, i.e. the planetary disk, we require $r_{\rm planet} \geq x_{\rm line}$.

For simplicity, a circular planetary disk of radius r is considered at a time of central meridian passage. The equivalent width W of the bump in the observed spectral line profile is due to the missing line absorption plus the missing continuum from the occulted stellar photosphere and is approximately

$$W_{\rm bump} \approx \frac{r^2}{1-r^2}\, W_{\rm line}. \tag{4.6}$$

The bump amplitude, or strength, as a fraction of the continuum flux is (see also Collier Cameron 1992)

$$R_{\rm bump} \approx \frac{W_{\rm bump}}{\Delta\lambda_{\rm bump}} \approx \; (1 - R_{\rm line}) \, \frac{x}{(r^2 + x^2 + \delta^2 + i^2)^{1/2}} \, \frac{r^2}{1-r^2}, \tag{4.7}$$

where δ denotes the width of the phase smearing due to a non-zero integration time, i the width of the instrumental profile, and $\Delta\lambda_{\rm bump}$ is given by

$$\Delta\lambda_{\rm bump} \approx \; \Delta\lambda_{\rm rot} \; (r^2 + x^2 + \delta^2 + i^2)^{1/2} \tag{4.8}$$

with

$$\Delta\lambda_{\rm rot} \approx \; \frac{\pi}{2} \, \frac{\lambda_0 \; v \sin i}{c} \; . \tag{4.9}$$

Here we assumed that the change of the planetary's disk velocity signature ($\Delta\lambda_{\rm bump}$) is very similar to that of a spot on the stellar surface rather than a, say, co-rotating feature like a prominence. Its intensity signature on the line profile will be fairly constant during the transit, only marginally modulated due to the stellar limb darkening, and thus significantly different than the signature from a cool spot. If we see a narrow and large-amplitude bump in the wings of a spectral line that is moving through the line profile with approximately constant strength, it must be a planet or any other body that appears dark against the stellar disk at the wavelengths of the observations (given that we observe a solar-type star and can exclude the signatures of high-order non-radial pulsations as seen in very hot O and B-type stars). The only confusion that may arise is from dark surface spots moving through the profile for a total duration of approximately $1/2\ P_{\rm rot}$.

If we now require $R_{\rm bump} \approx 3\sigma_{\rm pixel}$ for a three sigma detection, we may compute the smallest detectable planetary disk radius at the stellar central-meridian passage as a function of spectral line broadening and stellar rotation period. The orbital period of the planet, or brown dwarf, is implicitely adopted such that the transit in front of the stellar disk takes in about the same time than a surface feature to cross the stellar disk from limb to limb, i.e. approximately $1/2\ P_{\rm rot}$. Here is an example.

V471 Tau is an eclipsing K2-dwarf star in a close binary, its inclination of the stellar rotation axis as well as of the orbital plane is $i \approx 90°$. Its spectral lines are rotationally broadened by $v \sin i = 90\ {\rm km\,s^{-1}}$, the effective surface temperature is $T_{\rm eff} \approx 5000$ K, and the stellar rotation period is $P_{\rm rot} = 0.52$ days ($\Omega/\Omega_\odot \approx 50$). The star has V=9.7 mag and the companion (a white dwarf) is not seen at red wavelengths. Ramseyer *et al.* (1995) obtained a Doppler image of this outstanding object, and we adopt their original instrumental characteristics for our initial estimate and then proceed as if we were using the Keck+HIRES or VLT+UVES instruments or equivalent. Fixing the duration of the transit to approximately $1/2\ P_{\rm rot}$, the orbital period of an extra-solar planet would be expected to be between, roughly, 8 and 80 days, depending whether we assume its fractional distance from the parent star to be $R/a = 0.1$ or 0.01, respectively.

With Δt the total integration time, $\lambda/\Delta\lambda$ the spectrograph's resolving power and S/N the desired signal-to-noise ratio, the minimum detectable disk radius of a planet in units of the stellar disk according to Eqs. (4.4–4.9) is

- $\boxed{r_{\rm min} \approx 0.14}$ for $\Delta t = 30$ min, $\lambda/\Delta\lambda = 40,000$, S/N≈100:1, using Ca I 6717 Å,
- $\boxed{r_{\rm min} \approx 0.09}$ for $\Delta t = 15$ min, $\lambda/\Delta\lambda = 120,000$, S/N≈100:1, using Ca I 6717 Å,
- $\boxed{r_{\rm min} \approx 0.07}$ for $\Delta t = 15$ min, $\lambda/\Delta\lambda = 120,000$, but using the much weaker Fe I 6713 Å line (its $FWHM$ is ≈0.12 Å instead of 0.17 for Ca I).

If the integration time could be cut down to be negligible compared to the phase smearing due to the stellar rotation period, i.e. in the ideal case if $\Delta t \longrightarrow 0$, then $r_{\rm min} \approx 0.03 - 0.04$, solely depending upon the intrinsic width of the used spectral line. This would be the radius of the theoretically smallest detectable stellar surface element, or the smallest detectable disk size of the transiting planet.

As a comparison, a transit by Earth and Jupiter in front of the solar disk as seen from another star would give the following disk ratios:

$$\text{Earth at 1 AU}: r \approx 0.01$$
$$\text{Jupiter at 5 AU}: r \approx 0.1$$

Thus, for bright parent stars and with the largest telescopes and spectrographs at hand, we could already start to look for Jupiter-size planets in spectral line profiles!

It is a pleasure to thank Dr. John B. Rice of Brandon University for his many scientific and other highlights that drive our joint Doppler-imaging efforts. Thanks also to the CFHT Corporation for their continuous support of our *Gecko* observing runs, and to my colleague Dr. Yvonne Unruh for providing two of the spectra in Fig. 2 and for many fruitful discussions. Of course, there would be much less science without money, and it is crucial to thank the Austrian Science Foundation for their support through grants S7302-AST and S7301-AST. Finally, I compliment the meeting organizers, and especially its chair Rafael Rebolo, for inviting a talk on K–M stars in a brown-dwarf meeting!

REFERENCES

Barnes J. R., Collier Cameron A., Unruh Y. C., Donati J.-F. & Hussain G. A. J. 1998 *MNRAS*, **299**, 904

Berdyugina S. 1998 *A&A*, **338**, 97

Berdyugina S. & Tuominen I. 1998 *A&A* **336**, L25

Bruls J. H. M. J., Solanki S. K. & Schüssler M. 1998 *A&A* **336**, 231

Collier Cameron A. 1992 *Surface Inhomogeneities on Late-type Stars*, P. B. Byrne & D. J. Mullan (eds.), Springer, p. 33

Collier Cameron A. 1995 *MNRAS* **275**, 534

Collier Cameron A. & Unruh Y. C. 1994 *MNRAS* **269**, 814

Dempsey R. C., Neff J. E., Strassmeier K. G., Airapetian V. S. & Lim J. 1998 *ApJ*, in preparation

Deutsch A. 1958 *IAU Symp. 6, Electromagnetic Phenomena in Cosmical Physics*, P. Lehnert (ed.), Cambridge University Press, p. 209

Donati J.-F. 1998 *MNRAS*, in press

Donati J.-F. & Collier Cameron A. 1997 *MNRAS* **291**, 1

Donati J.-F., Semel M., Carter B. D., Rees D. E. & Collier Cameron A. 1997 *MNRAS* **291**, 658

Donati J.-F., Collier Cameron A., Hussain G. A. J. & Semel M. 1998 *MNRAS*, in press

Hatzes A. P. 1995a *Poster Proceedings, IAU Symp. 176, Stellar Surface Structure*, K. G. Strassmeier (ed.), University of Vienna, p. 90

Hatzes A. P. 1995b *ApJ* **451**, 784

Johns-Krull C. M. & Hatzes A. P. 1997 *ApJ* **487**, 896

Joncour I., Bertout C. & Bouvier J. 1994a *A&A* **291**, L19

Joncour I., Bertout C. & Menard F. 1994b *A&A* **285**, L25

Kürster M. 1993 *A&A* **274**, 851

Kürster M., Hatzes A. P., Pallavicini R. & Randich S. 1992 *7th Cambridge Workshop on Cool Stars, Stellar Systems, and the Sun*, M. S. Giampapa & J. A. Bookbinder (eds.), PASPC 26, p. 249

Kürster M., Schmitt J. H. M. M. & Cutispoto G. 1994 *A&A* **289**, 899

Maceroni C., Vilhu O., van't Veer F. & Van Hamme W. 1994 *A&A* **288**, 529

Montes D. & Martín E. L. 1998 *A&AS* **128**, 485

Neuhäuser R., Wolk S. J., Torres G., *et al.* 1998 *A&A* **334**, 873

O'Neal D., Neff J. E. & Saar S. H. 1996 *9th Cambridge Workshop on Cool Stars, Stellar Systems, and the Sun*, R. Pallavicini & A. Dupree (eds.), PASPC 109, p. 621

Petrov P. P., Gullbring E., Gahm G., *et al.* 1995 *Poster Proceedings, IAU Symp. 176, Stellar Surface Structure*, K. G. Strassmeier (ed.), University of Vienna, p. 217

Piskunov N. E. 1990 *IAU Colloq. 130, The Sun and Cool Stars: Activity, Magnetism, Dynamos*, I. Tuominen, D. Moss & G. Rüdiger (eds.), Springer, p. 309

Piskunov N. E. 1996 *IAU Symp. 176, Stellar Surface Structure*, K. G. Strassmeier & J. L. Linsky (eds.), Kluwer, p. 45

Piskunov N. E. & Rice J. B. 1993 *PASP* **105**, 1415

Press W. H., Teukolsky S. A., Vettering W. T. & Flannery B. P. 1992 *Numerical Recipes*, Cambridge University Press

Ramseyer T. F., Hatzes A. P. & Jablonski F. 1995 *AJ* **110**, 1364

Rice J. B. 1996 *IAU Symp. 176, Stellar Surface Structure*, K. G. Strassmeier & J. L. Linsky (eds.), Kluwer, p. 19

Rice J. B. & Strassmeier K. G. 1996 *A&A*, **316**, 164

Rice J. B. & Strassmeier K. G. 1998 *A&A*, in press

Rice J. B., Wehlau W. H. & Khokhlova V. 1989 *A&A* **208**, 179

Saar S. H., Piskunov N. E. & Tuominen I. 1994 *8th Cambridge Workshop on Cool Stars, Stellar Systems, and the Sun*, Caillault *et al.* (eds.), PASPC 64, p. 661

Spruit H. C. & van Ballegooijen A. A. 1982 *A&A* **106**, 58

Stout-Batalha N. M. 1997 *PhD. Dissertation*, University of California at Santa Cruz

Stout-Batalha N. M. & Vogt S. S. 1996 *IAU Symp. 176, Stellar Surface Structure*, K. G. Strassmeier & J. L. Linsky (eds.), Kluwer, p. 337

Strassmeier K. G. 1995 *Science with the VLT*, J. R. Walsh & I. J. Danziger (eds.), ESO Astrophysics Symp., Springer, p. 87

Strassmeier K. G. & Rice J. B. 1998a *A&A*, **330**, 685

Strassmeier K. G. & Rice J. B. 1998b *A&A*, in press

Strassmeier K. G., Rice J. B., Wehlau W. H., Hill G. M. & Matthews J. M. 1992 *A&A* **268**, 671

Strassmeier K. G., Welty A. & Rice J. B. 1994 *A&A* **285**, L17

Struve O. 1930 *ApJ* **72**, 1

Unruh Y. C. & Collier Cameron A. 1995 *MNRAS* **273**, 116

Unruh Y. C., Collier Cameron A. & Guenther E. 1998 *MNRAS* **295**, 781

Valenti J. A., Piskunov N. & Johns-Krull C. M. 1998 *ApJ* **498**, 851

Vogt S. S. 1981 *ApJ* **247**, 975

Vogt S. S. & Penrod G. D. 1983 *PASP* **95**, 565

Vogt S. S., Penrod G. D. & Hatzes A. P. 1987 *ApJ* **381**, 396

X–ray Emission from Cool Dwarfs in Clusters

By SOFIA RANDICH

Osservatorio di Arcetri, Largo E. Fermi 5, I-50125, Firenze, Italy

ROSAT has allowed the detection in X–rays of a large fraction of M dwarfs both in young open clusters such as the Pleiades and α Persei, and in the older Hyades. No decline of the average X–ray luminosity occurs between α Per and the Pleiades, while a rather steep decay is seen between the latter and the Hyades. The similarity of the Pleiades and α Per M dwarfs X–ray activity distributions simply reflects the similarity in their rotation distributions. It is more difficult to understand, instead, why the Hyades are, on average, significantly underluminous with respect to the Pleiades, since, due to the long spin-down timescales for M dwarfs, a large fraction of moderate or even rapid rotators are still present in the Hyades.

Although fully convective stars as active as stars with a radiative core have been observed, based on the Hyades, there might be an indication for a slight drop of the average X–ray emission level below the fully convective boundary mass, indicating a possible loss of efficiency of the mechanism of magnetic field generation.

Stars with masses down to 0.13 $M_\odot$ and 0.19 $M_\odot$ have been detected in the Pleiades and the Hyades, respectively. These detections, together with that of a 0.04 $M_\odot$ brown dwarf in the Chamaleon I star forming region and of very-low mass dwarfs in the field, support the idea that there is not a cut-off mass below which stars do not have coronae anymore.

---

1. Introduction

Einstein and *ROSAT* X–ray surveys of several open clusters have allowed a big step forward in our understanding of the age-rotation-X–ray activity paradigm and of the coronal properties of cool stars in clusters (e.g., Micela et al. 1990; Caillault 1996; Randich 1997; Jeffries 1998). Among the most important results is the finding that the decay of X–ray activity with age can be described by a power law $L_X \propto t^{-\alpha}$, as originally suggested by Skumanich (1972), but with α varying with both stellar mass and the age interval. Rotation is the crucial parameter in determining the level of X–ray activity; in particular, saturation of X–ray emission above a given rotation threshold, although not understood from a theoretical point of view, has been empirically shown to be one of the key ingredients to interpret the observed X–ray properties of stellar clusters and their evolution with age.

Most of the papers and reviews on *ROSAT* results for stellar clusters have concentrated on solar-type and K-type stars, dedicating relatively less attention to M dwarfs and, in particular, to very low mass stars. In this paper I "revisit" *ROSAT* observations of open clusters concentrating on the low mass end of the main sequence. I discuss the issue of the age–activity connection for M dwarfs; I address the issue of whether there is a change of the type of dynamo at the fully convective boundary mass; I finally present the results and the prospects offered by the AXAF satellite for the detection and the investigation of X–ray emission from very low mass stars and brown dwarfs (BDs).

2. X–ray activity *versus* age

2.1. *The comparison of XLDFs: results*

As it is the case at optical wavelengths, deep surveys are needed in order to detect and study low mass stars: X–ray observations of both field and cluster stars have indicated that, in quiescent state, the ratio of X–ray to bolometric luminosity, L_X/L_{bol}, does not exceed the "saturation" value of 10^{-3}, and thus optically faint stars are also X–ray faint, even if very X–ray active. Micela et al. (1990) detected only four out of the 39 Pleiades dM stars included in their *Einstein* field of view; the latest-type detected object had a B−V color = 1.52 or a mass of $\sim 0.5 M_\odot$. The implication was that either these stars constituted the high luminosity tail of an X–ray luminosity distribution function (XLDF) most of which fell below the sensitivity threshold, or they were exceptional objects (e.g., tidally locked binaries). *ROSAT* observations have confirmed that dM stars in clusters do have coronae and do emit in X–rays. Stauffer et al. (1994) detected 27 of the 38 Pleiades dwarfs with $1.8 < V - I_C < 3$ included in their PSPC fields; similar detection rates were obtained for the α Persei cluster by Randich et al. (1996) and Prosser et al. (1996), with the detection rate becoming as high as 75 % in the high sensitivity regions of the *ROSAT* images. As to the older Hyades, Stern et al. (1995) detected 29 % of the Hyades M dwarfs in the All-Sky Survey (RASS) data. Considerably higher detection rates were achieved in the deeper PSPC pointings of selected areas of the cluster: Pye et al. (1994) detected 23 out of the 32 M dwarfs present in their PSPC field of view, seven out 11 (64 %) M dwarfs were detected by Stern et al. (1994), and *all* M dwarfs in the field of view, apart from three most likely cluster non-members, were detected by Reid et al. (1995).

The detection of such a large number of M dwarfs in these clusters makes it possible to infer the average X–ray properties and to extend the age–activity relationship to stars considerably less massive than the Sun.

In Figure 1 I show the XLDFs for α Per (70 Myr –dashed curve), Pleiades (120 Myr –dotted curve) and Hyades (600 Myr –solid curve) stars with $(B-V)_0 > 1.42$ or $(V-I_C)_0 > 1.6$. For α Per I have used the merged raster scan (Randich et al. 1996) and deep PSPC pointings (Prosser et al. 1996) data; for the Pleiades I merged the data of Stauffer et al. (1994) and Micela et al. (1996); finally, for the Hyades the data of Pye et al. (1994) and Reid et al. (1995) were considered. Although their *ROSAT* fields cover a much smaller area than the RASS survey, these observations are significantly deeper and thus allow to overcome the problem of severe censoring. For all the clusters, only stars with masses larger than 0.3 $M_\odot$ were taken into account since *a)* very few fainter/later-type stars were covered by *ROSAT* pointings of the younger clusters; *b)* the incompleteness/contamination of optical catalogs for the younger clusters would anyway prevent a correct statistical analysis; *c)* the presence of a possible drop of the average X–ray activity below 0.3 $M_\odot$ could bias the comparison (see Sect. 3 below); and, *d)* below 0.3 $M_\odot$, stars in the younger clusters are significantly brighter in the optical than stars in the Hyades and thus, for the same activity level as indicated by L_X/L_{bol} ratios, they would have a higher L_X (i.e., more simply, for the same surface X–ray flux, they have a larger emitting area). The median and average log L_X values inferred from the various *ROSAT* surveys of the three clusters are summarized in Table 1. Note that the median L_X is not quoted for the Hyades sample of Stern et al. (1995; RASS data) since less than 50 % of the M dwarfs were detected. I mention in passing that a large number of M dwarfs were detected also in the 30–35 Myr old clusters IC 2602 and IC 2391 (Randich et al. 1995; Patten & Simon 1996); since these samples are X–ray selected (a membership list for cool stars in these clusters did not exist prior to *ROSAT* surveys) and are likely to be incomplete, I exclude them from the comparison.

The table indicates that the values inferred from the various surveys are consistent with each other. This means that each sample can be considered as representative of its parent population and, therefore, a statistical analysis based on the merged samples and/or on the less censored samples should not introduce any bias. Both the figure and

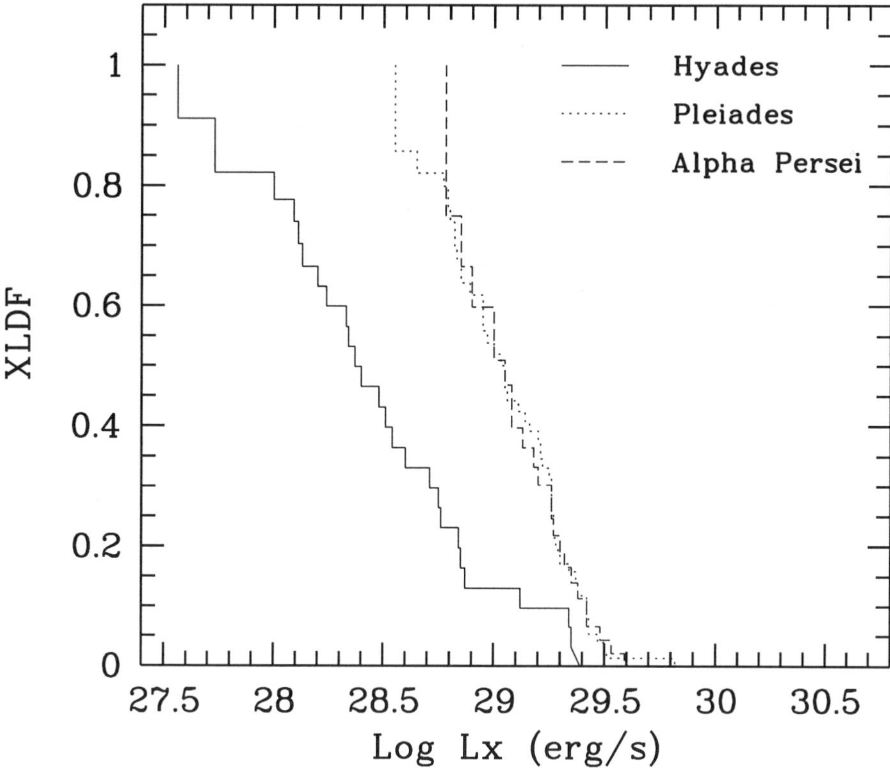

FIGURE 1. X–ray luminosity distribution functions for M dwarfs with masses $> 0.3\ M_\odot$ in the Hyades (solid curve), the Pleiades (dotted curve), and α Persei (dashed curve).

TABLE 1. Mean and median X–ray luminosities

Cluster (survey)	**mean**	**median**
	log $\mathbf{L}_X$ (erg/s)	
α Per (Randich et al. 1996)	28.95 ± 0.04	28.88
α Per (Prosser et al. 1996)	29.06 ± 0.04	29.01
Pleiades (Stauffer et al. 1994)	28.98 ± 0.04	28.99
Pleiades (Micela et al. 1996)	29.14 ± 0.08	28.85
Pleiades (merged)	29.03 ± 0.04	29.03
Hyades (Pye et al. 1994)	28.32 ± 0.14	28.32
Hyades (Reid et al. 1995)	28.53 ± 0.11	28.40
Hyades (Stern et al. 1995)	28.12 ± 0.04	—
Hyades (Pye + Reid)	28.41 ± 0.09	28.37

the table indicate that virtually no changes in the XLDF and in the mean/median X–ray luminosity occur between α Per and the Pleiades age (see also Randich et al. 1996; Prosser et al. 1996); on the contrary, the average level of X–ray activity declines by a factor ~ 4 between the Pleiades and the Hyades. The high luminosity tail of the Hyades XLDF extends to luminosities almost as high as the younger clusters XLDFs, which does not occur for solar-type stars; however, the bulk of the Hyades M dwarfs population is

X–ray underluminous with respect to the Pleiades and α Per as indicated by the mean and median L_X.

2.2. *The comparison of XLDFs: a possible explanation*

For slow rotators, the X–ray activity – rotation relationship has been often formulated as $L_X \propto v^n$, with n usually ranging between 1 and 4 (e.g., Pallavicini et al. 1981; Maggio et al. 1987; Micela et al. 1998). As I have mentioned in the introductory section, the level of X–ray emission then saturates above a given rotational velocity threshold: for $v_{rot} > v_{thr}$ (or $P < P_{thr}$), log $L_X/L_{bol} = -3$ and L_X does not increase any more with increasing rotation. The saturation relation has been found to be represented in the best and most uniform way by using the Rossby number (N_R –the ratio of the rotational period P over the convective cell turnover time τ) (e.g., Randich et al. 1996; Patten and Simon 1996). In this representation, and using the empirical expression for τ given by Noyes et al. (1984), the data for young clusters indicate that saturation occurs around log $N_R = -0.8 \pm 0.2$; since τ gets larger for lower mass stars, this means that they saturate at progressively larger periods/lower velocities. Stauffer et al. (1997b) estimated that the velocity threshold for a 0.4 $M_\odot$ star is around 5–6 km/s.

The flat age dependence of X–ray activity between α Per and the Pleiades is therefore not surprising and it is simply explainable as due to the fact that M dwarfs in these two clusters have similar rotation distributions and, in particular, most of them rotate faster than 5–6 km/s and are saturated.

More surprising, instead, is the steep decay of the X–ray activity emission level that is observed between the Hyades and the Pleiades. Due to the dependence of the spin-down timescales on stellar mass (the lower the mass the longer the timescale), a significant fraction of Hyades M dwarfs are still moderate (vsini = 12-15 km/s) or even fast (vsini = 20-30 km/s) rotators and should, in principle, have saturated X–ray emission with L_X levels comparable to what observed in the younger clusters. 25 % of the Pleiades M dwarfs in the sample of Stauffer et al. (1994) have log $L_X/L_{bol} < -3.2$, i.e., below saturation (a similar percentage holds for the Micela et al. 1996 sample); none of them has measured rotational velocities apart from two which have vsin $i \leq 9$ km/s. The assumption can thus be made that the small fraction of non–saturated stars in the Pleiades is constituted by slow rotators, or stars with $v_{rot} \leq 5 - 6$ km/s. On the other hand, about 80 % of the Hyades M dwarfs are non–saturated, which eventually reflects into the fact that the bulk of the Hyades population is X–ray underluminous with respect to the Pleiades. About 20 % of the non–saturated stars with measured rotational velocities, however, have detected velocities *above* the 5–6 km/s saturation threshold. Stauffer et al. (1997a) have in fact found that there seems to be a bifurcation of the saturation relation for the Hyades M dwarfs, with part of the stars with 6 km/s vsin i upper limits being saturated and another fraction saturating only at higher velocities. This two–branches relationship is shown in Figure 2 (from Stauffer 1997a). As an explanation, they suggested that stars that show saturation at 6 km/s are the descendants of Classical T Tauri stars (CTTS), while stars that show saturation at higher velocities would be the descendants of weak-lined T Tauri stars (WTTS); the two groups of stars, due to the different timescales for dissipation of their circumstellar disks arrive on the zero age main sequence (ZAMS) as slow/fast rotators and with large/small differential rotation. The difference in radial and latitudinal differential rotation would eventually result in a difference of the dynamo induced activity (both coronal and chromospheric).

It is not clear why such a difference in dynamo induced activity should appear only at the Hyades age, since differences in differential rotation between stars with long-lived and short-lived disks should be present already at the Pleiades age; nevertheless, based on the

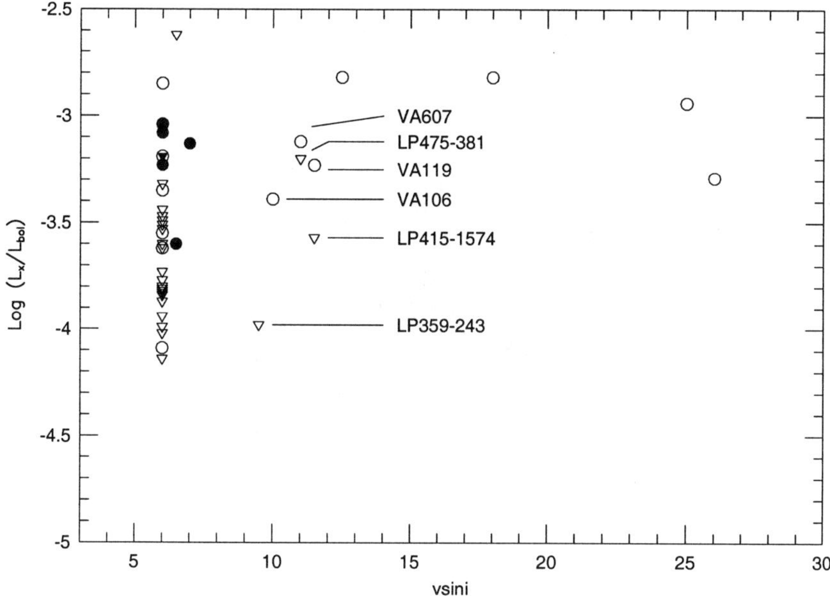

FIGURE 2. From Stauffer et al. (1997a), their Fig. 13. Log L_X/L_{bol} as a function of projected rotational velocity vsin i. Filled symbols correspond to SB2's. Downward-pointing triangles correspond to X–ray upper limits. All points at vsin i = 6 km/s are rotational velocity upper limits.

empirical findings, the hypothesis can be made that there are two reasons for the large decay of the average X–ray activity level between the Pleiades and the Hyades: the first one is the same as for solar–type stars, i.e., stars that are saturated at 120 Myr, at 600 Myr have slowed-down below 5–6 km/s; they are probably a large fraction of the stars with non–detected or non-measured rotation and they likely have v_{rot} of the order of a few km/s. The second reason is that the one branch rotation-activity relationship which presumably holds for young clusters (as at least it is suggested by the still relatively few measured rotational rates) has converted into a two-branches relationship, with part of the stars with moderate rotation lying below the saturation level. More rotational data are needed; they would allow us to determine whether a two–branches relationship does hold for young clusters or not and would provide a more accurate estimate of the fraction of moderately rotating/non–saturated stars in the Hyades.

3. Coronal activity below the fully convective boundary mass

As well known, stars are predicted to be fully convective around 0.3 $M_\odot$. According to models of stellar dynamos, a fundamental change in the mechanism of generation of the magnetic field at the fully convective mass and, therefore, a change in the coronal activity is expected (Giampapa & Liebert 1986; Rosner & Weiss 1992) . Namely, the shell dynamo cannot be at work in fully convective stars which lack a radiative core and a shear layer at the basis of the convection zone. A distributive dynamo or a fibril dynamo that would generate chaotic magnetic fields have been proposed has alternative processes that could operate in fully convective stars. The obvious question is whether a change

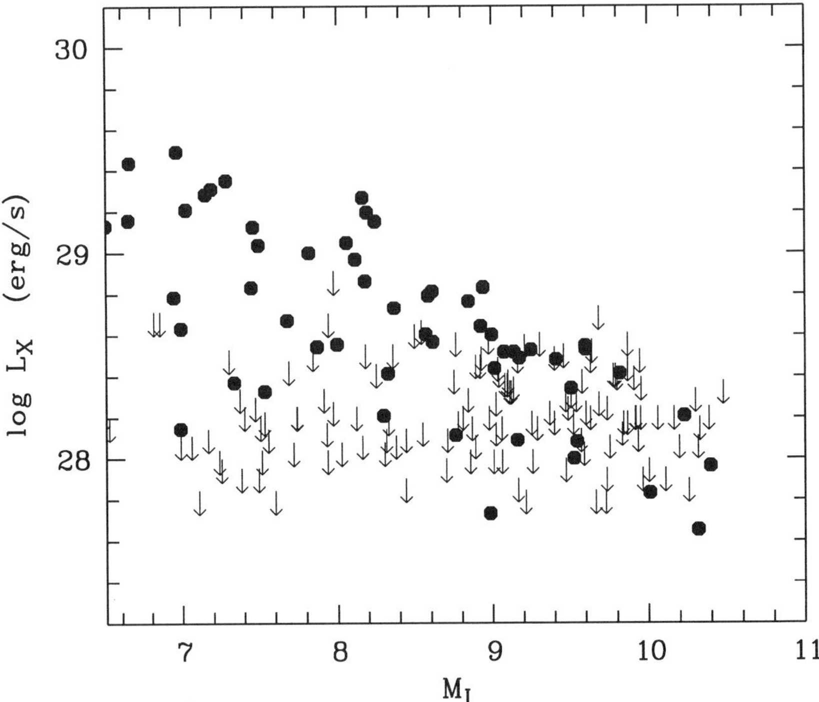

FIGURE 3. X-ray luminosity as a function of the *I* absolute magnitude for the Hyades M dwarfs. Filled circles and downwards arrows represent detections and upper limits.

in the activity level (and, in this context, in the coronal activity level) is evidenced by the observations.

The detection in X-rays of several field stars with masses below the fully convective boundary mass (see the review by Schmitt in this Volume and references therein) indicates that these stars do have coronae. Barbera et al. (1993), based on an extended *Einstein* survey, claimed to see a drop in the level of X-ray activity for field stars later than $\sim$ M5 and interpreted this as a proof of a reduced efficiency in magnetic field generation. On the contrary, more recent *ROSAT* observations of field dMe and dM stars do not evidence changes in the level of X-ray activity (as expressed in terms of log L_X/L_{bol}) from early to late-type M dwarfs (e.g., Fleming et al. 1993; Fleming et al. 1995; Giampapa et al. 1996). All these studies (and conclusions) were based on inhomogeneous samples as far as age and metallicity are concerned and on relatively small samples. Given the activity-age dependence and the possible metallicity-activity dependence, it is important to investigate in detail the coronal properties at (and below) the fully convective boundary for a co-eval and uniform metallicity sets of stars, in order to place better constraints on any change in the X-ray emission level. A study of this kind has been carried out for the Pleiades by Hodgkin et al. (1995), who detected a possible turnover of the mean log L_X/L_{bol} around an *I* absolute magnitude $M_I \sim 8-9$ (or a mass M $\sim$ 0.4–0.3 $M_{\odot}$). In the following, I am going to investigate this issue for the Hyades.

In Figure 3 I plot $\log L_X$ vs. M_I for all Hyades M dwarfs with $M_I \geq 7$, while in Figure 4 log L_X/L_{bol} vs. M_I is shown. According to the predictions of Baraffe et al.

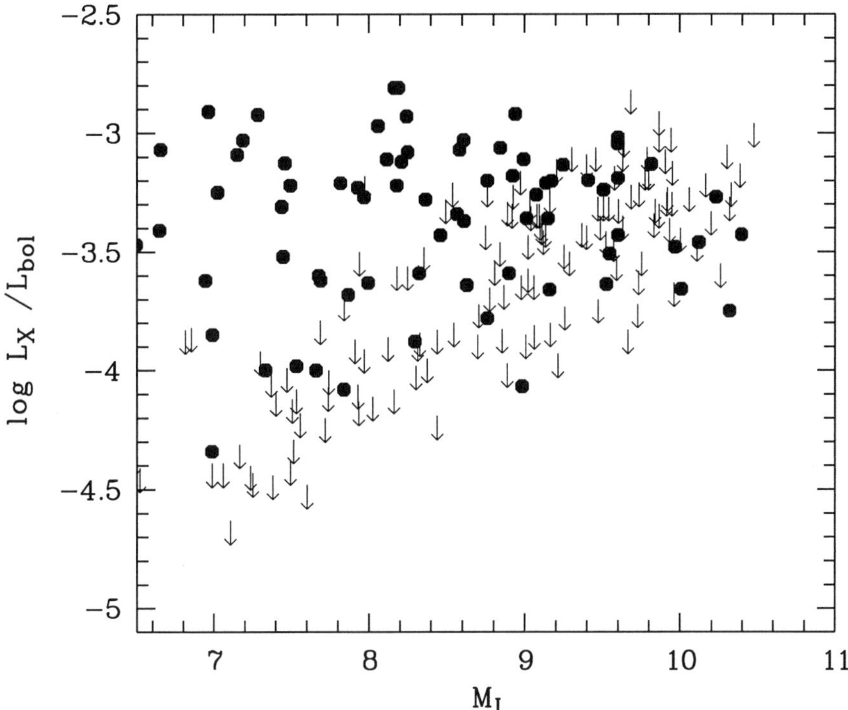

FIGURE 4. Same as Fig. 3, but log L_X/L_{bol} values are shown.

(1998), a 0.3 $M_\odot$ star at 600 Myr should have $M_I = 9.21$. Whereas the exact magnitude corresponding to the fully convective boundary mass is obviously model dependent, what is of interest here is to check whether appreciable variations of the X–ray properties occur starting at a given M_I. Fig. 3 shows that the maximum L_X decreases with M_I or mass; the maximum X–ray activity level is constrained by saturation and L_X^{max} is roughly equal to $10^{-3} \times L_{bol}$. It is therefore more appropriate to consider the L_X/L_{bol} as an activity indicator: Fig. 4 indeed confirms that the highest activity stars for most of the M_I range cluster around log $L_X/L_{bol} = -3$. In other words, Figs. 3 and 4 indicate that dynamo in fully convective stars can work as efficiently as in stars with a radiative core. However this simply means that, whatever dynamo mechanism is, it is bounded by the saturation condition. I believe that the fact that fully convective stars can be as active as more massive stars does not necessarily mean that there is not an average loss of efficiency in the magnetic field generation. A comparison of the average X–ray activity levels of stars above/below the fully convective boundary mass is instead needed for a more detailed investigation of this issue. Thus, I constructed cumulative distribution functions (CDF) of the log L_X/L_{bol} ratios for Hyades stars in four M_I magnitude bins: namely, $8 \leq M_I < 8.5$, $8.5 \leq M_I < 9$, $9 \leq M_I < 9.5$, and $9.5 \leq M_I < 10$.

The CDFs are shown in Figure 5. The comparison of the CDFs below log $L_X/L_{bol} \sim -3.5$ is not very meaningful, since they are dominated by upper limits; for higher log L_X/L_{bol} ratios, instead, the comparison of CDFs, in agreement with the results of Hodgkin et al. (1995) for the Pleiades, suggests that a drop of the average X–ray activity around the fully convective mass may indeed be present. The sample, however, is too heavily censored to regard this conclusion as definitive. More important, as stressed

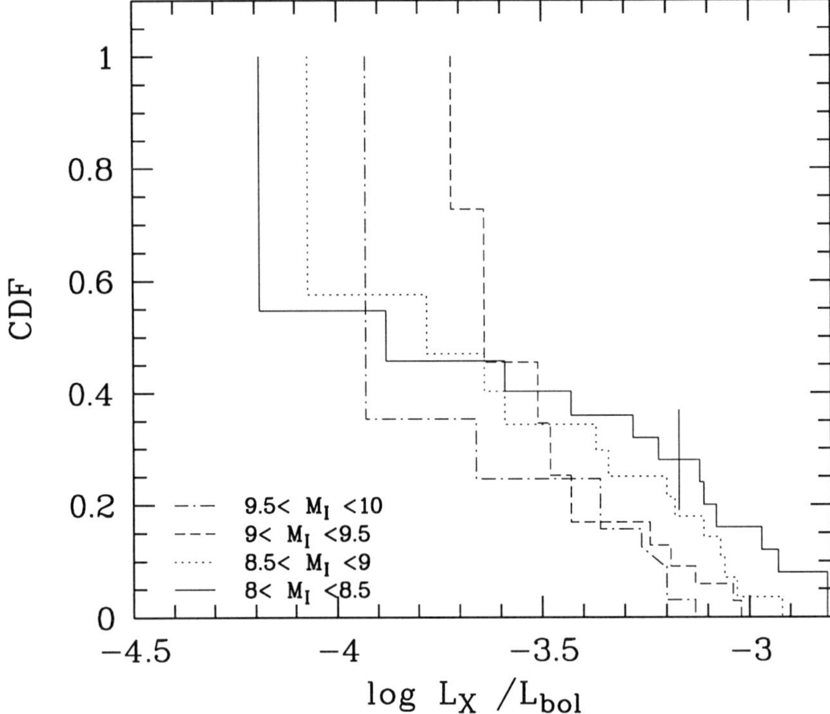

FIGURE 5. Cumulative distribution functions (CDF) of the log L_X/L_{bol} values for Hyades stars in four M_I bins. The typical error bar is also shown.

by Hodgkin et al. (1995), no conclusion can be drawn about the mass dependence of X–ray emission without taking into account rotation. How does the distribution of rotational velocities among fully convective stars compare with that of stars with a radiative core? And what about the L_X – rotation relationship for fully convective stars? These questions, again, point towards the need for extending the rotational database.

Finally, Figs. 3 and 4 show that Hyades stars have been detected fainter than $M_I = 10$ ($M \leq 0.25\ M_{\odot}$), but none of them has saturated X–ray emission. This introduces the subject of next section.

4. Very low mass stars and Brown Dwarfs

The topic of X–ray emission from very low mass stars or BDs in clusters is more a subject for the discussion of the prospects offered by future X–ray missions than for a detailed analysis of current results. As shown in the previous section, stars in the Hyades have been detected down to $M_I \sim 10.4$ or down to $\sim 0.2\ M_{\odot}$. This is an important result and, probably, one of the major achievements of *ROSAT* surveys of open clusters. Very low mass stars have been detected also in the younger clusters and in the field (see again the review by Schmitt in these Proceedings), but they are too few to try to attempt any detailed/quantitative discussion/comparison of their X–ray properties. In Table 2 I list a compilation of the lowest mass X–ray detected stars in five open clusters surveyed by *ROSAT* (masses have been inferred from M_I magnitudes using the models of Baraffe et al., 1998). The caveat should be kept in mind that the lowest mass for which a detection

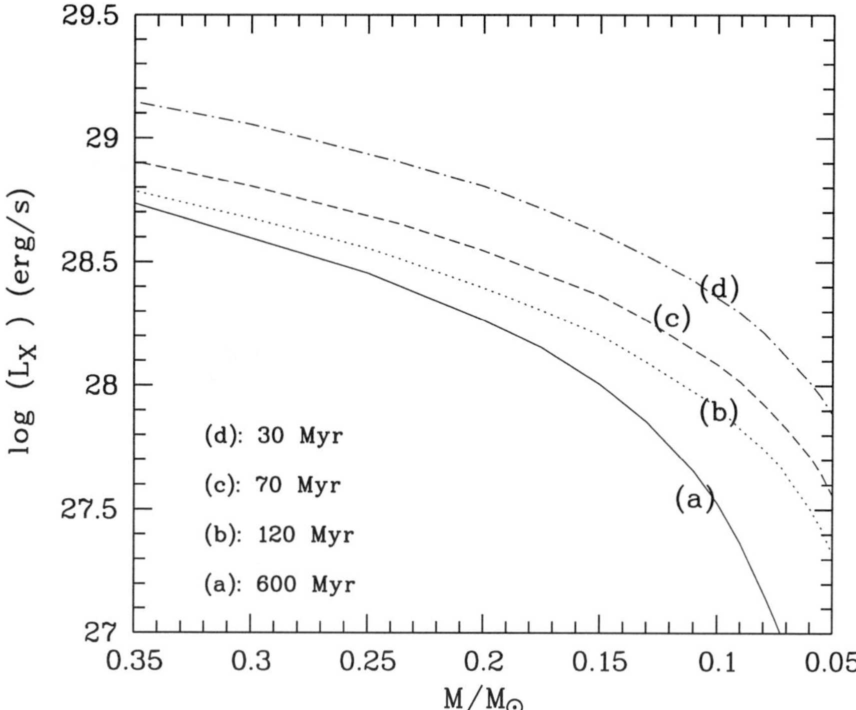

FIGURE 6. Predicted X–ray luminosity as a function of mass for 30, 70, 120, and 600 Myr. The predicted log L_X has been computed for saturation conditions, i.e., log $L_X/L_{bol}= -3$ and $L_X = 10^{-3}$ L_{bol}. The L_{bol} vs. mass relation from the models of Baraffe et al. (1998) has been assumed.

was obtained is most likely determined by the sensitivity of the observations rather than by any intrinsic physical reason. The table indeed suggests that there is not a threshold mass below which stars cannot support coronae any more. I mention in passing that the detection of AP 275 in α Per is uncertain; whereas a X–ray source has clearly been detected, the offset between the X–ray and optical positions is somewhat larger than what normally accepted. The X–ray source is weak, however, and a larger uncertainty is associated to its position. Moreover, no other possible known optical counterparts are present close to the X–ray source. If the detection of AP 275 is real, this would be the lowest mass star detected in a stellar cluster, with a mass very close to the substellar limit. Note that a very short rotational period (P $\sim$ 7.6 hrs) and a high vsin i (33$\pm$6 km/s) were measured for this object by Martín & Zapatero-Osorio (1997) and by Basri & Martín (1998), respectively. An additional, qualitative, feature to be noted in Table 2 is that almost all of the X–ray detected very low mass stars in the young clusters have a log L_X/L_{bol} ratio above the saturation value, indicating a possible flaring activity, whereas this is not the case for stars in the Hyades.

Finally, there is the question of whether it is (will) be possible to detect X–ray emission from brown dwarfs. Assuming that there is not a cut-off mass for supporting coronae, detection of BDs just depends on sensitivity. In Figure 6 I plot the expected log L_X as a function of mass using the L_{bol} vs. mass relationship of Baraffe et al. (1998) and assuming the saturation condition log L_X/L_{bol} =−3. The four curves correspond to 600 Myr (solid), 120 Myr (dotted), 70 Myr (dashed), and 30 Myr (dashed–dotted).

TABLE 2. The lowest mass X–ray detected stars

Cluster	star	M_I	Mass ($M_\odot$)	$\log L_X$ (erg/s)	$\log L_X/L_{bol}$	Ref.
IC 2602	R19	7.87	0.36	29.52	−2.70	Randich et al. (1996)
IC 2391	R28B	8.82	0.21	28.71	−3.16	Patten & Simon (1996)
α Per	AP 275 ?	10.74	0.09	28.69	−2.45	Prosser et al. (1996)
	AP 143	8.71	0.30	29.18	−2.64	Prosser et al. (1996)
Pleiades	HHJ 48	10.6	0.13	28.76	−2.50	Hodgkin et al. (1995)
Hyades	VA260	10.40	0.19	27.96	−3.43	Pye et al. (1995)

The saturation condition means that, as in the optical, older BDs will be X–ray fainter than younger ones and thus, higher sensitivity surveys are needed in order to detect them. The figure indicates that to detect e.g. a 0.075 $M_\odot$ object with saturated X–ray emission one would need sensitivity thresholds of $\log L_X \sim 27.1$, 27.7, 27.9, 28.16 erg/s in the four cases, which roughly correspond to the Hyades, the Pleiades, α Per and IC 2602/2391 ages. Such sensitivities were not reached by *ROSAT* surveys of the clusters listed in Tab. 2, and, indeed, it is not surprising that Neuhäuser et al. (this Volume) did not detect X–ray emission from none of the four Pleiades BDs for which they analyzed *ROSAT* data. Only under the assumption of an intense flaring activity, lower sensitivities than the ones listed above would be sufficient to detect BDs. Viceversa, should the log L_X/L_{bol} ratios for BDs be systematically below the saturation value, higher sensitivities would be required. Finally, note that significantly lower sensitivities are needed to detect very young and very bright BDs in star forming regions (see below).

AXAF could, in principle, lead to the detection of BDs in open clusters. I refer to the papers of Pallavicini and Linsky in this Volume for a more detailed discussion of this topic.

As a final remark, I would like to note that at the time this meeting was held no detections of BDs had been reported. Very recently, Neuhäuser & Comeron (1998) announced the detection in X–rays of *Cha H*α1, a newly discovered BD (estimated mass $= 0.04 \pm 0.01$ solar masses) in the Chamaleon I star-forming region. They derived a $\log L_X = 28.41$ erg/s and a $\log L_X/L_{bol} = -3.44$. They suggested coronal activity as the most likely reason for the X–ray emission from this object, implying that at least young BDs can support magnetic coronae.

5. Conclusions

ROSAT observations have allowed the detections of several M dwarfs in open clusters of various ages, confirming that very cool dwarfs in clusters do emit in X–rays.

As it occurs for solar-type stars, the average level of X–ray activity shows a decline between ~ 100 and ~ 600 Myr. On one hand this means that, for stars older than the Pleiades, activity can still be used as a proxy for age. On the other hand, this result is unexpected within the age–rotation–activity scenario. The spin-down timescales for M dwarfs are much longer than for solar-type stars and several Hyades M dwarfs have indeed been measured to have a moderate rotation, comparable to that of the Pleiades. The reason why Hyades M dwarfs are significantly X–ray fainter than the Pleiades could be linked to the fact that the rotation–activity relationship for Hyades M dwarfs and, specifically, the saturation relation, seems to be somewhat different from that observed for younger clusters; in particular, stars exist in the Hyades that show moderate rotation, but are non–saturated. As suggested by Stauffer et al. (1997a), this fact could be ultimately linked to the rotational structure and rotational evolution of low mass stars.

The detections of low-mass and very low mass stars in clusters (and in the field) and of a 0.04 $M_\odot$ BD in the Chamaleon I star–forming region suggest that there is not a cut-off mass below which stars cannot support coronae anymore.

Very active fully convective stars with log L_X/L_{bol} at and above the saturated value have been observed, indicating that dynamo in these stars can be at least as efficient as in more massive stars with radiative cores. However, the average X–ray activity level in the Pleiades and the Hyades seems to slightly decay below the fully convective boundary mass, suggesting a general loss in the efficiency of the magnetic field generation and a change in the type of dynamo. The knowledge of the rotation–activity relation for fully convective stars would help addressing directly this issue.

The bottom line is that, although the available rotational database has considerably grown in the last years, both in quantity and quality, more data are needed to help understanding the X–ray data.

I wish to dedicate this paper to the memory of Charles Prosser, who, with his hard and dedicated work, gave a fundamental contribution to the field of open clusters and who left the stellar astronomical community a most precious heritage, a database for members of several clusters.

I am indebted to Roberto Pallavicini for his valuable comments on the manuscript. I thank Isabelle Baraffe, for sending electronic files of her evolutionary tracks and John Stauffer for kindly allowing me to use Fig. 13 of his 1997 paper and for sending me the postscript file of that figure.

REFERENCES

Baraffe, I., Chabrier, G., Allard, F., & Hauschildt, P.H. 1998 *A&A* **337**, 403.

Barbera, M., Micela, G., Sciortino, S., et al. 1993 *ApJ* **414**, 846.

Basri, G., & Martín, E.L. 1998 *ApJL*, in press.

Caillault, J.-P. 1996 in *9th Cambridge Workshop on Cool Stars, Stellar Systems, and the Sun*, ASP Conf. Ser. 109, R. Pallavicini and A.K. Dupree eds., p. 325.

Fleming, T.A., et al. 1993 *ApJ* **410**, 387.

Fleming, T.A., Schmitt, J.H.M.M., & Giampapa, M.S. 1995 *ApJ* **450**, 401.

Giampapa, M.S., & Liebert, J. 1986 *ApJ* **305**, 784.

Giampapa, M.S., et al. 1996 *ApJ* **463**, 707.

Hodgkin, S., Jameson, R.F., & Steele, I.A. 1995 *MNRAS* **274**, 869.

Jeffries, R. 1998 in *Solar and Stellar Activity: Similarities and Differences* J. Butler and G. Doyle (eds.), in press.

Maggio, A., Sciortino, S., Vaiana, G.S., et al 1987 *ApJ* **315**, 687.

Martín, E.L., & Zapatero–Osorio, M.R. 1997 *MNRAS* **286**, L17.

Micela, G., Sciortino, S, Vaiana, G.S., et al. 1990 *ApJ* **348**, 557.

Micela, G., Sciortino, S, Kashyap, V. et al. 1996 *ApJS* **102**, 75.

Micela, G., Sciortino, S., Harndnen, F.R., Jr., et al 1998 *A&A*, in press.

Neuhäuser, R., & Comeron, F. 1998 *Science*, in press.

Noyes, R.W., Hartmann, L., Baliunas, S.L., Duncan, D.K., & Vaughan, A.H. 1984 *ApJ* **279**, 763.

Pallavicini, R., Golub, L., Rosner, R., et al. 1981 *ApJ* **248**, 279.

Patten, B.M., and Simon, T. 1996 *ApJS* **106**, 489.

PROSSER, C.F., RANDICH, S., STAUFFER, J.R., SCHMITT, J.H.M.M., & SIMON, T. 1996 *AJ* **112**, 1570.

PYE, J.P., HODGKIN, S.T., STERN, R.A., & STAUFFER, J.R. 1994 *MNRAS* **266**, 798.

RANDICH, S. 1997 *Mem. SAIt* **68**, 971.

RANDICH, S., SCHMITT, J.H.M.M., PROSSER, C.F., & STAUFFER, J.R. 1995 *A&A* **300**, 134.

RANDICH, S., SCHMITT, J.H.M.M., PROSSER, C.F., & STAUFFER, J.R. 1996 *A&A* **305**, 785.

REID, N., HAWLEY, S.L., & MATEO, M. 1995 *MNRAS* **272**, 828.

ROSNER, R., & WEISS, N.O. 1992 in *Solar Cycle* K.L. Harvey (ed.), p. 511.

SKUMANICH, A. 1972 *ApJ* **171**, 565.

STAUFFER, J.R., CAILLAULT, J.-P., GAGNÉ, M., ET AL. 1994 *ApJS* **91**, 625.

STAUFFER, J.R., BALACHANDRAN, S.C., KRISHNAMURTHI, A., ET AL. 1997a *ApJ* **475**, 604.

STAUFFER, J.R., HARTMANN, L.W., PROSSER, C.F., RANDICH, S., BALACHANDRAN, S.C., ET AL. 1997b *ApJ* **479**, 776.

STERN, R.A., SCHMITT, J.H.M.M., PYE, J.P., ET AL. 1994 *ApJ* **427**, 808.

STERN, R.A., SCHMITT, J.H.M.M., & KAHABKA, P. 1995 *ApJ* **448**, 683.

X-ray Variability in dM Stars

By G. MICELA AND A. MARINO

Osserv. Astronomico di Palermo G. S. Vaiana, Piazza del Parlamento 1, 90134 Palermo, Italy

The issue of the variability of coronal emission is a current problem of stellar physics. Properties of X-ray variability, amplitude and time scales give us information on the physics underlying the coronal emission. In this work we present results from a systematic analysis of X-ray variability of nearby M stars, mainly focused on medium and long term time scales. Taking advantage of archival data of ROSAT-PSPC observations, we explore time scales from days to months. Variability is present at all explored time scales and its amplitude is independent of stellar X-ray and visual luminosity. Results are compared with solar X-ray variability properties suggesting that in dM stars emitting Structures have a spatial distribution more homogenous that in the Sun. Analogous studies on dM stars of the Pleiades indicate that in this cluster the spread observed in the X-ray luminosity function can be explained in terms of variability.

1. Introduction

The study of temporal variations is a very powerful tool to characterize and study the properties of a population of X-ray sources. Studies of typical time scales and amplitude of the observed variability can provide useful information on dimensions and physical conditions of the regions where X-ray emission originates. Comparative studies of the variability properties within a homogeneous class of X-ray sources are useful to determine or constrain the mechanisms generating their X-ray emission. To pursue such studies, a large number of homogenous observations are required. In this work we present some results from a systematic study of X-ray variability on dM stars in the solar neighborhood and in the Pleiades.

The Sun is usually considered to be the prototype of low mass stars and often we extrapolate the knowledge we have on the Sun to the other stars. this approach is certainly very effective but in some cases it has to be applied with caution since the physical conditions of the stars can be different from those in the Sun. In M stars, for example, it is not clear which kind of dynamo is operating and if the solar model can be applied; in this context variability studies and the comparison with the variability properties of the Sun give us an effective tool to test the properties of coronae of dM stars. Previous systematic studies on variability of X-ray emission of dM stars, based on *Einstein* and Exosat observations, concerned mainly short term variability (Ambruster *et al.* 1987; Collura *et al.* 1988; Pallavicini *et al.* 1990).

In this work we have used as reference the solar data from Solrad experiment in the 8-20 Å band. Data have been obtained via Web at http://www.ngdc.noaa.gov/stp/stp.html. The energy bandpass is not exactly the same but it is very similar to that of typical stellar observations. These solar data cover 5 years of observations from March 1968 to February 1973, spanning a time interval comparable with that spanned by ROSAT. The Sun is the only star for which we have a good knowledge of X-ray variability on all time scales (e.g. Vaiana *et al.* 1973; Vaiana and Tucker 1974; Kreplin 1977; Zombeck *et al.* 1978; Withbroe *et al.* 1985), that extend from minutes to years. We know in detail the solar properties since we can relate directly the variations observed in the spatially integrated solar X-ray flux to the spatial distribution of such emission as revealed by high-resolution X-ray images. Since no equivalent studies are possible for stars in this context, variability studies allow us to test the properties and the origin of coronae of late-type stars.

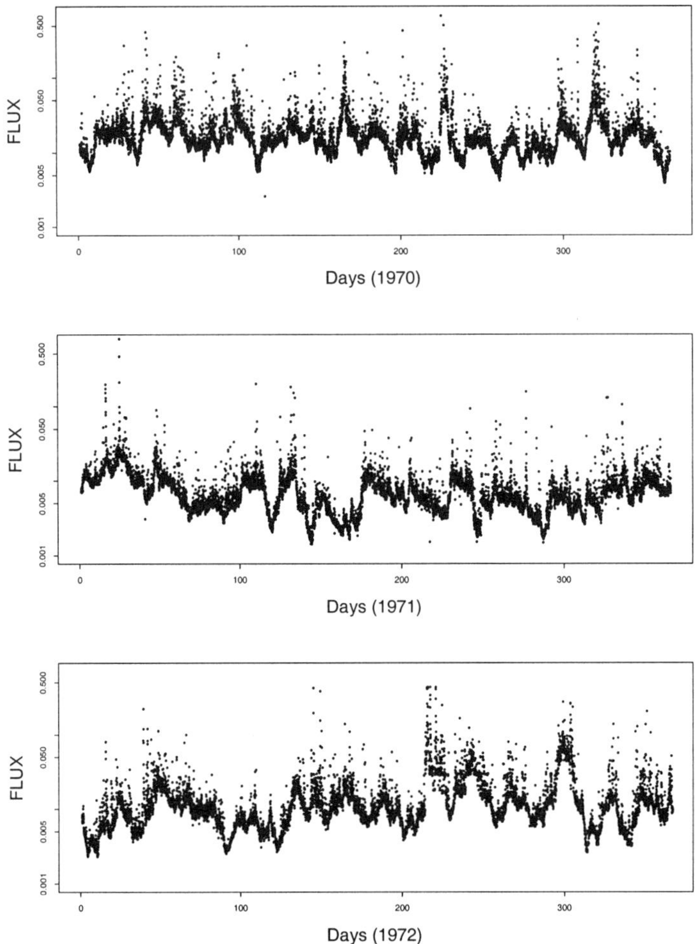

FIGURE 1. Light curve of the Sun as observed by SOLRAD in the 8-20 Å between 1970 and 1972, each point represents the solar emission integrated in one hour.

For nearby field dM stars we have analyzed ROSAT PSPC pointed observations, that were obtained between 1991 and 1994, while for the Pleiades dM stars we have studied *Einstein* IPC, ROSAT PSPC and ROSAT HRI observations.

In order to compare data obtained with a similar time pattern, for the Sun we have used the hourly data, that compare well with the typical length of ROSAT time orbits, typically of the order of few thousands of seconds. In Fig. 1 we report the light curve of the Sun, in three of the explored years by SOLRAD. These light curves show very clearly the rotational modulation that typically produces amplitude variations of less than a factor 5 and the occurency of flares with amplitude up to two orders of magnitude.

With these data we have computed the time distribution function for the Sun. The curve fits very well with a log-normal distribution with a variance of .09 and gives the fraction of time, in a period of 5 years, that the Sun spends at a given value of X-ray flux. The distribution indicates that the Sun spends more than 70% of the time with an X-ray flux within a factor two of its median value and 90% of the time within a factor three, and that only for 1% of the time the X-ray flux of the Sun is different from its median

TABLE 1. Results of the Kolmogorov-Smirnov test applied to the ROSAT PSPC light curves of dM field stars: In columns 2 and 3 are reported the number of cases for which we can reject the hypothesis that the source is constant at the confidence level indicated in column 1.

Confidence level	**Stars $M_v < 13$**	**Stars $M_v > 13$**
> 99%	31 (48%)	14 (67%)
95-99%	4 (6%)	2 (10%)
90-95%	6 (9%)	1 (5%)
<90%	24 (37%)	4 (19%)

value by a factor ten. Note, however, that we are studying five years of solar observations just after a solar maximum, hence the previous percentages refer to this period and not to the entire solar cycle. If a stellar population were composed by stars identical to the Sun we expect that the X-ray luminosity function would have the shape of the solar time distribution. To test which fraction of the variance of X-ray luminosity distribution of the late type field star population can be due to variability, we compare this distribution with the luminosity function observed by Schmitt *et al.* (1995) in the solar neighborhood for G, K and M stars. A comparison of the variances shows that variations such as those observed on the Sun can account at most for 40% of the observed spread for dK stars and for not more than 15% of the spread observed in dG and in dM stars. Hence variations such as those observed in the Sun in a period of 5 years can be responsible for only part of the spread observed in the X-ray luminosity functions of late type field stars. The remaining spread can be due to stellar parameters, such as rotation, but we cannot exclude further contribution from variability on a longer time scale.

2. Stellar observations

We have explored X-ray variability properties of nearby dM stars using the X-ray light curves of the PSPC pointed observations of M stars contained in the CNS3 catalog. Among them we have selected all the stars detected with more than 40 counts and observed at off-axis angle $\leq$ 48 arcmin. Our sample consists of 55 stars, for a total of 86 distinct observations. A detailed analysis of these data is reported in Marino, Micela, & Peres (in preparation).

Most of the lifetimes are shorter than 10ksec with about half of the observations shorter than 5 ksec. The elapsed times are typically less than 2 days with a tail up to 10-15 days and few stars observed one or two years apart.

To evaluate on statistical grounds the presence of variability, we have applied the Kolmogorov-Smirnov (KS) test on all the stars of our sample. The results are reported in Table 1. We see that most observations, both of the earliest and the latest stars, show significant variability. Furthermore variability does not seem to depend on the activity level.

To further explore the variability properties of our stars we have fragmented each observation into individual orbits, and evaluated the X-ray luminosity in each orbit, retaining those in which the source has been detected with at least 40 counts. In Fig. 2 we report the cumulative time distribution functions of the stars with at least 10 distinct luminosity values. As it is evident we are looking at stars with very different activity levels with X-ray luminosity spanning more than two orders of magnitude. All these stars

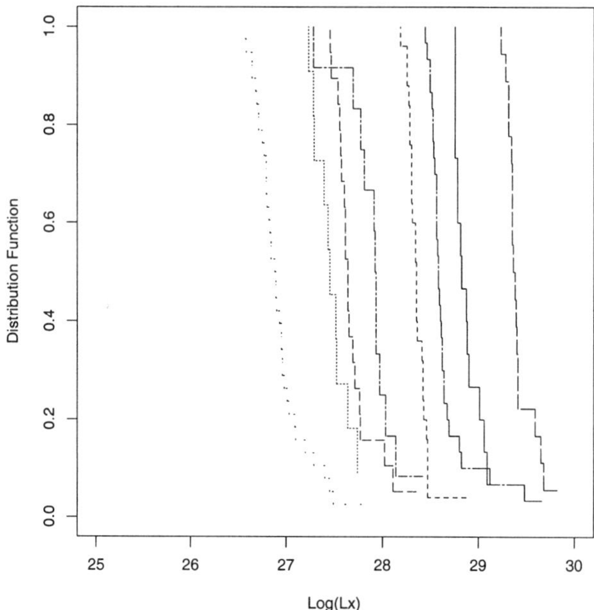

FIGURE 2. Cumulative X-ray time distribution functions of the stars with at least 10 distinct luminosity values. From the left to right: GL551, GL866, GL65, GL294, GJ630.1, GL873, GL388, GL490)

show similar variations independently of the luminosity level. The tail at high value of Lx are typically due to flares.

For each star we compute the distribution of amplitude variation as the ratio between the observed value of Lx and the minimum value observed for that star. Since variability does not seem to depend on the activity level, we compute the normalized cumulative distribution of amplitude variations for the whole sample. The final distribution is shown in Fig. 3. The plot gives us the fraction of time that a dM star spends with an X-ray activity level larger than a given quantity of its minimum value. So a star spends about half time with a L_x larger than 50% over its minimum value and 10% of the time with a Lx larger than 4 times its minimum value.

With the aim to compare the properties of our sample with those of the Sun we have create a set of 500 samples from the Sun distribution. For each sample we have simulated each of the 32 stars we have used to build the amplitude distribution of Fig. 3, drawing from the solar set the same number of observations we have in the PSPC sample. For example, for the 5 stars observed twice, we have randomly selected 5 pairs from the solar light curve (each pair represents a star), then for the 6 stars observed 3 times we have selected 6 sets of 3 points from the solar curve (each set simulates one of these 6 stars), and so on. The idea is to simulate the same pattern of observations typically used for stars, i.e. to simulate the data we would have obtained if we had observed the Sun with the same time pattern used to observe the stars. In this way we can study the Sun with the same bias due of the stellar observations to the time sampling. For each simulated sample we have computed the distribution of amplitude variation as done for the stars. The region occupied by these simulations, together with the observed dM distribution, is shown in Fig.4. We see that the high luminosity tail of our M stars, is consistent with

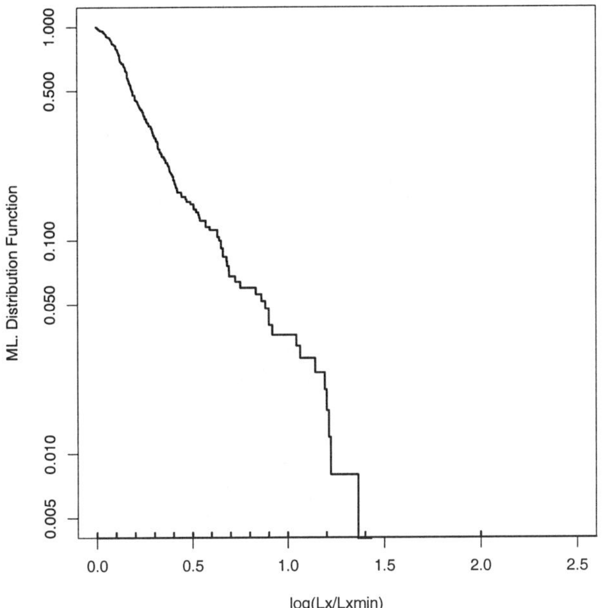

FIGURE 3. Normalized cumulative distribution of amplitude variations for all stars of our field sample.

the tail of the solar distribution that typically is due to flares. On the contrary the body of distribution is depressed with respect to the Sun, suggesting that flare distribution for dM stars is similar to that of the Sun, while the variations by a factor 2-5 that in the Sun are due mainly to rotation are less evident in dM stars.

This discrepancy between the distribution of the Sun and that of the nearby dM stars suggests that the spatial distribution of the regions originating the X-ray emission in dM stars is different (and more homogeneous) than in the Sun.

3. Pleiades dM stars

For the Pleiades we take advantage of the *Einstein* IPC, and of the pointed ROSAT PSPC and HRI exposures in the Pleiades direction. For the IPC observations, that were obtained in 1979/1981, we have used the X-ray luminosity reported in literature (Micela *et al.* 1990).

The PSPC observations were obtained in 1991 and 1992 and we have reanalyzed the images in a homogenous way with a Wavelet based algorithm. When possible we have analyzed separately the single temporal segments of the PSPC observations, taken at well separated times for a total of 7 different images.

The HRI exposures were obtained in 1994-1996, and also for these observations we have used a Wavelet based algorithm applied to each of the temporal segment of observations for a total of 9 different images. We note that the time distribution used for the Pleiades observations is different from that used for the dM field stars, hence the results we obtain for the Pleiades could be not immediately compared with those obtained for the field stars.

Using the same approach used for the dM field stars, we have obtained the distribution

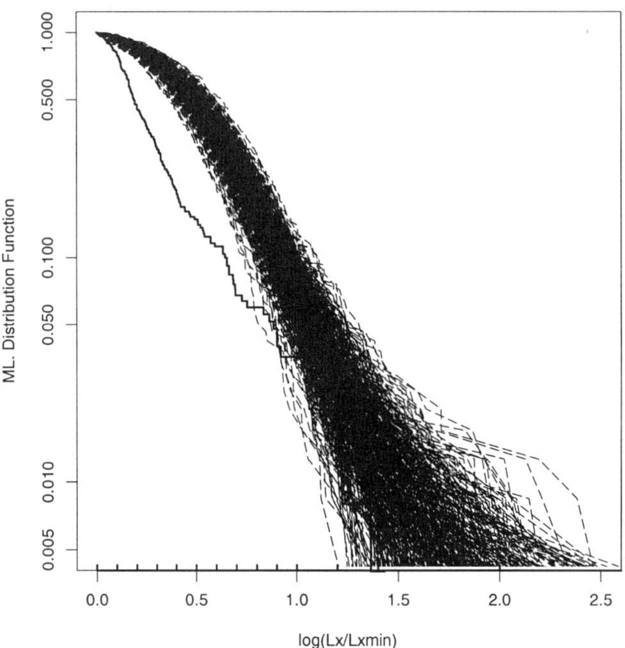

FIGURE 4. Time distribution of X-ray luminosity of nearby dM stars (solid line), compared with the region occupied by the simulations of the time distribution of the solar emission, when the Sun is sampled with the same pattern of the stars. Note the logarithmic scale in ordinate.

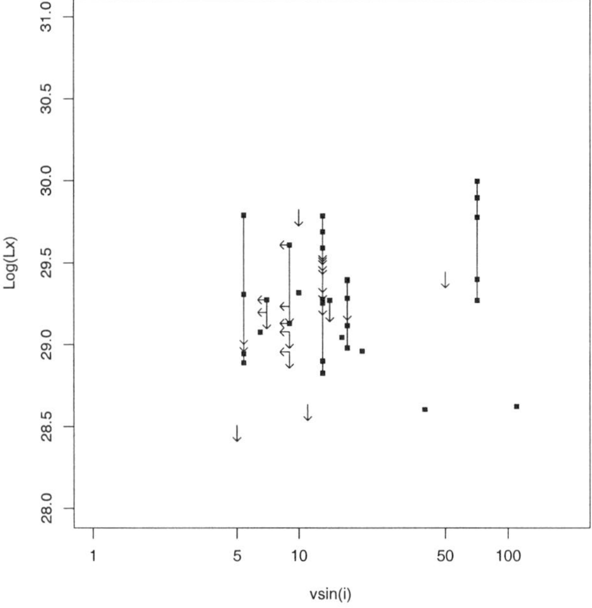

FIGURE 5. Scatter plot of L_x versus the rotation rate, as measured by the projected equatorial rotation rate ($v \sin(i)$); vertical line segments connect different measurements of the same star.

of the amplitude variations for the dM Pleiades stars. In this case if we compare this time distribution with the observed luminosity function for Pleiades dM stars we find that most of the observed spread in the luminosity function can be attributed to variability.

In Fig. 5 we report the scatter plot of L_x versus the rotation rate for the corresponding star, as measured by the projected equatorial rotation rate v $\sin(i)$; vertical line segments connect different measurements of the same star. From the figure we see that for Pleiades dM stars the spread in the activity-rotation relation can be entirely attributed to variability.

4. Summary

In this work we have presented results from a systematic analysis of X-ray light curves of dM stars both in solar neighborhood and in the Pleiades.

We found that X-ray variability is a common phenomenon in dM stars regardless of their "quiescent" activity level and visual magnitude, at least in the explored ranges.

The frequency of amplitude variations is consistent with the frequency of X-ray flares observed in the Sun, while rotational modulation is substantially suppressed in dM stars. This suggests that the spatial distribution of emitting structures in dM stars is different from that observed in the Sun.

The spread of the X-ray luminosity function in field dM stars cannot be explained by variability only, while it is entirely explained by variability in dM stars of the Pleiades.

We acknowledge financial support from ASI (Italian Space Agency), and MURST (Ministero della Università e della Ricerca Scientifica e Tecnologica. We acknowledge extensive discussions with R. Harnden, G. Peres, R. Rosner and S. Sciortino.

REFERENCES

AMBRUSTER, C.W., SCIORTINO, S., & GOLUB, L. 1987 *Ap. J. Suppl*, **65**, 273.

COLLURA,A., PASQUINI, L., & SCHMITT, J.H.M.M. 1988 *Astr. & Astroph.* **205**, 197

KREPLIN, R.W., DERE, K.P., HORAN, D.M. & MEEKINS, J.F. 1977 *The solar output and its variations*, Colorado University Press, 187.

MICELA, G., SCIORTINO, S., VAIANA, G. S., HARNDEN, F. R. JR., ROSNER, R., & SCHMITT, J. H. M. M. 1990 *Ap. J.* **348** 557.

PALLAVICINI, R., TAGLIAFERRI, G., & STELLA, L. 1990 *Astr. & Astroph.* **228**, 403

SCHMITT, J. H. M. M., FLEMING, T. A., & GIAMPAPA, M. S. 1995 *Ap. .J.* , **450** 401.

VAIANA, G.S., DAVIS, J.M., GIACCONI, R., KRIEGER, A.S., SILK, J.K., TIMOTHY, A.F. & ZOMBECK M. 1983 *Ap. J. (Letters)* **185**, L47-L51

VAIANA, G.S. & TUCKER, W.H. 1974, *X-ray Astronomy* eds. Giacconi, R., Gursky, H., Reidel, D. 169.

WITHBROE, G.L., HABBAL, S.R. & RONAN, R. 1985 *Sol. Phys.* **95**, 297.

ZOMBECK, M.V., VAIANA, G.S., HAGGERTY, R., KRIEGER, A. S., SILK, J. K. & TIMOTHY, A. 1978 *Ap. J.* **38**, 69.

The Coronae of AD Leo and EV Lac

By S. SCIORTINO[1], A. MAGGIO[1], F. FAVATA[2] AND S. ORLANDO[3]

[1]Osservatorio Astronomico di Palermo, Piazza del Parlamento 1, I-90134 Palermo, Italy

[2]Astroph. Div. - SSD ESA, ESTEC, Postbus 299, NL-2200 AG Noordwijk, The Netherlands

[3]Solar Sys. Div. - SSD ESA, ESTEC, Postbus 299, NL-2200 AG Noordwijk, The Netherlands

We present results of the analysis of X-ray observations of the active M dwarfs AD Leo and EV Lac. The PSPC spectra can be fitted with one- (EV Lac) or two-component (AD Leo) isothermal MEKAL models, and very low metallicity ($\sim$ 0.1 solar); during an intence flare the spectrum of EV Lac can be fitted only by adding a second component with log T $\sim$ 7.5. The SAX light-curves of AD Leo and EV Lac also show the occurrence of several flares. The fits of the SAX spectra require at least three thermal MEKAL components and best-fit coronal plasma metallicity below solar for AD Leo and only marginally below solar for EV Lac.

We have also fitted the SAX spectra of AD Leo and EV Lac with model spectra from constant cross-section static coronal loops. One-loop models fail to fit the observed spectra. A second loop component, that accounts for most of the plasma emission at high energy, is required to obtain an acceptable fit. We interpret the fit results as pointing toward the existence of various (at least two) dominant classes of coronal emitting structures: the dominant one is composed of hundreds of compact loops, with lower maximum temperature and length smaller than 0.1 the stellar radius, covering no more than 1% of stellar surface; the second one, responsible for the high energy emission, is composed at least of tens of quite elongated loops, covering a very small fraction of stellar surface.

1. Introduction

Einstein IPC X-ray spectra of solar-type stars, observed with sufficient photon statistics, usually require two emission components from an optically thin plasma to be adeguatelly modeled (cf. Vaiana 1983, Schmitt *et al.* 1990). In the case of dMe stars the fits usually result in one component at few million degrees, and a second one at few tens million degrees. While the need of two thermal components does not necessarily imply the presence of two distinct emitting "regions" at different temperatures in the stellar coronae, it certainly reflect the presence of a (complex) thermal structure in stellar coronae. This description has been confirmed and extended by the analysis of spectra gathered with the Position Sensitive Proportional Counter (PSPC) on board ROSAT. The recent analysis of ASCA, EUVE and SAX spectra (White 1996; Drake 1996; Favata 1998; Pallavicini 1998 and references therein cited) of several nearby and/or active stars has further reinforced a description of stellar coronal spectra in terms of scaled version of the solar emission.

There are several reasons that makes the study of the coronal emission of dMe stars a worth effort. First of all, their X-ray emission level makes them among the most intense coronal emitters, hence they are a good test-bed for any proposed mechanism for the heating of stellar coronae. Their internal structure, reflected by their cool photospheres, should minimize any contribution by non-magnetic (acoustic) heating processes making easier to pin point the characteristic features of the coronal magnetic heating mechanism (down to the so called micro-flare level) and their relation to any kind of dynamo activity, either based on a scaled version of the solar $\alpha-\omega$ dynamo or on the more elusive turbulent dynamo (cf. Durney *et al.* 1993) that has been suggested to be at work in the (almost) fully convective late dM stars (Stern *et al.* 1995).

To gain insight in the above questions we have obtained observations of the active dMe stars AD Leo and EV Lac with the SAX X-ray astronomy satellite.

AD Leo (GJ 388) is a single dM4.5e star at a trigonometric distance of 6.15 pc with an X-ray flux to the Earth $f_X \sim 3\ 10^{-11}$ erg sec^{-1} cm^{-2} (Barbera *et al.* 1993; Schmitt *et al.* 1995), EV Lac is a single dM4.5e star at a trigonometric distance of 5.08 pc and $f_X \sim 4\ 10^{-11}$ erg sec^{-1} cm^{-2} (Schmitt *et al.* 1995). They are both well known flare stars, extensively studied and are known to emit from infrared up to X-rays. Their space velocities make them members of young disk population based on the criterion of Eggen (1973). They are among the few dM stars for which, adopting the Zeeman broadening technique, it has been possible to derive the (mean) value of the surface magnetic field in the spots, with the associated area coverage (Saar 1990). The deduced magnetic fields are of the order of 10^3 Gauss (similar to the intensity of magnetic field in solar spots) with area coverage fraction above 70% (i.e. well above the typical spot coverage in the Sun). This is a clear evidence that some kind of dynamo action should be at work in these almost fully convective stars.

It is worth to note that the IPC spectrum of AD Leo cannot be adequately fitted, for an assumed solar composition of emitting plasma, either with 1-T or 2-T Raymond-Smith plasma emission models, or with a continuous power-law emission measure distribution (Schmitt *et al.* 1990).

2. The PSPC Observations of AD Leo and EV Lac

There is one single PSPC observation of AD Leo lasting about 27 ksec, while EV Lac has been observed six times with observations of 3 to 7 ksec each.

The entire spectral analysis, including the analysis of SAX spectra of Sect. 3, has been performed using the XSPEC V.10 package and the so-called MEKAL plasma emissivity model. PSPC spectra and light-curves extraction has been performed using PROS V2.4.

2.1. *AD Leo PSPC spectrum*

In order to make easy and more consistent the comparison with the results of the rest of our analysis, based on a different plasma emissivity code, we have re-analyzed the AD Leo PSPC spectrum whose analysis, based on the Raymond-Smith emissivity model (Raymond & Smith 1977; Raymond 1988), has been presented by Giampapa *et al.* (1996, hereafter G96). We find that: a) it is impossible to obtain a statistically acceptable two-component fit if we keep fixed the value of hydrogen column density to $N_H = 10^{18}$ cm^{-2} that is within a factor two from the value recently derived by the analysis of an EUVE spectroscopic observation of AD Leo (cf. Cully *et al.* 1997); b) our analysis based on the MEKAL plasma emissivity model results in higher temperatures ($\log T_L = 6.82$, $\log T_H = 7.22$) and in $EM_L/EM_H = 5.9$ much greater than the published ones (cf. G96) derived adopting the Raymond-Smith plasma emissivity code. It is worth noting that the individual emission measure EM_L and EM_H obtained with our fit are both significantly lower than those reported by G96 because the MEKAL plasma emissivity per unit emission measure is greater than the one predicted by the Raymond-Smith plasma emissivity.

2.2. *EV Lac PSPC spectra*

Since analysis of the EV Lac PSPC spectra has not been published yet, we have analyzed the available archive data. As a first step we have constructed the PSPC light-curves;

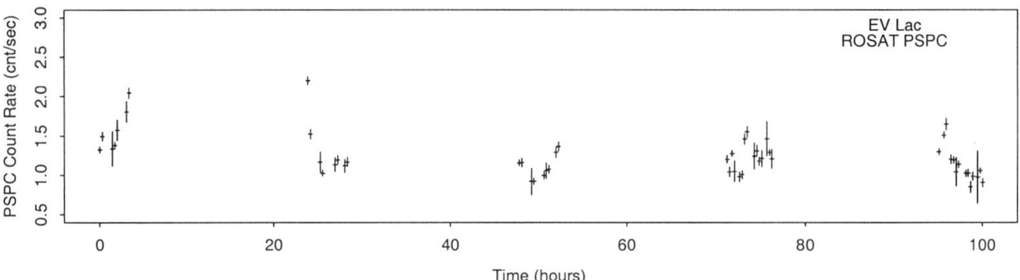

FIGURE 1. PSPC light-curve of EV Lac, accumulated in 1000 sec bin size, obtained combining 5 distinct observations taken in 5 consecutive days. Notice that the data are suggestive of a coronal variability up to a factor 2 over the time scale explored.

for the ROR's from 201583 to 201587 spanning 5 consecutive days, we have constructed a single light curve (cf. Fig. 1) to investigate the presence of intence flares.

Our fit results clearly shows that i) except for the strong flare (ROR 200984), the PSPC spectra of EV Lac can be described by best fit 1-T models with derived parameters (log T $\sim$ 6.8, Emission Meassure $\sim$ 5-7 10^{51} cm^{-3}) that are compatible within the error bars with a single value, ii) a coronal metallicity lower than solar ($Z/Z_\odot \sim 0.1$) is required to obtain good fits in all cases, iii) during the flare an high-temperature component (log T $\sim$ 7.6) has to be added to obtain a best fit, while the low-temperature component remains unchanged; iv) the limited spectral resolution and energy bandpass of the ROSAT PSPC, is unable to well constraint the value of the flare high temperature component.

3. The SAX Observations of AD Leo and EV Lac

The new SAX data where acquired with the Low Energy Concentrator Spectrometer (LECS, Parmar *et al.* 1997), which covers the energy range 0.1–10 keV with a spectral resolution $\Delta E/E \simeq 20\%$ at 1 keV and scaling as $E^{-1/2}$, and with the Medium Energy Concentrator Spectrometers (MECS's, Boella *et al.* 1997), which covers the energy range 1.5–10 keV with a spectral resolution $\Delta E/E \simeq 8\%$ at 6 keV and scaling as $E^{-1/2}$. The SAX observation of AD Leo was performed on April 23–34, 1997, and resulted in an effective exposure time of 43.3 ks for the three MECS units, and of 19.6 ks for the LECS†. The SAX observation of EV Lac was performed on December 7–8, 1997, and resulted in an effective exposure time of 33.9 ks for the two MECS‡, and of 9.7 ks for the LECS.

The source spectra were re-binned so to have at least 30 counts per energy bin, and bins with energies below 0.1 and above 7.0 keV were discarded in the case of LECS, while for the MECS spectra we have retained the bins with energies between 1.5 and 7.0 keV.

3.1. *The SAX X-ray Light-Curves*

The AD Leo X-ray light curves, accumulated in 400 sec bins, for the LECS and the three combined MECS detectors are shown in Fig. 2. The LECS light curve shows evidence of variability up to a factor 2 on a time scale of few thousands seconds, associated to the occurrence of flares (and possibly of other low-amplitude variability). The presence of variability is confirmed by the results both of the χ^2 and the Kolmogorov – Smirnov tests, while the same tests confirm that the background is consistent with being constant.

† The LECS exposure time is always significantly shorter than the MECS exposure time, because the LECS was operated during Earth dark time only, at the time of both observations.

‡ In the time separating the two observations one of the three MECS's has developed an unrecoverable problem and has been turned off.

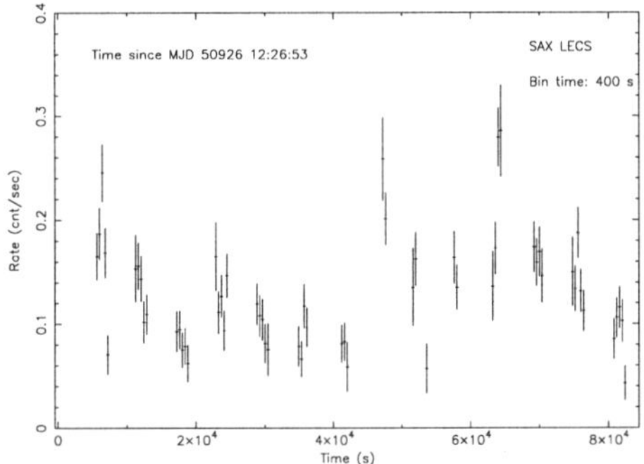

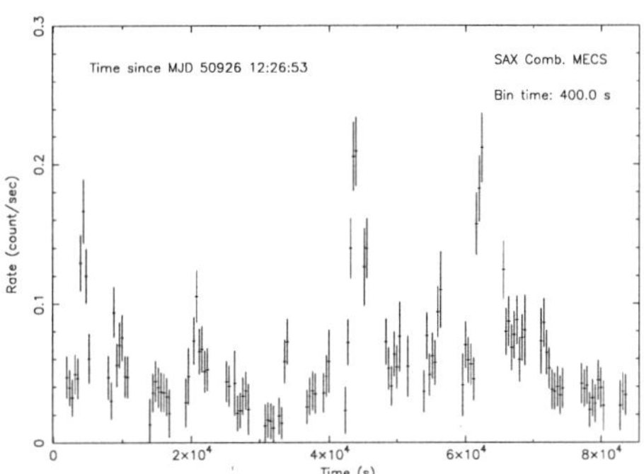

FIGURE 2. (a) [LEFT-PANEL] The light curve of the SAX/LECS observation of AD Leo accumulated in 400 second bins, shows that the coronal emission level of this star is highly variable. (b) [RIGHT-PANEL] The light curve of AD Leo accumulated in 400 sec bins, from the three combined MECS detectors onboard of SAX. The curve shows the occurrence of flares at about 6, 42, and 60 ks since the beginning of the one day long observation.

The MECS light-curve also shows evidence for intense source variability, again confirmed both by the a simple χ^2 test and a Kolmogorov – Smirnov test. From the inspection of Fig. 2 it is evident the occurrence of, at least, 3 flares during the 1 day long observation. The peak count rates of these flares are $\sim$ 4–5 times higher than the "quiescent"coronal emission level. The overall appearance of LECS and MECS light curves strongly indicates that a large fraction (30-50%) of the overall emitted X-rays occurs in flares.

As for AD Leo the SAX observation of EV Lac spans approximately one day. With the same procedure adopted for AD Leo we find that the background-subtracted LECS and combined MECS light-curves indicate the presence of variability, whose existence has been confirmed by applying the χ^2 and the Kolmogorov-Smirnov tests. The MECS light curve shows two intense flares, occurring respectively 20 and 60 ks since the beginning of the observation, plus some less extreme variability. The decay times of the two flares derived with formal fits are about 4.5 ks and 1.4 ks, respectively.

3.2. *Fit of SAX Spectra*

Given the limited counting statistics of the sources we have jointly fitted the available LECS and MECS spectra assuming a constant normalization among the four detectors fixed in such a way that a source of given flux yields in the LECS 70% of the count-rate in any of the three MECS (that instead are all equivalent).

As demonstrated by an extensive set of simulations presented by fmp+97, LECS spectra allow us to get fair diagnostic capabilities for the global metallicity of coronal plasmas, thanks to the wide spectral coverage. On the contrary the low statistics of the present SAX spectra, together with the spectral resolution above 0.6 keV lower than for the ASCA/SIS, makes unrealistic the analysis with variable individual abundances (as we will show in the following, acceptable fits can be found varying only the global metallicity in the fit).

3.2.1. *The SAX spectra of AD Leo*

We first attempted to fit the available spectra with a 2-T MEKAL model with the global abundance of metals in the emitting plasma, relative to the solar value, left as a free parameter in the fit. This fit yields a value of χ^2 of 185.1 with 145 d.o.f. (probability 1.4%) and does not result in an acceptable model description. In order to obtain an

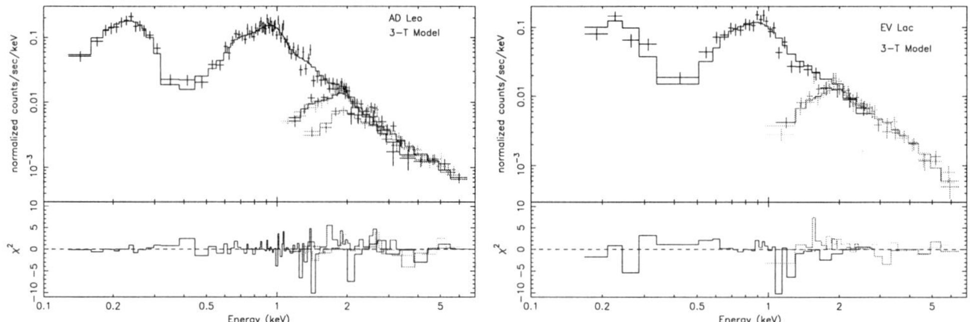

FIGURE 3. [RIGHT-PANEL] The observed SAX spectra of AD Leo together with the best-fit three-temperature MEKAL spectrum, yielding a global metallicity 0.25 times solar. Note that most of improvement of this fit with respect to the 2-T fit occurs in the spectral region around 0.6 keV. [LEFT-PANEL] The observed SAX spectra of EV Lac together with the best-fit three-temperature MEKAL spectrum, yielding a global metallicity 0.59 times solar. Again note that most of improvement in the 3-T fit occurs in the spectral region around 0.6 keV. Bottom panels y-axis shows the (signed) contribution of each bin to the total χ^2.

acceptable fit we have added a third thermal MEKAL component and have repeated the fit leaving the metallicity free to vary. The resulting best-fit (log $T_1 \sim 6.6$, log $T_2 \sim 7.1$, log $T_3 > 8.13$, with 90% confidence ranges on $Z/Z_{\odot} = 0.22$-0.35) model is shown in Fig. 3. The confidence intervals show that we can only set a lower-limit to the value of the high-temperature. The fit formally converges with a value of χ^2 of 163.0 for 145 d.o.f. corresponding to a probability level of $\sim$ 9.0%. While the value of metallicity is constrained to be lower than solar, available data do not constrain the value of the temperature of the hotter component nor allow us to discriminate between the (assumed) thermal and a non-thermal nature of this hot component.

3.2.2. *The SAX spectra of EV Lac*

We first attempted to fit the available spectra with a 2-T MEKAL model with metallicity left free to vary. This fit yields a value of χ^2 of 103.8 for 78 d.o.f. corresponding only to a probability of 1% to give an adequate description of EV Lac coronal spectrum. In order to improve our model description we have added a third thermal MEKAL component and have repeated the fit leaving the metallicity free to vary. The fit yields a value of χ^2 of 95.2 (for 78 degree of freedom) that corresponds to a null hypothesis probability of 2.9%, i.e. a marginally acceptable fit. The 3-T fit parameters are log $T_1 \sim 6.5$, log $T_2 \sim 6.9$, log $T_3 \sim 7.4$, with 90% confidence ranges on $Z/Z_{\odot} = 0.36$-1.11.

4. Loop modeling

In view of the complexity of the thermal structure of the coronae of AD Leo and EV Lac, we have attempted to fit the SAX spectra with a model synthesized from constant cross-section static coronal loops (Serio *et al.* 1981; Ciaravella *et al.* ,1996)). This modeling approach has been recently adopted to fit the PSPC spectra of F and G-type stars (Maggio & Peres 1997; Ventura *et al.* 1998).

As an extension to the works cited above we have constructed a grid of coronal loop models having as free parameter also the plasma metallicity, beside the loop length, l, and the plasma maximum temperature, T_{max}. This model has as further free parameter

TABLE 1. Summary of 2-loop model fits of SAX Spectra

Star	Z $(Z/Z_\odot)$	Log(T_{max1}) (K)	Log(l_1/R_*)	f_1	Log(T_{max2}) (K)	Log(l_2/R_*)	f_2	Prob (%)
AD Leo	0.23 [0.20-0.30]	7.12 [7.09-7.14]	-2.3 [< -1.0]	1.6e-4 [<3.1e-3]	8.5 [> 8.2]	-2.1 [unconstr.]	1.0e-10 [unconstr.]	7.3
EV Lac	0.45 [0.32-0.69]	6.90 [6.74-7.01]	-2.3 [<-1.4]	2.1e-4 [<1.7e-4]	7.53 [7.47-7.58]	-2.3 [<-1.2]	9.4e-7 [<1.2e-5]	1.7

1- The reported parameter uncertainty ranges are computed at 90% confidence level assuming a single interesting parameter.
2- Given the small source distance, and the result obtained by Cully *et al.* (1997), all the AD Leo fits were performed assuming $N_H = 10^{18}$ cm^{-2}.
3- Given the small source distance and the fits of PSPC spectra we have decided to adopt $N_H = 10^{19}$ cm^{-2}. We have checked that no acceptable fit can be obtained with a lower hydrogen column density.

the surface filling factor, f, i.e. the fraction of stellar surface covered by a given class of loops.

We have found that one-loop model cannot give an adequate description of the observed spectra. We have then added a second loop component to our model obtaining the fitting results shown in Fig. 4 and summarized in Table 1. As a matter of fact the "quality" of the best-fit 2-loop models, as measured by the probability level of the χ^2 test, is better than for the 2-T models, and slightly worse than for the best-fit 3-T models. Note that 3-T models and loop models have the same number of fitted parameters.

In order to properly interpret loop fit results it is necessary to recall that if the loop length is smaller than the pressure scale height, $H = 2kT_{max}/\mu g = 6000T_{max}/(g/g_\odot)$ (where g is the stellar gravity, and T_{max} is in units of K) the solution of the loop static equations imply the so-called RTV (Rosner *et al.* 1978) scaling law among T, l and the loop base pressure, p, namely $T_{max} = 1.4\ 10^3(pl)^{1/3}$.

Our fit results are consistent with loops having lengths much smaller than the pressure scale height, hence RTV scaling law applies. In this case, T_{max} (and hence the product pl) is the only parameter well constrained by the fit, while the other quantities, including the surface filling factor, may take a range of possible values while yielding the same quality of the fitting result. This is an *intrinsic* property of the model and *cannot* be overcome. Moreover, note that the overall X-ray luminosity emitted by a given loop can be expressed as $L_X \sim 2.16\ 10^{16}\ \gamma\ (R_*/R_\odot)^2\ T_{max}^{7/2}\ f\ l^{-1}$ erg/sec, where γ is the fraction of energy input (heating) that is emitted in the X-ray band. Since T_{max} is usually well determined, the overall observed luminosity constrains the quantity f/l; this is the way in which we have estimated the upper limits on f from the upper limits on l, as reported in Table 1. Since T_{max} and lp are related by the scaling laws, the above dependence implies that, for loop lengths shorter than the pressure scale height, l, p and f can occupy a range of admissible solutions (possibly limited also by the constraint $f < 1$). In other words, when $l < H$, the loop modeling cannot provide *one solution*, but instead a range of admissible solutions with different area coverage and base pressure (or length) of the emitting loop. In any case, the above fitting results with loop models allow us to infer some of the properties of the stellar coronae, which could not be otherwise derived.

As a further step in the interpretation of the fitting results, let us consider that the area coverage factor can be written as $f = N\alpha^2\beta^2/4$ where N is the number of loops of a given component, α is the loop aspect ratio, i.e the ratio between loop base radius

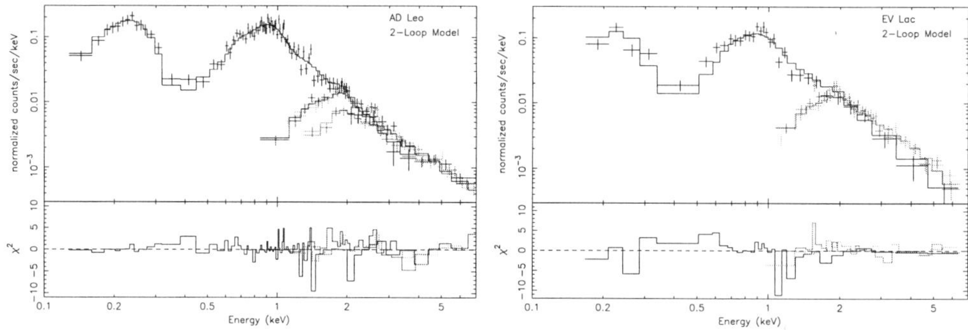

FIGURE 4. [LEFT-PANEL] The fit of the SAX spectra of AD Leo with a model of emission from a 2 coronal loop plasma model. For this fit the hydrogen column density has been taken fixed to log $N_H = 18$. [RIGHT-PANEL] The fit of the SAX spectra of EV Lac with a model of emission from a 2 coronal loop plasma model. Note that, similarly, to the 3-T fit case, the fit is statistically unsatisfactory mostly because the model does not reproduce the emission below 0.5 keV. For this fit the hydrogen column density has been taken fixed to log $N_H = 19$, however leaving free to vary the value of N_H improves only marginally the fit.

and length†, $\beta = l/R_*$ is the ratio between loop length and stellar radius. For each of the loop components we have derived the values of f and of β, hence we can deduce the value of N as a function of the loop aspect ratio. *Note that f/β is constant within the confidence regions determined by the fitting.*

Let us focus on the AD Leo low-T_{max} loop: in this case our analysis gives a best-fit solution $N\alpha^2 = 4f/\beta^2 \sim 25.5$, and a formal 90% confidence limit which translates in $N\alpha^2 > 1.24$. Hence, for $\alpha \sim 10^{-1}$ we would have a best-fit solution $N \sim 2500$, with a formal 90% confidence limit $N > 124$. Assuming instead the maximum loop aspect ratio, $\alpha = 1/\pi$, we would obtain $N > 12$. We interpret this findings as evidence that the lower T_{max} loops are indeed quite compact with length smaller than 0.1 stellar radius (and smaller than the pressure scale height) and that they cover a fraction not greater than 0.3% of stellar surface with tens to hundreds of loops.

Going to the high-T_{max} loops, we have $N\alpha^2 = 6.3\ 10^{-6}$, with the condition $N \geq 1$, this implies $\alpha \leq 2.5\ 10^{-3}$, i.e. the solution is compatible with elongated and thin loops covering a very small fraction of the stellar surface. While we cannot set a 90% confidence limit on β, hence on the value of α, yet if we assume as possible solution loops with $\alpha \sim 0.1$, i.e. typical solar loops, we can deduce that the loop length should be extremely small, $\leq 10^{-5} R_*$, in order to meet the condition $N > 1$. Similar arguments apply also for EV Lac.

Our findings indicate that the coronae of active dM stars can be described by (at least) two major emitting structures (let say two major loop classes) with well distinct maximum temperatures. The corona of AD Leo is globally "hotter" than the corona of EV Lac. In both stars the loops with the lower T_{max} are compact (aspect ratio < 0.3-1.5), with length smaller than 0.1 stellar radius, base pressure higher than typical solar loops, and area filling factor of 10^{-4}–10^{-3}. There should be, at least, tens-hundreds of such loops to explain the observed emission. The higher T_{max} loops tend to have a more elongated structure (aspect ratio in the 10^{-1}–10^{-3} range), while maintaining a length smaller than 0.1 stellar radius. Because their area coverage factor is quite small,

† For the given loop geometry, we have the constraint $\alpha < 1/\pi \sim 0.32$ to avoid overlap of the loop foot-points

only at most ten of them are contemporary present in the corona. Given their elongated structure we argue that their base pressure is even higher than the pressure of the lower T_{max} loops, and they are likely to be non-steady and continuously replenished in the corona. This picture is consistent with that derived from the analysis of AD Leo PSPC data (Giampapa *et al.* 1996).

While we have not done detailed analysis of any specific flare, yet our results seem to contradict the existence of AD Leo loops with length comparable to the stellar radius (Cully *et al.* 1997), while it is more consistent with the recent analysis of the AD Leo flare seen with the ROSAT PSPC that has been interpreted as occurring on a loop whose length was of the order of 0.1 stellar radius (Reale & Micela 1998).

SS, and AM acknowledge partial support from ASI and MURST.

REFERENCES

BARBERA M., MICELA S., SCIORTINO S., HARNDEN, F.R. J., ROSNER R. 1993, *ApJ* **414**, 846

BOELLA G., CHIAPPETTI L., CONTI G. *et al.* 1997, *A&AS* **122**, 327

CIARAVELLA A., PERES G., MAGGIO A., SERIO S. 1996, *A&A* **306**, 553

CULLY S., FISHER G., HAWLEY S. L., SIMON T. 1997, *ApJ* **491**, 910

DRAKE J. U. 1996, in R. Pallavicini, A. Dupree (eds.), Cool Stars, Stellar Systems, and the Sun. Ninth Cambridge Workshop, Vol. 109 of *ASP Conf. Series*, ASP, 203

DURNEY B. R., DE YOUNG D., ROXEBURG I. W. 1993, *Solar Phys.* **145**, 207

FAVATA F. 1998, in R. A. Donahue, J. A. Bookbinder (eds.), Cool Stars, Stellar Systems, and the Sun. 10th Cambridge Workshop, Vol. in press

FAVATA F., MAGGIO A., PERES G., SCIORTINO S. 1997, *A&A* **326**, 101

GIAMPAPA M. S., ROSNER R., KASHYAP V., FLEMING T. A., SCHMITT J. H. M. M. 1996, *ApJ* **463**, 707

MAGGIO A., PERES G. 1997, *A&A* **325**, 237

PALLAVICINI R. 1998, in L. Scarsi, H. Bradt, P. Giommi,, F. Fiore (eds.), The Active X-ray Sky: Results from BeppoSAX and Rossi-XTE, Vol. in press

PARMAR A. N., MARTIN D. D. E., BAVDAZ M. *et al.* 1997, *A&AS* **122**, 1

RAYMOND J. 1988, in R. Pallavicini (ed.), Hot thin plasmas in Astrophysics, Kluwer, Dorderecht

RAYMOND J., SMITH B. 1977, *ApJS* **35**, 419

REALE F., MICELA G. 1998, *A&A* **334**, 1028

ROSNER R., TUCKER W. H., VAIANA, G. S. 1978, *ApJ* **220**, 643

SAAR S. H. 1990, in J. O. Stenflo (ed.), Solar Photosphere: Structure, Convection and Magnetic Fields, Kluwer, Dorderecht, 427

SCHMITT J. H. M. M., COLLURA A., SCIORTINO S. *et al.* 1990, *ApJ* **365**, 704

SCHMITT J. H. M. M., FLEMING T. A., GIAMPAPA M. S. 1995, *ApJ* **450**, 392

SERIO S., PERES G., VAIANA G. S., GOLUB L., ROSNER R. 1981, *ApJ* **243**, 288

STERN R., SCHMITT J., KAHABKA P. 1995, *ApJ* **448**, 683

VAIANA G. S. 1983, in O. Stenflo (ed.), Solar and Stellar Magnetic Fields: Origins and Coronal Effects, Reidel, Dorderecht, 165

VENTURA R., MAGGIO A., PERES G. 1998, *A&A*, in press,

WHITE N. E. 1996, in R. Pallavicini, A. Dupree (eds.), Cool Stars, Stellar Systems, and the Sun. Ninth Cambridge Workshop, Vol. 109 of *ASP Conf. Series*, ASP, 193

Prospects of Future X-ray Missions for Low Mass Stars and Cluster Stars

By R. PALLAVICINI

Osservatorio Astronomico di Palermo, Piazza del Parlamento 1, I-90134 Palermo, Italy

The new generation of X-ray missions to be launched by the end of this century will provide excellent opportunities for the study of very-low mass stars and brown dwarfs as well as of cool stars in open clusters and star forming regions. AXAF and XMM will be highly complementary in this respect, with AXAF leading the field for the detection of very faint objects and the study of crowded regions, and XMM allowing medium to high resolution spectroscopy to fainter limits for a large number of stars in open clusters and nearby star forming regions. With the help of simulations of AXAF and XMM spectra, and estimates of the sensitivity limits for typical imaging and spectroscopic observations, I discuss the prospects offered by these two missions for the study of low-mass stars and cluster stars.

---

1. Introduction

The next few years will be a marvellous time for X-ray astronomy, with the launch of AXAF (Advanced X-ray Astrophysics Facility) in spring 1999 and of XMM (X-ray Multi Mirror Mission) and ASTRO-E (the new Japanese X-ray mission) in early 2000. These new powerful missions will produce a great leap forward in all fields of X-ray astronomy, from nearby stars to the most distant objects in the Universe. They will be far more sensitive than past and ongoing X-ray missions and will be equipped with new detectors (CCD and microchannel plate cameras, transmission and reflection gratings, and X-ray microcalorimeters) that will allow detection of fainter objects as well as detailed medium to high-resolution spectroscopy of the brightest sources. Significant advances are expected in the field of low-mass stars and brown dwarfs as well as in observations of late-type stars in stellar clusters and associations. In this paper, I will present a brief outline of the imaging and spectroscopic capabilities of AXAF and XMM, with emphasis on observations of very low mass stars and stellar clusters.

2. The Advanced X-Ray Astrophysics Facility (AXAF)

AXAF is a major NASA X-ray mission to be launched in spring 1999. Due to the superbe quality of its reflection optics, the emphasis of the mission will be on imaging capability with subarcsecond spatial resolution. Medium and high-resolution spectroscopy will also be possible with transmission gratings. The detector assembly features 4 focal plane detectors (2 mainly for imaging and 2 mainly for spectroscopy) that can be used alternatively. Objective grating spectrometers can be inserted in the optical beam behind the mirror assembly to obtain dispersed spectra. With this instrument assembly, it will be possible to obtain high-resolution images down to a sensitivity better than $\sim 1 \times 10^{-15}$ erg cm^{-2} s^{-1} that are recorded by either a high-resolution camera (HRC-I) or a CCD camera (ACIS-I). Medium resolution spectroscopy will be possible with the CCD camera, whereas high-resolution spectroscopy will be possible with the transmission gratings (HETG and LETG) and focal plane detector arrays (ACIS-S and HRC-S, respectively). AXAF will be launched in a highly eccentric orbit with a period of 64 hours, allowing uninterrupted observations of up to 200 ks. The lifetime of the mission is expected to range from a minimum of 5 years to more than 10 years. Detailed information on AXAF can be found on the WWW at the URL address *http://asc.harvard.edu.*

There are three main structural elements in AXAF: a) the High Resolution Mirror Assembly (HRMA); b) the 2 transmission gratings (a Low Energy Transmission Grating LETG and a High Energy Transmission Grating HETG) that can or cannot be inserted in the beam; c) the 4 focal plane detectors (the 2 High Resolution Cameras HRC-I anf HRC-S for imaging and spectroscopy, respectively, and the 2 AXAF CCD Imaging Spectrometers ACIS-I and ACIS-S for imaging and spectroscopy, respectively). The HRC and ACIS detectors are mounted on a focal plane platform that can be moved perpendicularly to the optical axis of the telescope. Thus the different detectors can be used alternatively depending on the science objectives of the observation. The two primary imaging detectors (HRC-I and ACIS-I) are bidimensional arrays whereas the two primary spectroscopic detectors (that will be used mainly when the transmission gratings are in the beam) are one dimensional arrays to record the dispersed spectra.

The two primary instruments for imaging observations are HRC-I and ACIS-I. The High Resolution Camera (HRC-I) is based on microchannel plates and is an updated version of similar instruments flown on *Einstein* and ROSAT. The FOV is 30×30 arcmin and the spatial resolution is better than 1 arcsec. The spectral band covered by the instrument extends from 0.1 to 10 keV, with limited spectral resolution (E/ΔE ~ 1 at 1 keV). The effective area is ~ 200 cm^2 at 1 keV and ~ 50 cm^2 at 6 keV. The limiting sensitivity for point sources is $\sim 2 \times 10^{-15}$ erg cm^{-2} s^{-1} in 3×10^5 s (5σ). The AXAF CCD Imaging Spectrometer (ACIS-I) is an array of 4 CCD chips, the FOV is $\sim 17 \times 17$ arcmin and the spatial resolution is better than 1 arcsec. The spectral band extends from ~ 0.5 to 10 keV, with an energy resolution E/ΔE of ~ 10 at 0.5 keV and ~ 50 at 5 keV. The effective area is ~ 600 cm^2 at 1 keV and ~ 200 cm^2 at 6 keV. The limiting sensitivity for point sources is $\sim 2 \times 10^{-15}$ erg cm^{-2} s^{-1} in 10^5 s (5σ). The low energy response of ACIS has been increased, albeit at a somewhat lower spectral resolution, by using back-illuminated chips rather than front illuminated ones.

High-resolution spectroscopy with AXAF is possible by using two sets of transmission gratings. The High Energy Transmission Grating (HETG) is generally used in combination with ACIS-S as detector. It allows covering simultaneously two energy ranges, a High Energy (HEG) and a Medium Energy (MEG) range, covering the overall spectral band ~ 0.4 to 10 keV ($\sim 1-30$ Å). The energy resolution E/ΔE ~ 1000 for the HEG and ~ 500 for the MEG at 1 keV, decreasing to higher energies. The effective area at 1 keV is ~ 20 cm^2 for the HEG and ~ 80 cm^2 for the MEG. The limiting flux for spectroscopy with the HETG is $\sim 5 \times 10^{-12}$ erg cm^{-2} s^{-1}. The Low Energy Transmission Grating (LETG) is generally used in combination with HRC-S. The spectral band covered extends from ~ 0.08 to 6 keV ($\sim 2 - 160$ Å) and the energy resolution E/ΔE ~ 2000 at 0.1 keV, decreasing to higher energies. The effective area at 1 keV is ~ 20 cm^2. The limiting flux for spectroscopy with the LETG is $\sim 5 \times 10^{-12}$ erg cm^{-2} s^{-1}.

AXAF will be the most sensitive X-ray imaging instrument in the next few years owing to the subarcsec quality of its mirrors. It will allow detection of the faintest sources, including the faintest members of nearby stellar clusters as well as possibly brown dwarfs. AXAF will also have the highest spectral resolution at low energies, with a resolving power exceeding 1000 at the lowest energy end. Note that the resolving power of grating spectrometers decreases with increasing energy, while the reverse occurs for non-dispersive detectors such as CCDs. The resolving power of the AXAF LETG and MEG becomes comparable to that of ACIS around 5 keV and 7 keV, respectively, whereas the resolving power of the HEG remains somehat higher than ACIS up to the highest energy end of the accessible spectral range. ACIS on the other end is much more sensitive than the grating instruments.

3. The X-Ray Multi Mirror Mission (XMM)

XMM is a major ESA X-ray mission to be launched in early 2000 (at about the same time of the Japanese mission ASTRO-E). XMM is highly complementary to AXAF, since its emphasis is on high throughput and spectroscopic capabilities, at the expenses of a lower spatial resolution ($\sim$ 15 arcec). XMM will have three different types of instruments that will be operated *simultaneously*: a) three CCD cameras (EPIC); b) two Reflection Grating Spectrometer (RGS); and c) an Optical Monitor (OM). The primary instruments for X-ray imaging are the CCD cameras (EPIC). Medium resolution spectroscopy will also be possible with EPIC (in much the same way as with ACIS) for sources brighter than $\sim 1 \times 10^{-13}$ erg cm^{-2} s^{-1}. Higher resolution spectroscopy will be possible with the Reflection Gratings (RGS) for typical sources brighter than $\sim 1 \times 10^{-12}$ erg cm^{-2} s^{-1}. Simultaneous optical/UV observations of the same fields imaged in X-rays will be obtained by the Optical Monitor (OM) for sources of up to B=24. XMM will be injected into a highly eccentric orbit with a period of 48 hours allowing uninterrupted observations of up to 140 ks (with a short gap at 70 ks). The expected lifetime is greater than 10 years, with a minumum of 2 years. Technical information about XMM can be found at the WWW site *http://astro.estec.esa.nl/XMM/xmm.html.*

The telescope assembly of XMM consists of 3 different telescope modules, each made of 58 nested grazing incident X-ray mirrors. Grating boxes are mounted behind two of the three modules and intercept $\sim$ 60% of the incoming photons. The intercepted light is diverted to the side and is collected by the two Reflection Grating Spectrometers (which use CCDs as detectors). The rest of the light goes to two identical MOS CCD cameras, which are two of the three CCD cameras that form EPIC. The third telescope module has no grating box behind it and the incoming photons are focussed on a third CCD camera, which is a new technology PN-type camera. Both the MOS and PN cameras use mosaics of CCD chips to cover the focal plane. The OM is mounted parallel to the X-ray mirror modules and images the same field.

The European Photon Imaging Camera (EPIC) is the basic XMM instrument for imaging and medium resolution spectroscopy. The two MOS cameras at the focus of the two partially obscured modules, and the PN camera at the focus of the free module, cover a FOV of 30×30 arcmin. The spectral band covered extends from $\sim$ 0.1 to $\sim$ 15 keV. The energy resolution is $\sim$ 100 eV, as typical for CCD cameras, with a resolving power E/ΔE increasing with energy from $\sim$ 10 to $\sim$ 60. The effective area is more than 1000 cm^2 for the free module. Taking into account the light loss due to the interception by the gratings, the total effective area of the EPIC camera is more than $\sim$ 2000 cm^2 at 1 keV, about a factor 4 higher than that of ACIS on AXAF. The limiting flux for source detection depends on the still uncertain confusion limit, but is expected to be of the order of 2×10^{-15} erg cm^{-2} s^{-1}. AXAF will go deeper in the longest exposures, probably as deep as $\sim 2 \times 10^{-16}$ erg cm^{-2} s^{-1}, due to the much higher spatial resolution. XMM however will be more sensitive than AXAF in the spectroscopic mode due to the higher throughput. The typical limiting flux for medium resolution spectroscopy with EPIC is $\sim 1 \times 10^{-13}$ erg cm^{-2} s^{-1}.

The Reflection Grating Spectrometer (RGS) on two of the three telescope modules covers the spectral band $\sim 5 - 35$ Å ($0.3 - 2.5$ keV) with a spectral resolution $\lambda/\Delta\lambda$ $\sim 300 - 600$ (at first order) over that range. The effective area is $\sim$ 100 cm^2 per module at 1 keV, i.e. about one order of magnitude lower than for EPIC. The limiting flux for high-resolution spectroscopy with the RGS is therefore $\sim 1 \times 10^{-12}$ erg cm^{-2} s^{-1}.

In terms of resolution the RGS on XMM is comparable to the LETG on AXAF over its narrower spectral band (the RGS does not extend to the very low energies where the LETG reaches the highest resolution). It has however a much larger effective area (nearly a factor 10) than the LETG. Similarly, the resolving power of EPIC is comparable to that of ACIS, but the effective area is lower. A significant improvement of resolving power at high energies with non dispersive devices will be possible with the microcalorimeters that will be flown on ASTRO-E. These will have a much higher resolving power than CCDs, even higher at energies greater than $\sim$ 2 keV than that of the MEG and HEG on AXAF. The effective area and resolution will be comparable to that of RGS around 1 keV, but dispersive spectrometers will remain the best choice for high resolution spectroscopy at energies lower than 1 keV.

4. Stellar spectroscopy with AXAF and XMM

The spectroscopic instruments on board AXAF and XMM will allow taking full advantage of the wealth of information contained in stellar X-ray spectra. For sources that are sufficiently bright, it will be possible to apply detailed diagnostic techniques of the type exploited so far for solar observations. These diagnostic techniques will provide quantitative information on temperatures, densities, elemental abundances, stellar and interstellar absorption, and flow velocities (both line broadenings and shifts). The diagnostic capabilities of AXAF and XMM for bright stellar sources have been reviewed recently by Linsky & Gagné (1998), Mewe & Güdel (1998) and Pallavicini (1998), to whom I refer for details. Here I only stress a few points and present examples of AXAF and XMM spectra to show the kind of data we are going to deal with in the next few years.

Typical X-ray fluxes for bright coronal sources (like the nearest RS CVn binaries and flare stars) are in the range of $\sim 10^{-11} - 10^{-12}$ erg cm^{-2} s^{-1}. For this type of sources, it will be possible to obtain both medium (ACIS, EPIC) and high (HETG, LETG, RGS) resolution spectra with typical exposures ranging from $\sim$ 10 to $\sim$ 50 ks (at high resolution, the weakest of these sources are observable only with XMM). With XMM, medium (EPIC) and high (RGS) spectra will be obtained simultaneously, while separate observations using different instruments are required in the case of AXAF. These sources will be the ideal case for application of diagnostic techniques of the type discussed above. Examples of simulated spectra of Capella (the brightest coronal source in the sky, apart from the Sun) for EPIC, the RGS and the LETG can be found in Linsky & Gagné (1998) and Mewe & Güdel (1998) both for the whole spectral range covered by these instruments and for individual spectral lines, such as the He-like triplets used for density diagnostics. For the latter ones (e.g. the O VII line at $\sim$ 22 Å) the ratio of the intercombination to forbidden line is a sensitive diagnostic of density for the range of densities typical of stellar coronae.

In order to show the spectroscopic capabilities of XMM and AXAF for various types of coronal sources, I have performed simulations of EPIC, RGS and LETG spectra using the XSPEC package and the MEKAL code (Mewe *et al.* 1995), which is appropriate for optically-thin thermal sources. A $2 - T$ model has been assumed in all cases. The EPIC simulations are for the PN-CCD camera and the free telescope module, while the RGS simulations are for one of the other two modules taking into account that only a fraction of the incoming light is intercepted by the grating array.

I have considered the following representative cases:

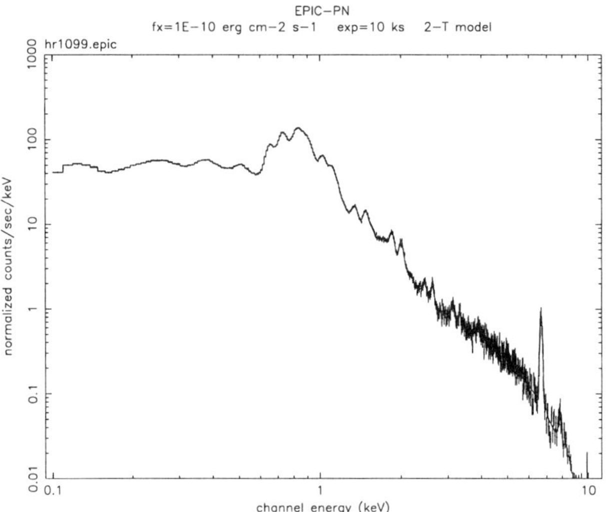

FIGURE 1. Simulated 10 ks EPIC-PN spectrum of HR1099.

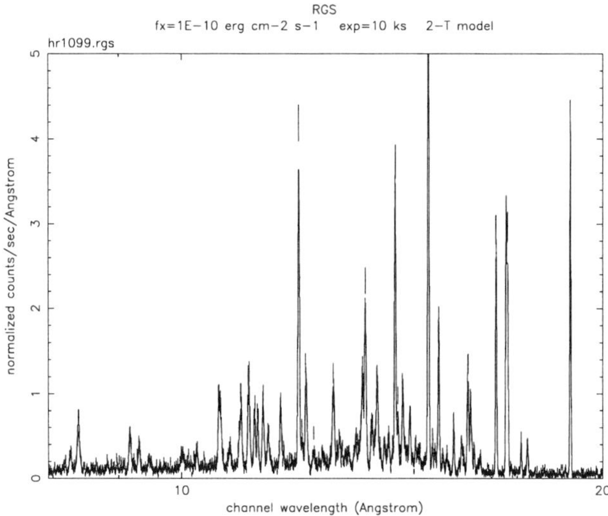

FIGURE 2. Simulated 10 ks RGS spectrum of HR1099.

• a very bright source (like the active binary HR1099) with a flux of $\sim 1 \times 10^{-10}$ erg cm^{-2} s^{-1}, $T_1 = 0.6$ keV, $T_2 = 2.5$ keV, $EM_2 = 2EM_1$ and $N_H = 6 \times 10^{18}$ cm^{-2}. The exposure time is 10 ks and the count rate is 83.9 cts/s in EPIC and 4.5 cts/s in the RGS (see Figures 1 and 2). Capella will give a spectrum of similar strength, but significantly softer.

• a bright source with a flux of $\sim 1 \times 10^{-11}$ erg cm^{-2} s^{-1}, $T_1 = 0.5$ keV, $T_2 = 2$ keV, $EM_2 = EM_1$ and $N_H = 1 \times 10^{19}$ cm^{-2}. The exposure time is 20 ks and the count rate is 10.9 cts/s in EPIC and 0.6 cts/s in the RGS. This source is representative of many nearby RS CVn binaries and dMe flare stars seen with XMM.

• a bright source (UX Ari) with a flux of $\sim 3 \times 10^{-11}$ erg cm^{-2} s^{-1}, an input spectrum

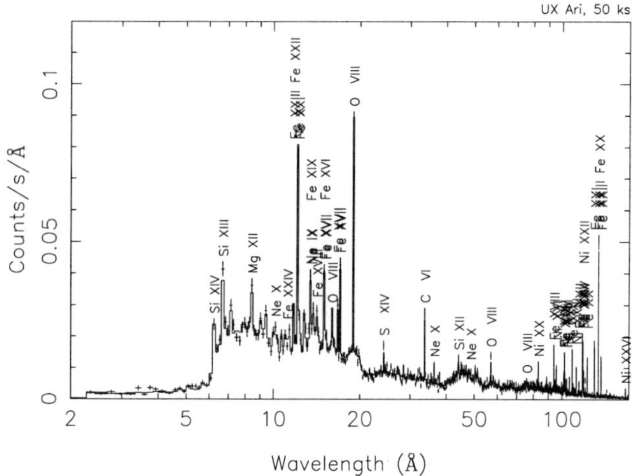

FIGURE 3. Simulated 50 ks LETG spectrum of UX Ari.

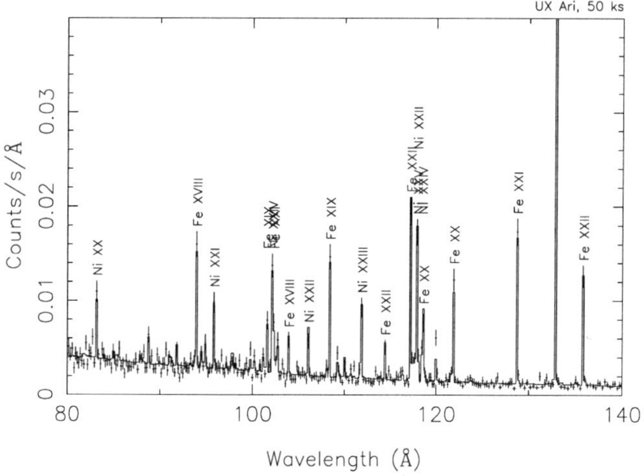

FIGURE 4. Expanded portion of the same spectrum as in Fig. 3.

similar to that of HR1099 but with reduced metallicity ($Z \sim 0.2$) with respect to solar. The exposure time is 50 ks and the count rate is 0.7 cts/s in LETG/HRC-S (see Figures 3 and 4). This source is representative of nearby RS CVn binaries seen with AXAF.

• a medium intensity source with a flux of $\sim 1 \times 10^{-12}$ erg cm^{-2} s^{-1}, $T_1 = 0.5$ keV, $T_2 = 2$ keV, $EM_2 = EM_1$ and $N_H = 1 \times 10^{19}$ cm^{-2}. The exposure time is 50 ks and the count rate is 1.1 cts/s in EPIC and 0.06 cts/s in the RGS (see Figures 5 and 6). This source is representative of many nearby solar-type stars as well as of the brightest sources in the Hyades and in nearby star forming regions.

• a weak source with a flux of $\sim 1 \times 10^{-13}$ erg cm^{-2} s^{-1}, $T_1 = 0.5$ keV, $T_2 = 2$ keV, $EM_2 = EM_1$ and $N_H = 3 \times 10^{19}$ cm^{-2}. The exposure time is 100 ks and the count rate is 0.1 cts/s for EPIC-PN (see Figure 7). No usable spectrum can be obtained for these weak sources with the RGS. In terms of flux, this source is representative of cool stars in

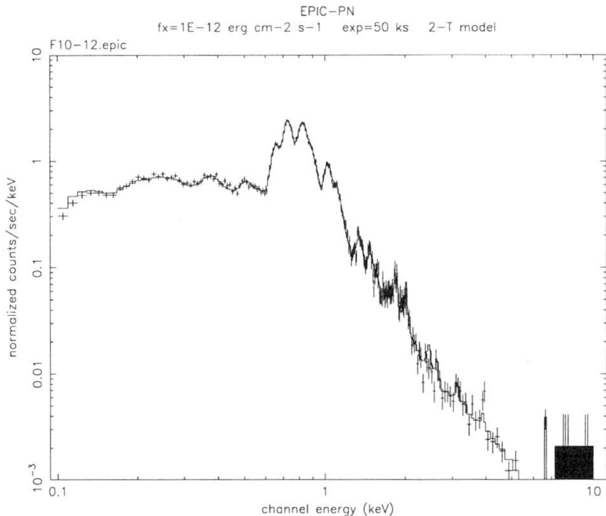

FIGURE 5. Simulated 50 ks EPIC-PN spectrum of a stellar source with a flux of $\approx 1 \times 10^{-12}$ erg cm^{-2} s^{-1}.

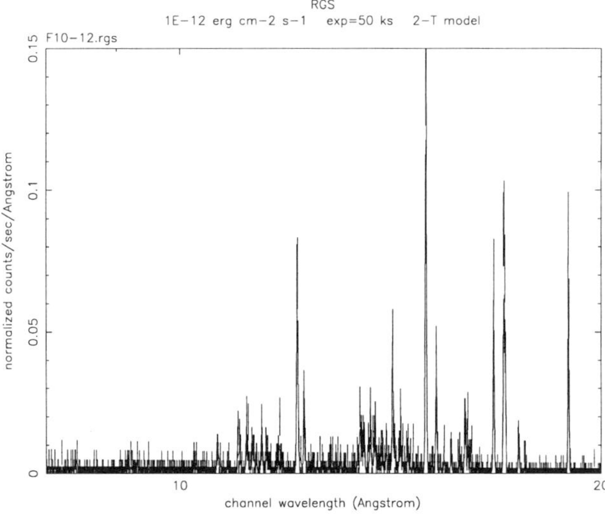

FIGURE 6. Simulated 50 ks RGS spectrum of the same source in Fig. 5.

nearby open clusters (e.g. the Pleiades) and in star forming regions, albeit the column density N_H may be substantially larger in some of the latter regions.

EPIC and ACIS observations will allow a complete spectroscopic characterization of all types of coronal sources, including shock-heated wind sources (hot stars), solar-type magnetically-dominated sources (cool stars), active binaries, flare stars, pre-main sequence objects, Pop II stars and some evolved stars. Moreover, it will be possible to obtain long interrupted observations for up to ~ 2 days (for studying, e.g., rotational modulation and/or eclipsing binary systems), as well as time-resolved spectra of flares at a much better sensitivity than possible at present with ASCA. However, in spite of the

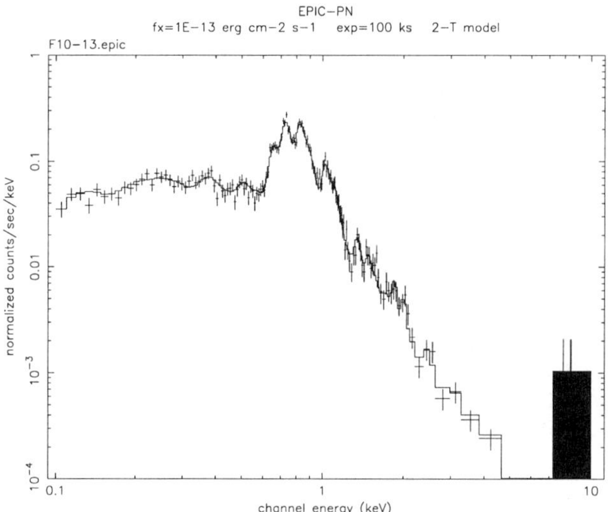

FIGURE 7. Simulated 100 ks EPIC-PN spectrum of a weak stellar source with a flux of $\approx 1 \times 10^{-13}$ erg cm^{-2} s^{-1}.

excellent diagnostics that will be possible for relatively bright sources, the simulations above show clearly than only limited high-resolution spectroscopy can be obtained for very low mass stars and cluster stars. This is the realm where imaging observations with AXAF and medium-resolution spectroscopy with XMM will be most appropriate.

5. Prospects for very-low mass stars and cluster stars

Very-low mass stars and brown dwarfs are expected to be faint X-ray sources, with typical luminosities of the order of the average solar luminosity ($L_x \sim 10^{27}$ erg s^{-1}). The available data, mostly from ROSAT, suggest that X-ray emission does not disappear in fully convective stars and that late M dwarfs are as efficient coronal emitters as other cool stars in terms of the ratio of X-ray to bolometric luminosity (Fleming *et al.* 1993, 1995; Schmitt 1998). A saturation limit appears to exist at $L_x/L_{bol} = -3$. If so, the intrinsic coronal emission of very-low mass stars and brown dwarfs must be quite low.

A simple calculation shows that at a limiting sensitivity of 2×10^{-15} erg cm^{-2} s^{-1}, that could be reached with both AXAF and XMM, a source with an intrinsic X-ray luminosity of 10^{27} erg s^{-1} will be detectable up to 65 pc. This limiting distance falls to 20 pc and 6 pc, respectively, for intrinsic X-ray luminosities a factor 10 and 100 lower. Conversely, the limiting distance for detection increases to 200 pc and 650 pc for sources with intrinsic luminosities a factor 10 and 100 larger. Detection of very faint low-mass stars, with $L_x \leq 10^{27}$ erg s^{-1} will thus be possible only in the solar neighbourhood, at distances less than $\sim$ 60 pc, i.e. at distance much closer than those typical of most open clusters and star forming regions. Low mass stars with X-ray luminosities in the range $L_x \sim 10^{28} - 10^{29}$ erg s^{-1} can be detected instead in many nearby open clusters and star forming regions within a distance of about 500 pc. This will allow extending to fainter magnitude the extensive surveys of nearby open clusters carried out by ROSAT (e.g. Randich 1997, 1998; Jeffries 1998). At 150 pc, a source with an X-ray luminosity of 10^{28} erg s^{-1} corresponds to a flux (neglecting interstellar absorption) of 3.7×10^{-15} erg cm^{-2} s^{-1}.

The possibility of detecting brown dwarfs has been discussed by several authors at this meeting (e.g. Linsky, Randich, Schmitt; see elsewhere in this volume). It relies on the assumption that brown dwarfs, if X-ray emitters, may be close to the saturation limit ($L_x/L_{bol} = -3$) and, therefore, younger BDs are brighter in X-rays. If this occurs, detection of brown dwarfs may be possible in the Pleiades and other nearby young clusters as well as in the field. For instance, brown dwarfs in the Pleiades are expected to have luminosities $\leq 5 \times 10^{27}$ erg s^{-1}, i.e. fluxes $\leq 3 \times 10^{-15}$ erg cm^{-2} s^{-1} (Randich 1998). The detection of brown dewarfs in X-rays will put important constraints on our understanding of dynamo action amd magnetic activity in fully convective stars.

More generally, AXAF and XMM can contribute significantly to the study of late-type stars in open clusters and associations, as well as to the study of pre-main sequence objects in star forming regions (SFRs). ROSAT has demonstrated that open clusters, whose members have all the same age and chemical composition but different masses, are ideal targets to study the evolution of angular momentum in late-type stars, and the dependence of coronal activity on mass and age. Although ROSAT has observed nearly all open clusters at distance $\leq$ 500 pc, and has surveyed extensively all nearby star forming regions, very little is known as yet on the variation of coronal spectra with age. The CCD cameras on XMM and AXAF can provide medium-resolution spectroscopic data for many open clusters and SFRs observed with ROSAT, and can thus complement in a very effective way the imaging and low-resolution data provided for these sources by the latter mission. For a few bright stars (e.g. in the Hyades and in Taurus-Auriga) it will also be possible to obtain high-resolution spectra with the RGS. Additional avantages provided in this area by the new missions are the high spatial resolution of AXAF imaging observations (which will allow resolving even the most crowded regions like the Trapezium in Orion) and the high-energy response (up to 10 keV) which will allow the detection of embedded sources such as those recently discovered by ASCA and ROSAT in ρ Oph and R CrA (Koyama *et al.* 1996; Grosso *et al.* 1997).

The typical fluxes for late-type stars in nearby open clusters and SFRs are of the order of $\sim 1 \times 10^{-13}$ erg cm^{-2} s^{-1}, which corresponds to $\sim$ 0.1 cts/s in EPIC for typical coronal spectra. Assuming that we need at least 5,000 cts to get a good spectrum, typical exposure times of $\sim$ 50 ks are required for CCD spectroscopy of cluster sources. With these exposure times, it will be possible to obtain spectra for many sources in nearby open clusters, spanning the age range from $\sim$ 30 to 700 Myr (e.g. the Hyades, the Pleiades, α Persei, IC 2602 and IC 2391, Praesepe, Coma, NGC 6475, etc.). EPIC and ACIS spectra can also be obtained in similar exposure times for many CTT and WTT stars in nearby SFRs like Taurus-Auriga, Chamaeleon, ρ Ophiuchi, R CrA, etc. With the RGS, it should be possible to obtain reasonably good spectra at least for a few bright sources in clusters and associations with typical fluxes $\geq 1 \times 10^{-12}$ erg cm^{-2} s^{-1} in exposure times of 50 to 100 ks.

6. Conclusions

The powerful X-ray missions that will be launched in the next couple of years will provide new important results on coronal emission from very-low mass stars and (possibly) brown dwarfs, as well as for stars in open clusters and star forming regions. AXAF and XMM will be highly complementary in this respect. The superior image quality of AXAF will allow extending to lower luminosities the census of low-mass stars in clusters and associations, and may allow the detection of brown dwarfs both in the field and in nearby young clusters. Very crowded regions like the Trapezium will be resolved in a multitude of individual sources. The high throughput of XMM, on the other hand, will allow medium resolution (CDD) spectroscopy for a large number of stars in clusters and SFRs, and high resolution (RGS) spectroscopy for a few of them. It will be possible for the first time to study the variations with age of coronal parameters like temperature, rather than simply the age dependence of X-ray luminosity. The high energy response and good sensitivity of XMM up to (at least) 10 keV will allow investigating very young embedded PMS objects, with ages as young as 10^5 years and even less. In this way, it will be possible to investigate the origin of coronal emission in fully convective stars (as in late M dwarfs and in brown dwarfs) as well as at very early stages in the course of stellar evolution (as in PMS embedded sources). It will also be possible to explore for the first time the variation with age of the spectroscopic properties of stars, thus providing important constraints for our understanding of coronal heating mechanisms.

I thank Antonio Maggio and Gianpiero Tagliaferri for assistance with the XMM simulations presented in this paper and Manuel Güdel for the AXAF simulations. I acknowledge financial support from ASI (Italian Space Agency) and MURST (Ministero dell'Università e della Ricerca Scientifica e Tecnologica). I thank the European Union (EC) for making possible the organization of the Three-Islands Euroconferences and Rafael Rebolo and IAC for organizing the first meeting of this series.

REFERENCES

FLEMING, T.A., GIAMPAPA. M.S., SCHMITT, J.H.M.M. & BOOKBINDER, J.A. 1993 *ApJ* **410**, 387.

FLEMING, T.A., SCHMITT, J.H.M.M., & GIAMPAPA, M.S. 1995 *ApJ* **450**, 401.

GROSSO, N., MONTMERLE, T., FEIGELSON, E.D., ANDRÉ, P., CASANOVA, S. & GREGORIO-HETEM, J. 1997 *Nature* **387**, 56.

JEFFRIES, R. 1998 in *Solar and Stellar Activity: Similarities and Differences* (eds. J. Butler & G. Doyle), PASP Conference Series, San Francisco, in press.

KOYAMA, K., HAMAGUCHI, K., UENO, S., KOBAYASHI, N. & FEIGELSON, E.D. 1996 *PASJ* **48**, L87.

LINSKY, J.L. & GAGNÉ, M. 1998, in *10th Cambridge Workshop on Cool Stars, Stellar Systems, and the Sun* (eds. R.A. Donahue & J. A. Bookbinder), PASP Conference Series, San Francisco, in press.

MEWE, R. & GÜDEL, M. 1998, in *10th Cambridge Workshop on Cool Stars, Stellar Systems, and the Sun* (eds. R.A. Donahue & J. A. Bookbinder), PASP Conference Series, San Francisco, in press.

MEWE, R., KAASTRA, J.S., LIEDHAL. D.A. 1995 *Legacy* **6**, 17.

PALLAVICINI, R. 1998, in *10th Cambridge Workshop on Cool Stars, Stellar Systems, and the Sun* (eds. R.A. Donahue & J. A. Bookbinder), PASP Conference Series, San Francisco, in press.

RANDICH, S. 1997, in *Cool Stars in Clusters and Associations: Magnetic Activity and Age Indicators* (eds. G. Micela, R. Pallavicini & S. Sciortino), Memorie Soc. Astron. It., vol. 68, p. 971.

RANDICH, S. 1998, this volume.

SCHMITT, J.H.M.M. 1998, this volume.

AUTHOR INDEX

Author index